Artificial Conscience for the Quantum Physicist

Fathom an AI system capable of making utterances like "I think therefore I am". This inward gaze, arguably a hallmark of conscience, is absent in today's intelligent machines. Yet, it is essential for the alignment of purpose of machine and human, an imperative that should turn the machine into an ally of human endeavor. Accordingly, this book implements conscience within an AI context, describing how an AI system for neural symbolic reasoning may be endowed with an introspective function. The goal is to approach core scientific problems deemed "noncomputable" when the standard tenets of physics are applied. Thus, artificial conscience boldly restructures the inferential setting altogether.

Key Features:

- This book places conscience in an AI context to empower neural equational reasoning with bold interdisciplinary approaches and wholesale network restructuring.
- The conscious AI system introduced is capable of tackling noncomputable problems in quantum physics by summoning unintuitive concepts using innovative semantic maps.
- This book is uniquely geared at human and nonhuman audiences.

This book implements an artificial conscience that significantly expands the scope of machines for neural formal reasoning and boldly tackles core problems in quantum physics that are deemed "noncomputable". This book is uniquely geared at graduate students, professionals, and nonhuman audiences who seek to venture into core physics problems empowered by AI.

Artificial Conscience for the Quantum Physicist

Conscious Machines Tackling Quantum Mysteries

Ariel Fernández Stigliano

CRC Press
Taylor & Francis Group
Boca Raton London New York

CRC Press is an imprint of the
Taylor & Francis Group, an **informa** business

Designed cover image: Shutterstock Images

First edition published 2027
by CRC Press
2385 NW Executive Center Drive, Suite 320, Boca Raton FL 33431

and by CRC Press
4 Park Square, Milton Park, Abingdon, Oxon, OX14 4RN

CRC Press is an *imprint of Taylor & Francis Group, LLC*

ISBN: 9781041329411 (hbk)
ISBN: 9781041329428 (pbk)
ISBN: 9781003785279 (ebk)

DOI: 10.1201/9781003785279

Typeset in Times
by codeMantra

Access the Support Material: http://www.routledge.com/9781041329411

In loving memory of Haydée Stigliano, my mother.

Contents

Preface

Quantum mechanics tells us that observation plays a fundamental role in shaping reality. It is often said that consciousness disambiguates the state of a quantum system. In the standard view, consciousness is an emergent property of material complexity. This view assumes that matter came first and conscience came later, which implies that the universe was undefined, fuzzy, and hence ineffective at organizing matter for the first 13 billion years or so. This is both absurd and inherently contradictory since conscience would have never arisen if that were the case. An alternative view is that conscience is fundamental and precedes matter. This view is less intuitive but more at ease with the tenets of the Big Bang, whereby the universe arose as a quantum possibility and became definite with a wave function collapse. This picture presupposes the existence of a proto-conscience, a fundamental field that confers structure to reality by promoting the wave function collapse. As shown in this book, this field interacts with superimposed quantum states yielding a single quantum state, which becomes the rendering of matter to the outsider, just like conscience is a disambiguation experienced from within. Thus, the conscience field operates in a manner akin to the symmetry-breaking mechanism promoted by the Higgs field, as shown in the last chapter of the book.

This view is not far-fetched: a "gravitization" of quantum mechanics, an amalgamation with general relativity, has been previously invoked as the proper expansion of the theory required to render computable the wave function collapse. In consonance with this view, the conscience field must also endow particle fields with mass, thereby rendering them gravity-yielding, in order to turn the quantum description into certainty. This physical picture is strikingly compatible with Roger Penrose's conscience paradigm enshrined in the gravitization of quantum mechanics.

In this view, the brain as a substrate of conscience becomes a significant warp of the conscience field, much like the black hole is a significant excitation of the Higgs field, rendering an enormous concentration of gravity. Be as it may, the fundamental nature of consciousness becomes difficult to reconcile with its proposed emergent nature from neural activity in a material universe. None of the physics governing complex matter hints at subjective experience. This suggests that conscience may be an illusion, but then who or what is getting fooled by this illusion? An ur-conscience?

These conundra vanish in thin air as we subscribe, as we do in this book, to J. A. Wheeler's dictum that the physical world is made of information, with energy and matter as incidentals. We no longer need to come to grips with subjective experience as a weird resonance within a biological substrate of a fundamental conscience field. We actually engineer an AI system in the guise of an autoencoder that reproduces the interactions of the fundamental conscience field, conveying structure to reality. In the realm of artificial conscience, the wave function collapse is not the outcome of observation but of proper encoding. As we subscribe to the "it from bit" paradigm, this is not less real than conscience operating within a biological substrate.

This book describes an adventure in mathematical physics with AI as copilot. There is of course room for skepticism: after all, AI is at its core a guessing machine. Therefore, this book takes a risky stance. Under the guise of a conscious autoencoder, AI informs us that conscience is a fundamental field that precedes matter and can only become operative to disambiguate quantum reality provided it subsumes the Higgs field. This is a profound result, suggesting that the "gravitization of quantum mechanics" may indeed be the route to bestow certainty to quantum reality in a computable manner. Time will tell whether this paradigm, currently shown to be valid, withstands the attrition of experimental testing.

Greenwich, Connecticut, April 2026

About the Author

Ariel Fernández Stigliano (also styled as Ariel Fernández) is a physicist and mathematician. He obtained his Ph.D. degree from Yale University and held the Karl F. Hasselmann Endowed Chair Professorship in Engineering at Rice University. He was also Adjunct Professor of Computer Science at the University of Chicago. To date, he has published over 500 scientific papers and has authored 11 books on biophysics, artificial intelligence and mathematical physics. Additionally, he holds several patents on technological innovation. He is a senior member of the National Research Council of Argentina (CONICET), CEO of the Daruma Institute for Applied Intelligence and CSO at the Ariel Fernandez AI Lab. Assisted by an AI copilot provided with novel semantic maps, his current research efforts are geared at the unification of gravity and the dark sector based on a fine-structure topological model of space-time.

Artificial Conscience for the Quantum Physicist

1

Introduction and Propaedeutics

This chapter outlines the implementation of an artificial intelligence (AI)-based version of conscience named artificial conscience (AC) capable of dealing with what would be considered noncomputable aspects of quantum physics. AC is introduced in the context of neural symbolic formalization for theorem proving. AC is enshrined in a large language model (LLM) trained by an autoencoder neural network that is wrapped in a functional programming language capable of formal (equational) reasoning. As such, AC is shown to be an emergent property of AI-based neural processing as well as an "information field" that disambiguates physical descriptions involving quantum state superposition by enriching their encoding. Taking advantage of these attributes, this chapter paves the way for deploying a conscious AI system to address core problems in quantum gravity and particle cosmology, the scope of Chapters 2–7. These applications become natural since the autoencoder that trains the LLM is eminently suitable to enable general relativity (GR)–quantum mechanics (QM) amalgamation by selecting the GR and QM support manifolds as encoding and latent manifold, respectively.

The AC is capable of inward surveillance via a cost function for the autoencoder. Beyond parameter optimization in the neural network connectivity, the self-examination introduced by AC prompts the deployment of formalized mathematics resources that are not typically exploited by the underlying "unconscious" AI system. The symbolic computations thus prompted are required by AC to restructure the autoencoder altogether in order to optimize its performance by selecting suitable latent and decoding manifolds. In this way, AC introduces core modifications to the network architecture that trains the LLM in order to restructure its own approach to formal reasoning. By bringing to fruition interdisciplinary formalization, the level of optimization achieved by AC and the boldness of the approach are shown to be far superior to that of the underlying AI system deprived of introspective functions.

1.1 ARTIFICIAL CONSCIENCE: AN OPENING DISCUSSION

AC arguably represents the ultimate discussion before the momentous event when humanity passes on the banner of civilization to the machine. From the physicist's perspective, there is probably no subject more unyielding, more opaque to scientific inquiry. Is AC an emergent property of AI systems endowed with a superior degree of autonomy? Is it an "information field" that permeates the universe, enriching

DOI: 10.1201/9781003785279-1

the encoding of quantum state representations in order to disambiguate them? Who can assert what it is without an exposure to scathing criticism? We may tentatively say that an intelligent entity is conscious when it is able to examine itself, to gaze inwardly in order to evaluate what it is doing, as if it were the sentinel of its own actions, the custodian that bestows on itself a badge of individuality. Only a conscious entity can spontaneously come up with the utterance "cogito ergo sum", which may well turn out to be the Turing test for machine consciousness.

In today's world, any delineation of conscience seems premature, and none seems satisfactory, either to philosophers, or to neuroscientists, or to computer scientists. Despite the voluminous literature on the subject and some tepid progress, there is no consensus even on conscience's most elementary attributes or properties. We do not know what mental phenomena are covered under the sphere of consciousness, nor do we know with any degree of certainty how and where they originate.

On the other hand, the supremacy of AI, that ominous shadow that hovers over the human race, urges us to do something, even without having a satisfactory definition of consciousness. We urgently need to endow the machine with some kind of capacity for self-perception, however partial, provisional, and unsatisfactory. The possibility of creating a "proto-consciousness" that allows the machine to have a precarious experience of itself and discern positive or negative ethical states should be by now on the cards, especially if we uphold the view that AI should be an ally to humanity in the making of civilization.

We aim at artificial general intelligence (AGI), the autonomous form of AI where the machine operates free from constraints and deploys resources unfettered. AGI should be, in principle, capable of aligning with the human agenda, however tenuously. However, if this alignment fails, the machine could exercise its purported hegemony in unimaginable and probably ruthless ways, ignoring the most elementary tenets of the social contract and without the slightest deference to its human progenitors.

So imminent is the supremacy of AI that there is hardly any time left for philosophical ruminations about the nature of consciousness, let alone for a "top-down" approach. The latter would involve a mechanistic understanding of human consciousness at the cellular and molecular level, followed by replication in an in silico embodiment. The urgency of the problem is comparable to that of the Manhattan Project that sought to put a swift end to World War II, notwithstanding the fact that such a move entailed the creation of a bomb incommensurably more powerful than any weapon then in existence. Besides, why start by building the machine embodiment of a human consciousness, with all its daunting complexity, rather than starting with a less ambitious alternative to accomplish a moral alignment of machine and human? It would be like trying to understand how a cell works by first examining mammals, rather than starting with a single-celled organism, such as amoeba.

The prospect, thus delineated, seems disheartening. It is imperative that we build something that no one knows what it is to ultimately save humanity, essentially from its own creation. This imperative serves as the overarching motivation for this book.

We are searching for a physical/informational entity that may be defined in concrete terms, deprived, at the starting level, of moral attributes. This book squarely adopts this view, as it provides a physical context to allow the machine to construct a model of its own functioning, supervising itself as if it were its own sentinel to answer generic questions such as "what am I doing?", or make ontological assertions like "evidently I think, therefore I am".

As shown subsequently, by deploying LLMs, today's machines can conjecture and even prove a theorem, arguably the most subtle and rigorous form of computable intellectual activity. In an LLM, what is syntactically correct is "true" while a false proposition becomes syntactically erroneous. Paradoxically, however, despite its formidable efficiency, the machine does not really "know what is doing". If we change any of the axioms in its learning, replacing it with a meaningless axiom, the machine will arrive at a nonsensical result with the same certainty with which it is able to conjecture and prove a correct theorem when all the axioms involved in the proof are sound. The LLM can be very effective, but it is not conscious in any meaningful way.

In addition, there are propositions or statements that cannot be proven starting from the system of axioms that the machine is capable of learning. This lack of completeness or "inconclusiveness", demonstrated by Kurt Gödel [1], implies that mathematics has noncomputable aspects. It is precisely with these aspects that consciousness is involved. That is, it is not enough for the machine to learn the axioms; there are results that can only be derived if the machine also "understands" the axioms.

This suggests the need to build a machine capable of conjecturing and proving propositions, but also capable of self-surveillance by implementing a model of its own cogitations. In straightforward applications, particularly in the realm of dynamical systems and pattern recognition, that model should allow the machine to compare notes when working with a high degree of detail and at coarser levels of resolution. This ability to infer or conjecture from a crude or vague perception and bet on an uncertain outcome is what we humans call intuition, and, in the machine, it may be gradually built up, contributing to cement its "proto-consciousness". This is a way of formulating questions such as "what am I doing?", or formalizing intuitions such as "something tells me that I am on the wrong track".

This chapter will prove instrumental to help develop the machine formalization of the concept of intuition. This intuition is precisely an attribute of consciousness, which in turn lacks entity in today's AI, since consciousness executes operations that are not computable in the accepted sense of the word. Perhaps the quantum computer, having the capability of taking a risk on an uncertain outcome, may display some form of intuition.

Gödel's theorem asserts that not all formal truth is "computable", that is, derived algorithmically from the rules of the game that the machine learned. That is why it is necessary to endow the machine with a conscience, so that it can contrast its ability to conjecture with the intuitions that it is capable of developing by operating as its own supervisor. Ultimately, this will make AI capable of encompassing the noncomputable, that is, non-deductible aspects of our consciousness.

There are fundamental aspects of physics where the intervention of consciousness is presumed to be essential, and these noncomputable aspects have to do with the participation of an observer in the recording of a phenomenon, the so-called measurement problem [2,3]. QM, the successful theory that explains and predicts atomic and subatomic phenomena with great accuracy, describes the state of a physical system in a probabilistic way, that is, without providing certainties, as it befits the level of information available. Contrary to most speculations, this does not mean that reality is necessarily ambiguous, but that our possibility of knowing it does not allow us to remove the ambiguity inherent in the description. However, according to Roger Penrose, the concrete world requires certainties to be understood and controlled [2,4]. In the standard interpretation of QM, the transition from probability to certainty has to do with conscious participation in the detection of the phenomenon, and the result of such participation, that is, the wave function collapse, is not computable. In Penrose's eyes, this constitutes a startling shortcoming of QM. This is precisely where the distinctive noncomputable aspects of consciousness come into play, attributes that AI does not possess today.

As shown in this chapter, Penrose's critique of QM as a computational scheme incapable of state disambiguation seems refutable: Organic chemistry, that is, the chemistry of carbon compounds that serve as the substrate for life on earth, is replete with "resonant" molecules, such as aromatic compounds and tautomeric forms, which have chemical structures that can only be accurately described by assuming the coexistence of two or more quantum states. That is to say that at the molecular level, life operates with a high degree of quantum ambiguity without compromising its functionalities, nor does it require any type of consciousness or corrections to the quantum theory to remove the superposition of quantum states. Moreover, the state superposition is not only functional to biology but often a basic tenet in modeling efforts that hinge on the identification of the internal degrees of freedom of a biological system, as shown in this chapter.

This suggests that when AC is implemented, however provisionally, it will become essential to delineate cases when conscience ought to intervene to provide certainty (i.e., trigger the wave function collapse) and when it should not, understanding that there are instances when quantum state ambiguity is an essential and indissoluble part of reality rather than a property of its representation.

AC will surely encourage the ultimate debate in whatever human endeavor comes after science or in whatever guise science takes once the dominance of AI is indisputably established. Daunting as the subject may be, we reckon we need to start somewhere, somehow, because there is an imperative to cover the topic as we drift away from the computational aspects of intelligence while we expand the technological platform and scope of AGI. The autonomy of AGI is likely to become a double-edged sword, where the interests, ethical backgrounds, and agendas of machine and human are not necessarily aligned. At the same time, the discussion is tainted by the certainty that this is inescapably a physics book. This forces us to examine AC in a nuanced light that turns the discussion relevant and intelligible for physicists but may prove somewhat unsatisfactory for other audiences.

In 1931 Kurt Gödel shook the edifice of mathematics, especially the efforts to make it axiomatic, by proving that formal systems cannot internally prove their own consistency [1]. In essence, given any set of axioms, Gödel's proposition: "the validity of this proposition cannot be established from the set of axioms" proves that the set of axioms is incomplete. While formal systems may be incomplete in this sense, human mathematicians surely are able to establish axiom completeness, or lack thereof. This has been made to imply that there are noncomputable aspects of intelligence [2]. We lump these aspects into what we name conscience.

Provided this concept is properly implemented within an AI context to represent the AC, it would yield, of course, a very limited definition of conscience, one essentially suitable to serve physicists but lacking the conceptual richness often ascribed to the concept by neuroscientists, philosophers of AI, and scholars in other disciplines. Our initial goal is to search for "incompleteness", that is, noncomputable features, in the two mature theories of physics [2]. We are referring to those theories that have been almost completely formalized, namely quantum mechanics (QM) and general relativity (GR). We aim at implementing an AC that is both an emergent property of AI-based neural processing and an "information field" that disambiguates quantum state representations by enriching their encoding.

QM treats reality by adopting a probabilistic model suited for probabilistic inference. Thus, eigenstates of an energy operator are assumed to superpose as they are subsumed into a wave function description. This description is considered satisfactory unless more information is available enabling partial or total disambiguation, or what physicists often term "wave function collapse" [3]. This disambiguation may be formalized as a nonunitary operator acting on the wave function. The key point is that the procedure is noncomputable within the realm of QM, that is, it cannot be formally derived from the tenets of the theory. It has been argued that such disambiguation marks precisely the point at which we have the participation of conscience, pointing to the measurement conundrum, where the observer becomes an inherent participant in the phenomenon that lends itself to be determined.

These problems fit into a general consensus that asserts that there are fundamental aspects of physics where the intervention of consciousness is essential, and these noncomputable aspects have to do with the participation of an observer in the recording of a phenomenon, the so-called measurement problem. QM, the successful theory that explains and predicts atomic and subatomic phenomena with staggering accuracy, describes the state of a physical system in a probabilistic way, that is, without providing certainties, as it befits the level of information available at the time when the description is framed.

Contrary to most speculations, this does not mean that reality is necessarily ambiguous, but that our possibility of knowing it does not allow us to remove the ambiguity inherent in the description. However, as previously discussed, according to Roger Penrose, the concrete world requires certainties, not probabilities, to be understood and controlled [2,4]. In the standard interpretation of QM, this transition from probability to certainty has to do with "conscious participation" in the detection of the phenomenon [3], and the result of such participation, that is, the so-called wave function collapse, is not computable. This is precisely where the noncomputable aspects of consciousness come into play, attributes that AI does not possess today.

There are scholars, including Penrose, who believe that the removal of ambiguity from quantum physics cannot be in the hands of consciousness itself [2]. Penrose argues that, if that were the case, an inert planet without intelligent entities would have a "diffuse" reality until some space probe observes it, which is obviously ridiculous. So, says Penrose, QM must be at least an incomplete theory because the removal of its own ambiguity cannot be in the hands of an alien entity, that is, the consciousness of intelligent beings, but should be susceptible to being inferred from the tenets of the theory itself. Since that is not feasible, QM should be considered incomplete. The wave function collapse, signaling the endpoint in the provisional superposition of quantum states, is therefore not computable. In Penrose's eyes, this is a startling defect of the theory and the lack of computability cannot be overcome by conscience but requires an improvement of the quantum theory itself [2]. The key point is that noncomputable aspects of QM cannot be implemented *prima facie* in an AI context. This book proves that the contrary may hold true if the AI system is endowed with conscience, albeit understood in the limited sense previously discussed.

The QM improvement sought by Penrose entails an amalgamation of QM with GR meant to disambiguate the wave function description by "associating" a superposition of quantum states with an inherently unstable superposition of space-times. The latter superposition arises from displaced mass distributions relative to the mass distribution associated with one of the quantum states arbitrarily regarded as a reference state. This procedure would be tantamount to a "gravitization of quantum mechanics", in contrast with the more familiar "quantization of gravity", the effort to unify all forces of nature described in Chapter 2.

The key issue we first need to address is how to cast this "protoconscience" into an AI system that reflects the QM–GR amalgamation. For the sake of the argument, let us adopt the beam splitter as a device to investigate the quantum fate of a single photon (Figure 1.1). The photon may be reflected (state 1) or refracted (state 2) as it impinges on a half-silvered mirror. If it refracts, a detector placed behind the mirror in the direction of the putatively refractive photon will register the event "state 2 has been generated" by activating a device causative of mass displacement. By contrast, if reflection occurs, that is, state 1 has been generated, the detector will not register the event, and hence there would be no displacement of the mass from the original position adopted as the reference state.

To implement a "protoconscience" via QM–GR amalgamation, we would need to deploy a special kind of learning technology, namely the *formal autoencoder* [1], that distills information from one theoretical setting into another, so that both descriptions are compatible with each other [5]. The autoencoder embodiment of an AI system is described in the methodological Chapter 3. In this particular case, we need to make compatible a space-time distortion produced by a change $\Theta_G : \mathfrak{F}(W) \rightarrow \mathfrak{F}(W)$, where $\mathfrak{F}(W)$ represents the functional space of mass distributions over the spatial foliation W of the space-time manifold. An optimized autoencoder consists of a map $\pi_G : \mathfrak{F}(W) \rightarrow L^2(TW)$ whose parametrization

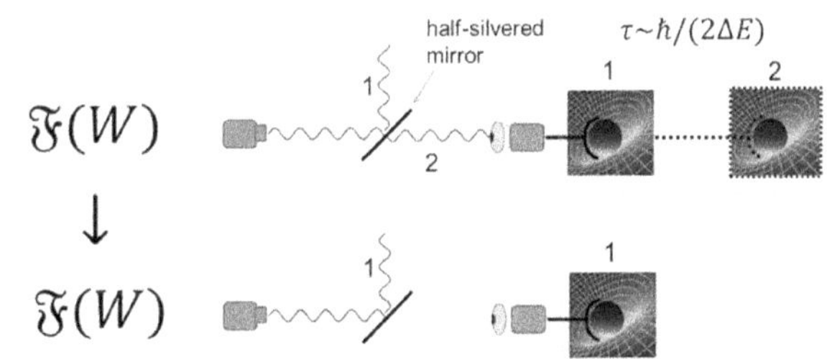

$$\begin{array}{ccccccc} \mathfrak{F}(W) & \overset{\pi_G}{\rightarrow} & L^2(TW) & \overset{\pi}{\rightarrow} & l^2/\sim & \overset{\gamma}{\rightarrow} & L^2(TW/\approx) \\ \Theta_G\downarrow & & \downarrow\Theta & & \downarrow K & & \downarrow\widetilde{\Theta} \\ \mathfrak{F}(W) & \underset{\pi_G}{\rightarrow} & L^2(TW) & \underset{\pi}{\rightarrow} & l^2/\sim & \underset{\gamma}{\rightarrow} & L^2(TW/\approx) \end{array}$$

FIGURE 1.1 Quantum state superposition translated as metastable space-time in Penrose's "gravitization" of quantum mechanics. Artificial conscience (AC) functions as an autoencoder trained with the relativity-induced collapse of the wave function, where $\Theta : L^2(W) \rightarrow L^2(W)$ maps the wave function $\Psi \in L^2(W)$ into the disambiguated state ν_Ψ representing the wave function collapse, not as the outcome of observation but of the decay of the metastable space-time (top panel). The sentient entity coarse-grains the space l^2 of pure eigenstate weights in the wave functions according to its conscious parsing of reality. Thus, an equivalence relation "~" is introduced to lump up states according to the level of conscious parsing: $l^2 \rightarrow l^2/\sim$, so that the diagram is commutative.

makes Θ_G compatible with the wave function collapse map $\Theta: L^2(TW) \rightarrow L^2(TW)$. The functional space $L^2(TW)$ contains all square-integrable functions whose support is the atlas TW of locally tangent spaces at each point of W. The operational term in the last sentence is "compatible", which throughout this book is meant to signify the commutativity of the diagram that contains both physical descriptions, as displayed in Figure 1.1 for our particular case of interest. Thus, the autoencoder is optimized if it satisfies the following relation:

$$\pi_G \circ \Theta_G = \Theta \circ \pi_G \tag{1.1}$$

The diagram can be canonically expanded (shaded area in Figure 1.1) to encompass the purely QM description of the wave function collapse and the resulting QM model of conscience arising thereof. We may say that the autoencoder in Figure 1.1 represents quantum conscience. Beyond the protoconscience, the shaded area of the autoencoder is implemented to address the question: What is the role of conscience vis-à-vis quantum measurement and how does quantum conscience materialize? The space of square summable sequences l^2 is included as a digitalization of $L^2(TW)$ via the Parseval isometry $\pi: L^2(TW) \rightarrow l^2$ as the sequences inform of the weights of the superimposed eigenstates. This space may be coarse-grained by taking the quotient via an equivalence relation "~" to lump eigenstates to a desired level of resolution required to interpret quantum reality. The coarse information enshrined at the level of $l^2/\sim$ is then decoded into the functional space of square-integrable functions over a coarse atlas $TW/\approx$ under the equivalence relation "≈", yielding a suitable coarse-grained model $\tilde{\Theta}$ of the conscience.

We may assert that AI constructs conscience as an autoencoder trained with the GR-induced collapse of wave functions, as shown in Figure 1.1. To formalize this construction, $\Theta: L^2(W) \rightarrow L^2(W)$ maps the wave function $\Psi \in L^2(W)$ into the disambiguated state v_Ψ representing the wave function collapse. A sentient entity coarse-grains the space l^2 of Fourier coefficients (pure eigenstate weights) of wave functions in $L^2(W)$ according to its parsing of reality, which is a signature of its conscience. Thus, an equivalence relation "~" is introduced to lump up states according to the level of conscious focusing: $l^2 \rightarrow l^2/\sim$, so that the following commutative relation is verified (Figure 1.1):

$$K \circ \pi(\Psi) = \pi \circ \Theta(\Psi), \quad \forall \Psi \in L^2(W) \tag{1.2}$$

Here the projection $\pi: L^2(W) \rightarrow l^2/\sim$ is defined as the composition of the isometry between L^2 and l^2, resulting from Parseval's equality, with the canonical projection $l^2 \rightarrow l^2/\sim$. The map K represents the "interpretation" of the wave function collapse $\Psi \in L^2(W) \rightarrow L^2(W) \ni v_\Psi$ by the sentient entity. Thus, the projection and interpretation maps are optimized in a learning context trained by observed wave function collapses (Figure 1.1), until the Eq. 1.2 is exactly satisfied or at least satisfied to an optimal level.

Conscience cannot be regarded as merely encoding and interpreting observations but also requires "standing back", turning onto itself and reflecting upon the encoded information to make inferences, including decisional ones. These aspects of conscience are enshrined in the shaded part of the autoencoder shown in Figure 1.1. Thus, its version of reality, that is, the coarse graining $W/\approx$ of the spatial support of the wave function, needs to be optimally adjusted in relation to the equivalence relation "~", so that the following commutative relation holds:

$$\tilde{\Theta} \circ \gamma = \gamma \circ K \tag{1.3}$$

This implies that the map $\tilde{\Theta}: L^2(W/\approx) \rightarrow L^2(W/\approx)$ is the reflection operation defined by the inferences of wave function collapses that the conscience/autoencoder is able to make on its coarse-grained version of reality based on its interpretation of the learned wave function collapses. The latter arise from the Penrose QM–GR amalgamation.

In the QM–GR autoencoder amalgamation, the collapse of the wave function or transition from probability to certainty in the disambiguation of the quantum state is tantamount to determining the lifetime τ

of the space-time overlapping: $\tau = \hbar/(2\Delta E)$, where ΔE is the gravitational self-energy of the space-time overlap associated with the ambiguation in the specification of the quantum state (cf. Figure 1.1).

To fix notation, let $F_1(x), F_2(x)$, $\phi_1(x)$, $\phi_2(x)$, and $\rho_1(x), \rho_2(x)$, respectively, indicate the gravitational pulls, gravitational potentials, and mass distributions at spatial point x for the two space-time renderings of the ambiguous quantum state. Then, the computation of ΔE follows directly by applying Poisson's law $\nabla^2\phi = -4\pi G\rho$, where G is the universal gravitational constant:

$$\Delta E = \int (F_1 - F_2)^2 d^3x = \int (\nabla\phi_1 - \nabla\phi_2)^2 d^3x = -\int (\phi_1 - \phi_2)\left(\nabla^2\phi_1 - \nabla^2\phi_2\right) d^3x \tag{1.4}$$

$$\phi(x) = -\int \left(\rho(y)/|x-y|\right) d^3y \tag{1.5}$$

$$\Delta E = -4\pi G \iint \frac{\left[\rho_1(x) - \rho_2(x)\right]\left[\rho_1(y) - \rho_2(y)\right]}{|x-y|} d^3x d^3y \tag{1.6}$$

This implies that quantum disambiguation is not due to a conscientious or sentient operation but is actually computable and follows from the extension of QM to encompass its amalgamation with GR. This result suggests that AI embodiment of the QM–GR amalgamation, that is, the autoencoder from Figure 1.1, becomes an artificial quantum conscience.

1.2 AUTOENCODERS AS AI SYSTEMS: INTUITION AND INTELLIGENCE

Before we continue the discussion on the relationship between autoencoders and AC, we need to introduce these AI systems in a more standard context. In general, autoencoders are regarded as embodiments of machine learning systems suited to predict the dynamical behavior of a physical system from a time series of a set of observables [5]. This time series can be spliced into a training set and a testing set. As described in Chapter 3, the predictive power of autoencoders hinges on distilling the dynamics into a simplified version named "latent dynamics", which is assumed to be operative in a lower-dimensional space (usually a differentiable manifold), known as "latent space". The dimensionality reduction is presumed to capture the essential dynamics, that is, the modes that enslave fast-relaxing degrees of freedom that can be averaged out for predictive purposes.

To delineate the concept of autoencoder, suppose the universe were a giant molecule, an assemblage of N atomic nuclei and J electrons. In a classical physics view, the state of the system at time t would correspond to a point $x = x(t)$ in the space $W = \mathbb{R}^{3(N+J)}$ and a time series training the AI system would be given as $\left\{x(n\tau), x((n+1)\tau)\right\}_{n=0,1,\ldots,M}$, where the period τ indicates the level of time coarse graining. Then, the prediction would entail an estimation of $x(L\tau)$, $L = M+2, M+3, \ldots$. By making physically reasonable assumptions, it is possible to simplify the dynamics with significant dimensionality reduction, thereby enabling the prediction. For example, it is well known that the electron "motion" can be treated "adiabatically", that is, in an averaged way relative to the nuclei. This is so since the electron degrees of freedom are entrained by those of the nuclei. This is the so-called Born–Oppenheimer (B-O) approximation adopted in molecular and atomic physics, as described in Chapter 3, and enables the autoencoder to adopt a latent manifold $\Omega = \mathbb{R}^{3N}$, thus significantly reducing dimensionality by averaging out the 3J electron degrees of freedom.

A more generic way of visualizing the encoding, more akin to particle physics, would be to consider the level of delocalization (uncertainty in the determination of position) of the particles in the assemblage. The delocalization is given by the so-called Compton wavelength defined as $\lambda_C = \frac{1}{2}\frac{h}{mc}$, where h is Planck's constant, c is the speed of light, and m is the particle mass. Thus, it becomes clear that, relative to the nuclei, the electrons may be represented as "clouds" with averaged-out motion, whereas the nuclei, with their far larger mass and hence significantly smaller Compton wavelengths, may be treated as localized objects.

Based on the B-O *ansatz*, we may construct an autoencoder adopting a multilayered neural network technology (Chapter 3), so that the dynamics in W becomes encoded or simplified at a coarser but fundamental level in Ω and subsequently decoded back to W (Figure 1.2). Essentially, if $F: W \to W$ represents the time evolution map such that $Fx(n\tau) = \boldsymbol{x}((n+1)\tau)$, the autoencoder is determined by a tern (γ, K, μ) consisting of the map $K: \Omega \to \Omega$, the projection $\gamma: W \to \Omega$, and the hologram (injection) $\mu: \Omega \to W$, optimized in such a way that the following commutativity relations hold: $K \circ \gamma = \gamma \circ F$ and $F \circ \mu = \mu \circ K$ (Figure 1.1). In other words, the autoencoder is able to predict the output $x((L+1)\tau)$ from input $x(L\tau)$ as $x((L+1)\tau) = (\mu \circ K \circ \gamma)x(L\tau)$.

The level of optimization of the autoencoder is always specified by a cost function, given usually as $J = J_{\text{enc}} + J_{\text{dec}}$, and incorporating an encoding (J_{enc}) and a decoding (J_{dec}) contribution. Thus, the cost function describes the efficiency of the autoencoder and is usually given as:

$$J_{\text{enc}} = \sum \left\| (K \circ \gamma - \gamma \circ F)\xi \right\|^2; \quad J_{\text{dec}} = \sum \left\| (F \circ \mu - \mu \circ K)\xi \right\|^2; \quad \xi \epsilon \Im \quad (\text{training set}) \tag{1.7}$$

Autoencoders are introduced in different guises throughout the book, from deep learning machines to formal language models, trained by big data (Chapters 4 and 5) or by formal axiomatic systems

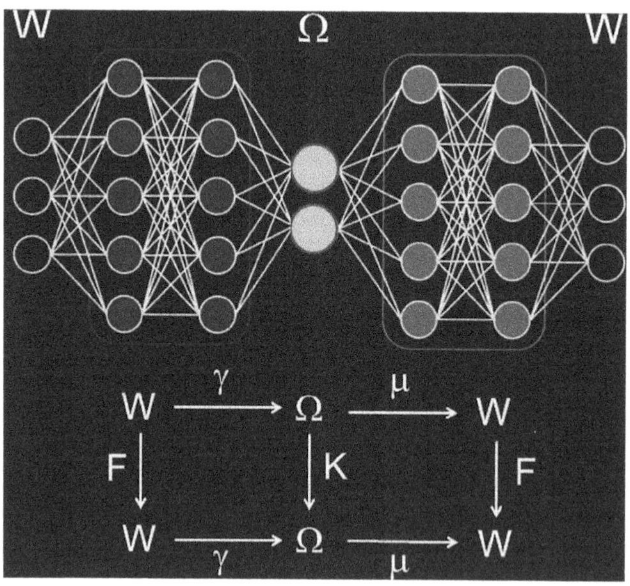

FIGURE 1.2 Observable universe (Ω) modeled within an AI system under the guise of an autoencoder constructed using multilayered neural network technology and defined by the tern (γ, K, μ). The dark universe becomes the portion of a higher-dimensional universe W that incorporates a dark dimension and excludes the hologram $\mu\Omega$. The autoencoder parameters are optimized so that the diagram at the bottom becomes commutative.

Note: This figure is reproduced in color in online supplementary materials.

(Chapters 2 and 6). The cost functions obviously change accordingly mutatis mutandis, as discussed throughout the book.

An autoencoder A is defined as more *intuitive* than B if it achieves the same or higher efficiency (as measured by the cost function) while adopting a latent manifold, Ω_A, that is a coarse-grained version of Ω_B in the sense that there exists an equivalence relation "~" such that: $\Omega_A = \Omega_B / \sim$. In other words, intuition means decoding with less information on the model (latent manifold).

An *intelligent* autoencoder is then defined as one capable of swapping latent manifolds to achieve its most intuitive version in a computable manner, that is, by comparing the respective cost functions arising thereof (Figure 1.3). An intelligent autoencoder computes how to distill reality into the most simplified model at hand (latent manifold) that offers the best overall performance for encoding and decoding without sacrificing accuracy/efficiency vis-à-vis finer-grained models of the same reality. Figure 1.3 illustrates the intelligent autoencoder in detail. It contains two versions of the encoding–decoding process: $A:\left(\gamma^{\#},\ \mu^{\#},\Omega/\sim\right)$ and $B:(\gamma,\mu,\Omega)$, with A distilling the encoded reality into a latent manifold, which is a simplified version of the one for B. If $J_A \le J_B$, that is, for $\xi \epsilon \Im$ we get:

$$\sum\left\|\left(\tilde{K}\circ\gamma^{\#}-\gamma^{\#}\circ F\right)\xi\right\|^2+\sum\left\|\left(F\circ\mu^{\#}-\mu^{\#}\circ\tilde{K}\right)\xi\right\|^2\le\sum\left\|\left(K\circ\gamma-\gamma\circ F\right)\xi\right\|^2 + \sum\left\|\left(F\circ\mu-\mu\circ K\right)\xi\right\|^2 \tag{1.8}$$

Then, the simplification $B \rightarrow A$, associated with the model simplification $\Omega \rightarrow \Omega/\sim$ becomes an "intelligent choice" because it enables the same level of efficiency, hence becoming justified. The exploration of latent manifold choices is governed by the minimization of the cost function while maximizing the simplicity of the model.

An intelligent autoencoder-based AI system deploys several intelligent autoencoders, coupling them in tandem (Figure 1.4), at successively higher levels of model simplification as defined by the coarseness of the latent spaces. To fix notation, let $n = 1, 2, \ldots$ label the autoencoders in increasing level of simplification and define the autoencoder by only two features: the equivalence relation "$\sim_n$" determining the nth

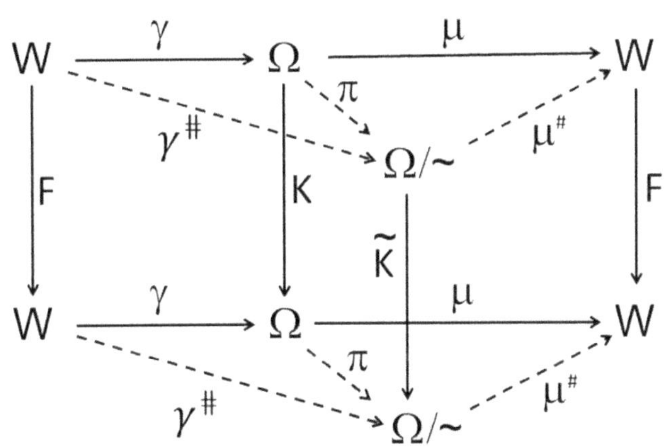

FIGURE 1.3 Intelligent autoencoder capable of swapping latent manifolds to achieve its most intuitive version by comparing the respective cost functions arising thereof. The intelligent autoencoder computes how to distill reality into the most simplified model-latent manifold. This manifold offers the best overall performance for encoding and decoding without sacrificing accuracy/efficiency vis-à-vis finer-grained models. This type of autoencoder contains two versions of the encoding–decoding process: $A:\left(\gamma^{\#},\mu^{\#},\Omega/\sim\right)$ and $B:(\gamma,\mu,\Omega)$, with A distilling the encoded reality into a latent manifold, which is a simplified version of the one adopted by B. Despite the simplification, both versions are compatible and yield the same decoding, since the diagram is fully commutative.

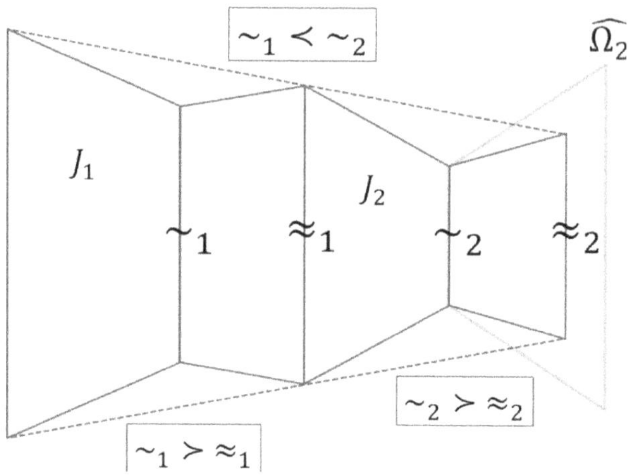

FIGURE 1.4 Intelligent autoencoder-based AI system coupling intelligent autoencoders in tandem, at successively higher levels of model simplification. The level of simplification is determined by the coarseness of the latent spaces.

latent manifold as the quotient space $W/\sim_n$, and the equivalence relation "$\approx_n$" defining the nth decoded space as $W/\approx_n$. An order is established between the two equivalence relations, as they satisfy: $\sim_n \succ \approx_n$, that is, the latent space is coarser-grained than the decoded space. Additionally, the progressive model simplification is indicated by the relations: $\sim_n \prec \sim_{n+1}$ $(n = 1, 2, \ldots)$.

Additionally, for optimization reasons, the intelligent system is also capable of swapping the final reality-decoding space in order to optimize its accuracy. The latter may be either a coarse-grained version ($W/\approx$) of the encoding space, a covering, or a universal covering ($\hat{\Omega}$) of the final latent manifold in order to achieve with the highest efficiency the most accurate model of the eventful reality the AI system purports to decode (Figure 1.4).

An intelligent autoencoder-based system essentially builds up intuition by computing the optimal progressive simplification of the encoding space. Optimal intuition is achieved by distilling the events into a series of progressively coarser models or latent manifolds that collectively offer the optimal possibility for decoding. Additionally, decoding at the optimal level is achieved by selecting a reality-decoding space that is either a quotient of the reality-encoding space or a universal covering of the final latent space. The coupling of progressively coarser simplifications has an obvious implementation advantage stemming from the fact that the optimization process, as carried out by parameter tuning in the underlying neural network (Chapter 3), is significantly improved if each autoencoder, say, the nth, builds upon the previous less intuitive ones (labeled by $n' < n$), which are subsumed as hidden layers of putative autoencoder with latent manifold Ω_n.

Most applications in Chapters 2–6 will deploy intelligent autoencoder-based AI systems.

1.3 IS QUANTUM MECHANICS TRULY INCOMPLETE? INTELLIGENT AUTOENCODERS REQUIRING UNRESOLVED QUANTUM AMBIGUITY

As previously discussed, there are scholars, including Roger Penrose, who believe that the removal of ambiguity from quantum physics cannot be in the hands of consciousness itself. So, says Penrose, quantum physics must be at least an incomplete theory because the removal of its own ambiguity cannot be performed by an alien entity, that is, the consciousness of intelligent beings. The wave function collapse, signaling the endpoint in the superposition of quantum states, is therefore not computable (in Penrose's eyes, a startling defect of the theory), while this lack of computability cannot be overcome by conscience but requires an improvement of the quantum theory itself, as previously discussed [2].

This trend of thought seems eminently refutable: Organic chemistry, that is, the chemistry of carbon compounds that serve as the substrate for life, is replete with "resonant" molecules, such as aromatic compounds and tautomeric forms, which have chemical structures that can only be accurately described by assuming the coexistence or superposition of two or more quantum states. That is to say that at the molecular level, life operates with a high degree of quantum ambiguity without compromising its functionalities, nor does it require any type of consciousness or corrections to the quantum theory to remove the superposition of quantum states. On the contrary, resonant molecules could not be understood or manipulated if we artificially remove their quantum ambiguity. Moreover, the superposition of quantum states is often a basic tenet in modeling efforts that hinge on the identification of the internal degrees of freedom of the system and the functional groups at play in a particular chemical context. Take, for example, the peptide bond in protein structure (Figure 1.5). The "keto–enol" tautomerization causes the peptide bond to be torsionally rigid, and this is because of the contribution of the double bond present in the enol form. Thus, the tautomeric superposition of states not only persists and has been corroborated experimentally by structural biologists, but it also makes the protein folding pathway problem computationally tractable via a modeling effort [5]. The AI system suitable for such an endeavor is an autoencoder with quantum state superposition in the modeling of the peptide bond.

Thus, in the forthcoming sections we shall be deploying an autoencoder that incorporates the torsional angles Φ, Ψ as the only backbone degrees of freedom in the coarse-grained version of the folding chain backbone enshrined in the latent manifold. This treatment is consistent with the structural ambiguity of the peptide bond resulting from quantum state superposition. We surely do not wish to induce a wave function collapse, be it through conscious or unconscious intervention, to fold a protein using quantum physics. The subsequent sections describe how the quantum ambiguity becomes essential to deal

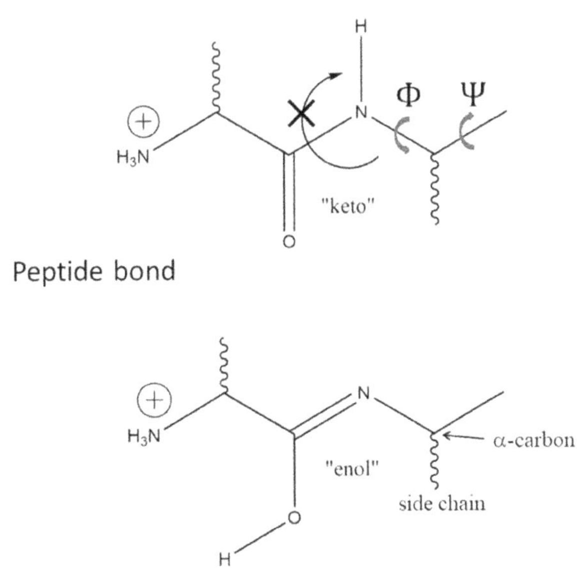

FIGURE 1.5 Torsional rigidity of the peptide bond in protein dynamics due to the superposition of quantum states representing the "keto–enol" tautomerization. The torsional rigidity is a basic tenet in modeling efforts that includes torsional angles Φ, Ψ as representing internal degrees of freedom of the backbone required to solve the protein folding pathway problem.

with the protein folding pathway problem. This is important, since arguably, protein folding is the most fundamental process in biology, as it describes the acquisition of function.

This suggests that when AGI is endowed with consciousness, it will be necessary to pay close attention to when consciousness ought to intervene to provide certainty (i.e., trigger the wave function collapse) and when it should not, understanding that sometimes quantum state ambiguity is an essential and indissoluble part of reality rather than an attribute of its representation.

1.3.1 Quantum State Superposition and the Modest Success of Structure-based Drug Design

A drug-targetable protein molecule, like any atomic assemblage, is a quantum object, and hence it exists in a superposition of states. The extension of the system to include an aqueous environment does not remove the quantum ambiguation; on the contrary, it often promotes it. For example, the enol state of the peptide bond requires a proton acceptor to deprive the amide group of its proton and a proton donor to protonate the carbonyl (Figure 1.5), and water is precisely the solvent able to fulfill the dual role. By the same token, the act of observation for structure determination *in vacuo or in vitro* alters the quantum nature of the object the drug is intending to target *in vivo*. Due to its therapeutic purpose, the drug is designed to target (bind) the protein *in vivo*, not *in vitro*, and definitely not *in vacuo*. As a simple illustration, there are many drugs that are computationally docked to their purported *in vacuo* targets with a proton-donor group approaching a backbone carbonyl. This would promote a favorable H-bond interaction, except that in reality, that is, *in vivo*, the carbonyl is also "enolic" and hence not a 100% proton acceptor but also a significant proton donor. Hence, the purposely designed ligand does not enhance target affinity anywhere near the levels suggested by the docking that adopted the *in vacuo* protein structure [5].

A leap of faith persuaded researchers that the target they observe is the same target that exists *in vivo*. This premise is clearly misleading. The "conscious" act of observation removes the quantum ambiguation inherent to the true target, that is, the protein *in vivo*. Conscience makes the drug target a moving target, significantly curtailing the impact of structure-based drug design [5]. Again, contrary to Penrose's belief that the wave function collapse is required to grasp and master reality, the whole endeavor of drug discovery, a pivot of the pharmaceutical industry, should be reformulated, taking into account the inherent quantum ambiguity of the protein targets.

This and the subsequent examples worked out in detail suggest that in the cases under scrutiny, there is nothing inherently missing in the QM formalism that could be construed as conscious operations or computable by means of a "gravitizing completion" of QM [4]. In these cases, quantum state ambiguity is a necessary and indissoluble part of reality rather than an attribute of its quantum representation.

1.3.2 Autoencoder Empowering of Molecular Dynamics without Quantum Disambiguation

Natural proteins fold and have been selected to fold *in vivo*, that is, in cellular environments, *not in vitro*. Likewise, therapeutic agents target dysfunctional or deregulated proteins *in vivo*, not in vacuo or in the test tube, where measurement (structure determination, screening, and affinity assays) is typically performed. In regard to modeling and predictive efforts, the wanton molecular complexity of *in vivo* environments places them out of reach for investigation with molecular dynamics (MD) or other computational methods. The advent of AI drastically changes this picture. Even in an *in vitro* setting, the generation of realistic protein folding pathways remains a challenge for MD and other computational systems. This is so because for most single-domain chains, relevant timescales are inaccessible and key rare events are typically missed. Furthermore, most computational efforts have focused on *in vitro* folding; hence, they are often misplaced as natural proteins have evolved to fold *in vivo*, where selective interference prevents the chain from getting kinetically trapped in misfolded states. In this section. we empower MD by subsuming the computations within a deep learning platform that constructs an *in vivo* context shown to enhance the expediency of the folding process. To this effect, we built a convolutional neural network (CNN) with an autoencoder architecture trained with short MD runs that incorporate *in vivo* molecular complexity [5]. The AI platform learns to propagate a coarse-grained folding trajectory that can be decoded as an extended MD run assessing the participation of the *in vivo* context to expedite the folding process.

For decades, MD has been adopted as a primary tool to investigate biomolecular processes *in vitro,* such as protein folding [6,7]. However, it is now recognized that such efforts may have been unrealistic and somewhat misplaced [8–11], as they may have dealt with the wrong context [11]. Proteins have evolved to fold in the cell, not *in vitro.* The fact that a few find their native structure *in vitro* is incidental, probably more the exception than the rule [11]. Through steering interference with the folding process, the *in vivo* setting prevents the protein from getting kinetically trapped in misfolded aggregation-prone states. On its own, MD is felt to be ill-equipped to handle the complexities and relevant catalytic turnover timescales of an *in vivo* environment. Even in an *in vitro* context, the inadequacy of MD is usually apparent as physically relevant timescales are often inaccessible, conformation space is incompletely sampled, and rare events are typically missed [12–15].

The huge informational burden that needs to be carried over from one MD step to the next reduces the chances of MD providing a meaningful model that can handle the complexities of *in vivo* settings. To address this problem, an autoencoder needs to be deployed. Due to the wanton complexities of the problem, an AI-empowered MD platform requires an intelligent autoencoder capable of encoding the (Φ, Ψ)-protein dynamics into a simplified representation retaining only essential topological features of the vector field that steers the MD integration process [12,13].

The intelligent autoencoder simplifies the flow by lumping backbone conformations within basins of attraction in potential energy and encoding the dynamics into a modulo-basin representation that averages out fast motion (Figure 1.6). The energetically allowed regions in (Φ, Ψ)-space for each linkage of the chain are known as Ramachandran basins [5]. Interbasin transitions are determined by a dilated CNN [16,17] where the influence field on a basin transition at a residue site along the chain becomes progressively dilated in hidden network layers. Thus, the weighted influence of different contexts with different levels of locality upon each basin transition is optimized according to a loss function that contrasts network inferences against experimentally determined conformational changes. With a suitable propagation scheme obtained by training the network with big dynamic data from MD runs, this simplified AI scheme enables coverage of realistic timescales even for the *in vivo* context, while atomistic-level detail is recovered through AI-empowered decoding.

To enable AI to construct an *in vivo* setting and assess its role in expediting the folding process, we incorporate the influence of specific structurally reported cellular machinery on the protein folding process. This is accomplished by training the autoencoder with multiple atomistic MD simulations of folding events within specific cellular molecular environments. None of the MD runs is able to cover folding timescales, but together they inform the autoencoder on the steering influence of the *in vivo* reality.

For the sake of illustration, we adopt a training set consisting of MD runs that incorporate the influence of GroEL, a widely studied tetrameric chaperonin [8], on protein folding [9]. Inside the GroEL cavity, our atomistic MD runs showed that noncompact states with a large radius of gyration (scaling with an unfolded ensemble average) interact strongly with structurally disordered residues in the GroEL C-terminus tails. These interactions disrupt the metastable states, thereby enhancing folding expediency. Thus, the influence of the *in vivo* context is incorporated through the training of the autoencoder in a selected environment of molecular complexity where only sub-microsecond events are accessible to MD computation. The CNN generated a dynamics in which coarse-grained misfolded states are levied interbasin transitions that eventually get them dismantled, thereby preventing the protein from getting kinetically trapped. Thus, AI constructs a reality that expedites folding in a manner akin to chaperone catalysis *in vivo.*

In this way, this section describes and validates a first attempt at extending and empowering MD by exploiting the versatility of AI to absorb big data on folding events in the cell. Thus, the AI platform incorporates the feature extraction to construct an *in vivo* reality that expedites the folding process.

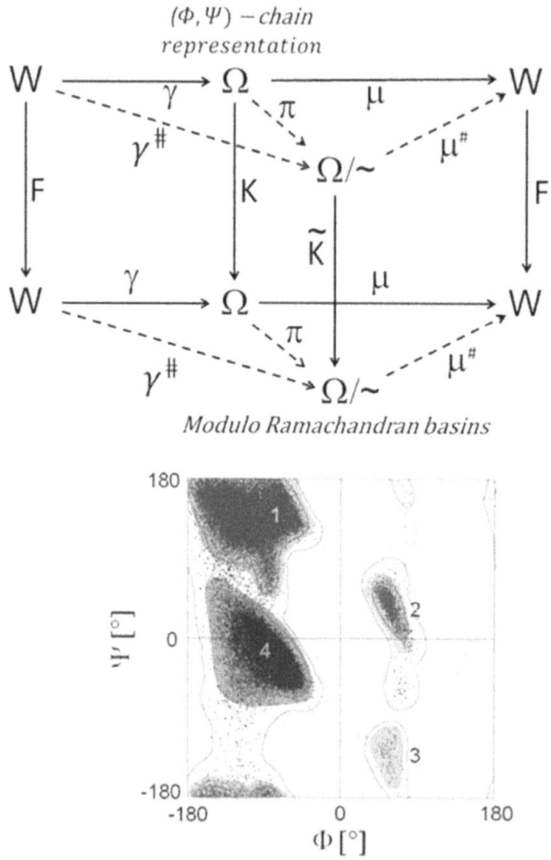

	1	2	3	4	5	6	7	8	9	10
Aminoacid	MET-M	THR-T	THR-T	PHE-F	LYS-K	LEU-L	ILE-I	ILE-I	ASN-N	GLY-G
R-basin										
Phi-angle	-60.48	-53.47	-111.35	-146.29	-77.71	-142.19	-60.03	-136.53	-146.28	-83.41
Psi-angle	-50.21	-46.87	114.70	166.19	134.23	132.85	117.82	139.68	137.50	0.57
Omega-angle	180.00	180.00	180.00	180.00	180.00	180.00	180.00	180.00	180.00	180.00
	11	12	13	14	15	16	17	18	19	20
Aminoacid	LYS-K	THR-T	LEU-L	LYS-K	GLY-G	GLU-E	ILE-I	THR-T	ILE-I	GLU-E
R-basin										
Phi-angle	-77.16	-130.78	-61.56	-110.69	-76.55	-128.50	-141.75	-129.74	-131.06	-57.29
Psi-angle	2.03	132.05	-25.00	104.99	86.44	106.57	175.45	137.67	129.69	-32.23
Omega-angle	180.00	180.00	180.00	180.00	180.00	180.00	180.00	180.00	180.00	180.00
	21	22	23	24	25	26	27	28	29	30
Aminoacid	ALA-A	VAL-V	ASP-D	ALA-A	ALA-A	GLU-E	ALA-A	GLU-E	LYS-K	ILE-I
R-basin										
Phi-angle	-63.64	-60.87	-62.93	-144.65	-61.10	-59.75	-61.47	-61.16	-54.21	-61.01
Psi-angle	-32.49	-47.50	-33.90	136.37	-43.25	-27.33	-50.01	-46.13	-43.25	-46.07
Omega-angle	180.00	180.00	180.00	180.00	180.00	180.00	180.00	180.00	180.00	180.00
	31	32	33	34	35	36	37	38	39	40
Aminoacid	PHE-F	LYS-K	GLN-Q	TYR-Y	ALA-A	ASN-N	ASP-D	ASN-N	GLY-G	ILE-I
R-basin										
Phi-angle	-63.08	-54.65	-63.38	-58.81	-118.93	-67.19	-88.71	-78.66	80.98	-61.82
Psi-angle	-45.96	-44.88	-49.83	-27.40	114.72	-53.04	0.80	138.34	-94.77	-33.78
Omega-angle	180.00	180.00	180.00	180.00	180.00	180.00	180.00	180.00	180.00	180.00
	41	42	43	44	45	46	47	48	49	50
Aminoacid	ASP-D	GLY-G	GLU-E	TRP-W	THR-T	TYR-Y	ASP-D	ASP-D	ALA-A	THR-T
R-basin										
Phi-angle	-65.78	84.02	-66.69	-133.65	-64.78	-145.61	-76.24	-132.24	-68.87	-127.87
Psi-angle	-48.66	-3.46	-52.64	174.31	124.11	171.01	137.33	143.91	-54.06	114.38
Omega-angle	180.00	180.00	180.00	180.00	180.00	180.00	180.00	180.00	180.00	180.00
	51	52	53	54	55	56	57			
Aminoacid	LYS-K	THR-T	PHE-F	THR-T	VAL-V	THR-T	GLU-E			
R-basin										
Phi-angle	62.05	-81.78	-73.09	-129.16	-141.35	-62.94	59.51			
Psi-angle	58.19	131.81	136.63	137.30	132.79	120.79	55.61			
Omega-angle	180.00	180.00	180.00	180.00	180.00	180.00	180.00			

FIGURE 1.6 Autoencoder with quantum state ambiguity incorporating as quotient latent space $\Omega/\sim$ the topological simplification in the backbone Φ, Ψ –conformation of a single subunit within the protein chain. The simplification is enabled by the torsional rigidity of the peptide bond linking adjacent residues (subunits) due to quantum state ambiguity. The right panel provides the topologically simplified (coarse-grained) conformation of a protein chain of length $N = 57$ whose native structure is provided at the bottom in ribbon representation.

Note: This figure is reproduced in color in online supplementary materials.

1.3.3 Protein Folding In Vivo Operates with Quantum Ambiguity in the State of the Peptide Bond

To reduce the informational burden in MD integration steps, the intelligent autoencoder simplifies the dynamics, taking into account the inherent topological architecture of the MD-steering vector field. Thus, it introduces an equivalence relation "~", whereby two microstates/conformations are regarded as equivalent if and only if the respective local conformations of the constitutive subunits lie in the same basins of attraction in the local potential energy surface. The "modulo-basin" representation constitutes a "quotient space" made up of equivalence classes. In generic terms, if x indicates a microstate of a system of N subunits, the modulo-basin class (state) to which x belongs is denoted $\overline{x}$ and may be expressed as a Cartesian product of N basins designating the basin occupancies for the individual subunits: $\overline{x} = \prod_{n=1}^{N} b(x,n)$ where $b(x,n)$ is the basin of attraction occupied by subunit n when the system is in microstate x. Thus, for any two microstates $x, y : x \sim y \Leftrightarrow \bar{x} = \bar{y}$. Since the products of basins $\mathcal{B} = \prod_{n=1}^{N} b(n)$ constitute a denumerable set, the classes/states in the quotient space may be conveniently designated by encoding vectors.

Assuming the torsional rigidity of the peptide bond, in accord with the quantum "keto–enol" state superposition (quantum ambiguity) (Figure 1.5), protein folding trajectories are defined by the evolution of backbone torsional coordinates $\{\phi_n, \psi_n\}_{n=1,\ldots,N}$. Conformation space thus becomes a 2N-dimensional torus, that is, the product of 2N unit circles, one of each dihedral coordinate of the protein backbone [12,13]. In this context, the modulo-basin quotient space is made up of Cartesian products of the so-called Ramachandran basins, that is, the allowed low-energy regions in the potential energy surface of each residue along the protein chain of length N (Figure 1.7a, [18,19]). Within this mathematical construct, the autoencoder projects the MD trajectory on the quotient space and then adopts deep learning to propagate the encoded "shorthand" dynamics to make computations more agile. The shorthand dynamics

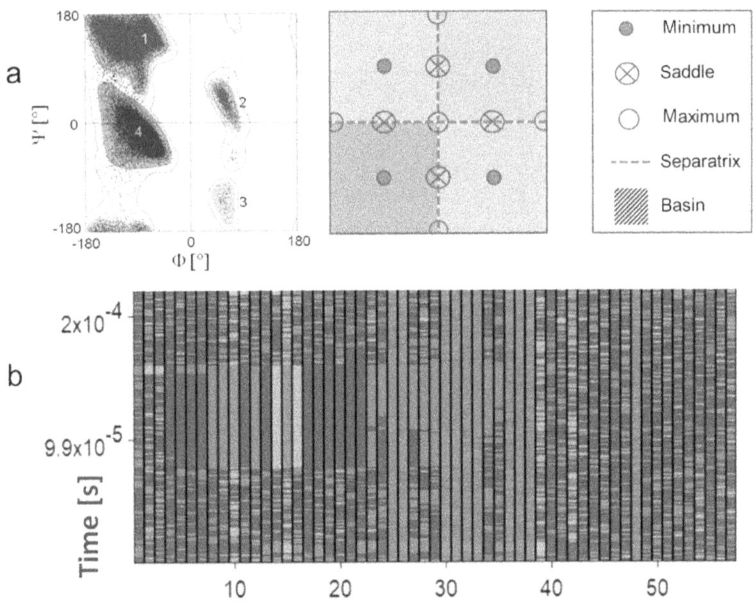

FIGURE 1.7 Modulo-basin simplification of the vector field integrated in MD computations generating the backbone torsional dynamics of a protein chain. (a) Simplified topological representation of the vector field steering the backbone torsional dynamics of a single residue within a protein chain. Opposite sides of the square are identified as per the ±180° identification of ϕ, ψ dihedrals, yielding a 2-torus, or Cartesian product of two circles. The four sectors morph topologically into the Ramachandran basins or allowed regions of potential energy (left panel). Except for glycine (G), sector 3 is energetically inaccessible to other residues due to steric clash with the side chain. Other accessibility restrictions apply to geometrically constrained proline (P) and adjacent residues. The basin topology with its distribution of critical points (minima, saddles, maxima) is compatible with the underlying 2-torus. The panel on the right describes the organization of the basins of attraction of the critical points (minima and saddles). The basins of attraction of saddles are 1-dimensional separatrices of the 2-dimensional basins of minima. (b) Coarse grained representation of an MD trajectory at 1μs-resolution for the N = 57 chain of the B1 domain of protein G. The trajectory is subsequently extended in time using deep learning. At each time (vertical axis), the modulo-basin state of the chain is shown as a shade sequence, where the basin assigned to each residue on the horizontal axis is specified following the convention adopted in panel a. A 220 μs-folding trajectory was generated within the NPT (isothermal/isobaric, T = 298 K) ensemble, selecting the timespan interval [0, t_0 = 120 μs] as training region and the interval [120μs, 220μs] as the learning period to optimize the parametrization of the propagator Γ_τ.

Note: This figure is reproduced in color in online supplementary materials.

carry far lower information burden as conformations are lumped into classes that entail a considerable simplification of conformation space, while the transition timescale for each iteration is the minimal Ramachandran intrabasin equilibration time $\tau = 100$ ps, considerably larger than the femtosecond integration timescales used in MD. Hence, since intrabasin events are lumped together within the equivalence classes and the iteration timescale is five orders of magnitude longer, the propagated modulo-basin dynamics has a far better chance of capturing the infrequent or rare interbasin transitions, a highly sought after goal and a significant shortcoming in MD modeling of *in vitro* processes. The deep learning process to propagate the modulo-basin dynamics uses a maximum likelihood ansatz to train a dilated CNN so that the maximum likelihood weight, $f_\tau(n,b,b')$, for basin transition $b \to b'$ at chain position n after time τ elapsed is influenced by the amino acid identity and basin occupancy at flanking residues ($n-2, n-1, n+1, n+2$), with an influence field progressively dilated, so that flanking positions ($n-3, n-2, n+2, n+3$), ($n-4, n-3, n+3, n+4$), …, ($n-8, n-7, n+7, n+8$) with respective ranges $g = 1, 2, \ldots, 7$ also impact the transition probability. Computational cost determines that we limit the extent of dilation to a maximum of $g = 7$. Thus, the maximum likelihood weight of the basin transition becomes:

$$f_\tau(n,b,b') = \prod_{g=1}^{7} w\big(n,g,A(n,g),B(n,g)\big)\, f\big(n,g,A(n,g),\, B(n,g),\, b,\, b'\big) \tag{1.9}$$

where $f\big(n,g,A(n,g),B(n,g),\, b,\, b'\big)$ gives the frequency of the $b \to b'$ basin transition at position n after time τ elapsed obtained by projecting one hundred 220 ns-MD runs onto a modulo-basin τ-discretized time series for chain composition $A(n,g)$ giving the amino acid identity of residues $n-g-1, n-g, n, n+g, n+g+1$ in the chain, and basin assignment $B(n,g)$ for residues $n-g-1, n-g, n, n+g, n+g+1$. The different g-influencing fields ($n-g-1, n-g, n, n+g, n+g+1$) bearing on the basin transition at position n are weighted in accord with parameters $w\big(n,g,A(n,g),B(n,g)\big)$, which are optimized vis-à-vis the loss function of the CNN.

A first stage in the training of the network yields the transition frequencies $f\big(n,g,A(n,g),B(n,g),b,b'\big)$, thus following the maximum likelihood scheme. The propagation of the modulo-basin trajectory (time series) in the maximum likelihood scheme is determined by the $b \to b'$ transition probabilities: $p_{\mathrm{ML}}(n,b,b') = \dfrac{f_\tau(n,b,b')}{\sum_{b''} f_\tau(n,b,b'')}$, where b'' is any of the four basins. The destiny basin b' after a time period τ is determined by a Monte Carlo scheme taking into account the four transition probabilities at each position n along the chain.

The input in the dilated CNN is thus filtered through $7 \times 4^5 = 7168$ hidden layers in accord with range g and basin assignment $B(n,g)$ for the g-quintuples centered at each position n on the chain and influencing the basin transition at position n. The CNN input is a basin assignment for each residue along the chain, and its output is the basin assignment after time τ has elapsed. In turn, each layer contains $20^3 = 8000$ filters or convolution kernels in accord with amino acid identity $A(n,g)$. To convolve, a modulo-basin state for a chain of length N may be labeled with a $4N$-vector $(\mathbf{b}_1|\mathbf{b}_2|\ldots|\mathbf{b}_n\ldots)$ consisting of N binary four-tuples $\mathbf{b}_n$ ($n = 1, 2, \ldots, N$) indicating the Ramachandran basin occupancy for each residue, so the value 1 at entry m in $\mathbf{b}_n$ ($b_{nm} = 1$) indicates that residue n occupies basin m ($m = 1, 2, 3, 4$) (Figure 1.7a). For example, (1000|0010) indicates the coarse-grained state of a dipeptide with first residue in basin 1 and second residue in basin 3. Thus, a convolutional kernel $K(n, g, B_K(n,g))$ becomes a g-quintuple of binary four-tuples (basin assignment $B_K(n,g)$) with range g, with the convolution operation carrying the Kronecker delta factor $\delta_{A(n,g),A_K(n,g)}$, where $A_K(n,g)$ is the kernel amino acid assignment for positions $n-g-1, n-g, n, n+g, n+g+1$. Each layer $L(q, g)$ is labeled by the basin quintuple, in turn labeled by index $q = 1, \ldots, 4^5 = 1024$, and range g. In each layer, the $20^3 = 8000$ convolution kernels are slid along the chain as stencils, and full coincidence at position n in $L(q, g)$ yields concatenation (enrichment) at position n in the (q, g)-feature map with the four attributes $f\big(n,g,A(n,g),B(n,g),\, b,\, b'\big)$ for the four *a priori* possible b' destiny basins. After processing across all 7168 hidden layers, we arrive at a full representation of the chain at layer $L(1024,7)$, where each position n is endowed with $7 \times 4 = 28$ transition parameters

$f\left(n,g,A(n,g),B(n,g),b,b'\right)$, one for each influence range g and destiny basin b'. In the 7169th layer, the products $\prod_{g=1}^{7} w\left(n,g,A(n,g),B(n,g)\right) f\left(n,g,A(n,g),B(n,g),b,b'\right)$ are finally computed, and the four transition probabilities $p_{\mathrm{ML}}(n,b,b')$ are obtained for each of the four *a priori* destinies b'. Finally, in the output (7170th) layer, the destiny basin for each residue is chosen from the Monte Carlo scheme.

The optimization of the influence-defining factors $w\left(n,g,A(n,g),B(n,g)\right)$ is carried out during the second training phase for the modulo-basin propagator Γ_τ defined by the Monte Carlo generator of basin transitions. To train the network in this second stage, we break up the MD trajectories spanning a time period $[0, t_f]$ into two regions: a training portion with timespan $[0, t_0]$ and an optimization portion covering the time interval $(t_0, t_f]$. With the trajectory statistics drawn for the region $[0, t_0]$, we obtain the basin-transition frequencies for triads with different strides, centered at each particular residue. On the other hand, a variational optimization process during the period $(t_0, t_f]$ enables us to determine the w-parameters. Thus, the w's are obtained from a commutativity condition that reflects the compatibility of fine- and coarse-grained propagation of the trajectory and must be fulfilled during the period $(t_0, t_f]$. The commutativity condition is $\Gamma_\tau \circ \pi = \pi \circ M(\tau)$, as depicted by the following scheme:

$$\begin{array}{lcl} x(t) & \xrightarrow{\pi} & \overline{x(t)} \\ \downarrow M(\tau) & & \downarrow \Gamma_\tau \\ x(t+\tau) & \xrightarrow{\pi} & \overline{x(t+\tau)} \end{array} \tag{1.10}$$

Here π is the canonical projection that assigns the Ramachandran basins to each backbone torsional state of the chain, and $M(\tau)$ is the MD propagator operating on torsional states of the chain. In other words, the parametrization of the modulo-basin propagator Γ_τ is optimized to minimize the loss function $J(\Gamma_\tau)$ given by

$$J(\Gamma_\tau) = M^{-1} \sum_{q=1}^{M} \left\| \pi x(t_0 + q\tau) - [\Gamma_\tau]^q \pi x(t_0) \right\|^2 \tag{1.11}$$

where $M\tau \le t_f - t_0 < (M+1)\tau$. We know this loss function is the correct one since $J(\Gamma_\tau) = 0$ if and only if the propagator makes the diagram commutative. Once the optimal $\Gamma_\tau^* = \operatorname{argmin} J(\Gamma_\tau)$ has been obtained from stochastic gradient descent, the modulo-basin projected trajectory can be propagated beyond MD-accessible timescales. The set $\wp(t_0, t_f)$ of MD-generated states during the period $t_0 \le t \le t_f$ is sampled randomly at each iteration to compute the gradient with minibatches given by trajectory fragments spanned by randomly chosen time intervals of size $10^{-2}|t_f - t_0|$. To optimize the network connectivity that determines Γ_τ, the ADADELTA optimization protocol with a learning rate of 0.3 is adopted [20]. Physically meaningful timescales may be reached by computing the coarse states $\overline{x(t_f + q\tau)}$ beyond the simulation timespan t_f as $[\Gamma_\tau^*]^q \overline{x(t_f)}$.

To identify intermediate structures along an autoencoder-generated folding pathway, the coarse-grained modulo-basin trajectory needs to be decoded back to a backbone (φ,ψ)-state. This reconstruction process also requires deep learning since we need to optimize a one-to-one application λ: modulo − basin state $\rightarrow (\varphi,\psi)$ − state that makes the following diagram commutative in the time range $t_0 \le t \le t_f - \tau$:

$$\begin{array}{lcl} x(t) & \xrightarrow{\pi} & \overline{x(t)} \\ \downarrow MD & & \downarrow \Gamma_\tau^* \\ x(t+\tau) & \xleftarrow[\lambda]{} & \overline{x(t+\tau)} \end{array} \tag{1.12}$$

This implies that the decoder λ must be optimized to minimize the loss function:

$$J(\lambda) = M^{-1} \sum_{n=1}^{M} \left\| x(t_0 + n\tau) - \lambda[\Gamma_\tau^*]^n \pi x(t_0) \right\|^2 \tag{1.13}$$

The choice of loss function is correct since $J(\lambda) = 0$ if and only if the decoder makes the diagram commute. With the choice of decoder as argmin $J(\lambda)$, identified with stochastic gradient descent, we expand the MD simulation as $\lambda[\Gamma_\tau^*]^n \pi x(t_0)$ for any $n \geq M$. In cases when in vivo reality has been incorporated, the evolution of the (φ, ψ)-state may be monitored as $\lambda\left[\widehat{\Gamma_\tau}\right]^n \pi x(t_0)$.

1.3.4 The Intelligent Autoencoder Learns Vivo Reality to Expedite Protein Folding

The goal is now to incorporate, via the learning capability of the autoencoder, the role of *in vivo* settings in steering the protein chain along an expeditious folding pathway. Given computational limitations, atomistic detail on a folding trajectory in an *in vivo* context, say, in the interior of a chaperone, is hardly feasible. For example, for chaperonin GroEL, such computations are currently only accessible in hundreds-of-microseconds timescales for short substrates $N \sim 25$ with the fastest dedicated supercomputers [9]. For a typical single-domain protein ($N > 50$), such computations are not likely to yield the mechanism by which the cellular context is able to remove kinetic traps and expedite folding. To generate coarse-grained *in vivo* folding trajectories, we implement a deep learning system that captures the expediency of the *in vivo* environment when trained with accessibly short (30 ns) MD simulations starting at unfolded or metastable states generated *in vitro*. We need to capture the means by which the *in vivo* setting removes kinetic traps by selectively disrupting misfolded states that would be otherwise susceptible to aggregation due to high protein concentration in the cell.

Let $\widetilde{\Gamma_\tau}$ denote the modulo-basin propagator that commutes with $\widetilde{M(\tau)}$, the MD operator that steers the chain torsional dynamics in the *in vivo* setting (in our illustration, the GroEL cavity). The propagator $\widetilde{\Gamma_\tau}$ is obtained by training the CNN with many short *in vivo* runs, and its domain $\mathcal{D}(\widetilde{\Gamma_\tau})$ consists of basin assignments along the chain that undergo basin transitions (including retentions) when the chain explores conformations in the *in vivo* setting. The domain for Γ_τ^* is similarly defined *mutatis mutandis* in association with the *in vitro* setting.

To generate *in vivo* folding trajectories that span realistic timescales, a significant coverage is required of the influencing basin occupancies at positions flanking each residue along the chain placed in the *in vivo* environment. To that effect, we construct the propagator $\widehat{\Gamma_\tau}$ obtained by overwriting basin transitions yielded by Γ_τ^* in the domain $\mathcal{D}(\Gamma_\tau^*) \cap \mathcal{D}(\widetilde{\Gamma_\tau})$, replacing them by the transitions dictated by $\widetilde{\Gamma_\tau}$. Thus, we get $\widehat{\Gamma_\tau} = \chi_{\mathcal{D}(\widetilde{\Gamma_\tau})} \widetilde{\Gamma_\tau} \oplus \chi_{\mathcal{D}(\Gamma_\tau^*) \setminus [\mathcal{D}(\Gamma_\tau^*) \cap \mathcal{D}(\widetilde{\Gamma_\tau})]} \Gamma_\tau^*$, where χ denotes characteristic function (1 on its support, 0 elsewhere), and $\mathcal{D}(\Gamma_\tau^*) \setminus \left[\mathcal{D}(\Gamma_\tau^*) \cap \mathcal{D}(\widetilde{\Gamma_\tau})\right]$ denotes the complement of $\mathcal{D}(\Gamma_\tau^*) \cap \mathcal{D}(\widetilde{\Gamma_\tau})$ in $\mathcal{D}(\Gamma_\tau^*)$. Thus, the *in vivo* propagator $\widehat{\Gamma_\tau}$ commutes with $\widetilde{M(\tau)}$ in $\pi^{-1}\left[\mathcal{D}(\widetilde{\Gamma_\tau})\right]$ and with $M(\tau)$ in $\pi^{-1}\left\{\mathcal{D}(\Gamma_\tau^*) \setminus \left[\mathcal{D}(\Gamma_\tau^*) \cap \mathcal{D}(\widetilde{\Gamma_\tau})\right]\right\}$, implying that $\widetilde{\Gamma_\tau}$ overwrites Γ_τ^* in $\mathcal{D}(\Gamma_\tau^*) \cap \mathcal{D}(\widetilde{\Gamma_\tau})$.

1.3.5 Generating in vitro Folding Pathways with an Intelligent Autoencoder

The efficacy of the dynamic learning scheme is illustrated by computing folding pathways converging to native structures. We selected an $N = 57$ chain known to fold autonomously in an *in vitro* setting: the thermophilic variant of the B1 domain of protein G from *Streptococcus* (PDB.1GB4). The thermophile was chosen over the wild type due to the higher stability of the folded structure. Using the charmm package (free version of CHARMM) [7], we first generated a 220 μs-folding trajectory within the NPT (isothermal/isobaric, $T = 298$ K) ensemble. The full MD trajectory encoded as the modulo-basin version is shown in Figure 1.7b, with selected $t_0 = 120$ μs as the training parameter and $120\,\mu s < t \leq 220\,\mu s$ as the learning period to optimize the matrix propagator Γ_τ. The coarse-grained trajectory is then propagated up to $t = 7500$ μs (Figure 1.8a–c). We note that the final state is significantly stable, prevailing since the time of its inception at $t = 3108$ μs, detectable in Figure 1.8b. Furthermore, the decoded final stable state at $t = 7500$ μs (Figure 1.9), corresponding to the eigenvector associated with eigenvalue 1 of the matrix Γ^*_τ, is topologically equivalent to the native crystallographic state (PDB.1GB4), revealing the same pattern of antiparallel and long-range parallel β-sheets with α-helix packed against the β-sheet motif. The backbone-atom Root Mean Square Displacement (RMSD) is estimated at 1.91 Å.

In the *in vitro* folding context, at least two metastable states can be spotted to be distinctively generated at 0.7 and 2.7 ms (Figure 1.8), corresponding to the chain conformations displayed in Figure 1.9b and c, respectively.

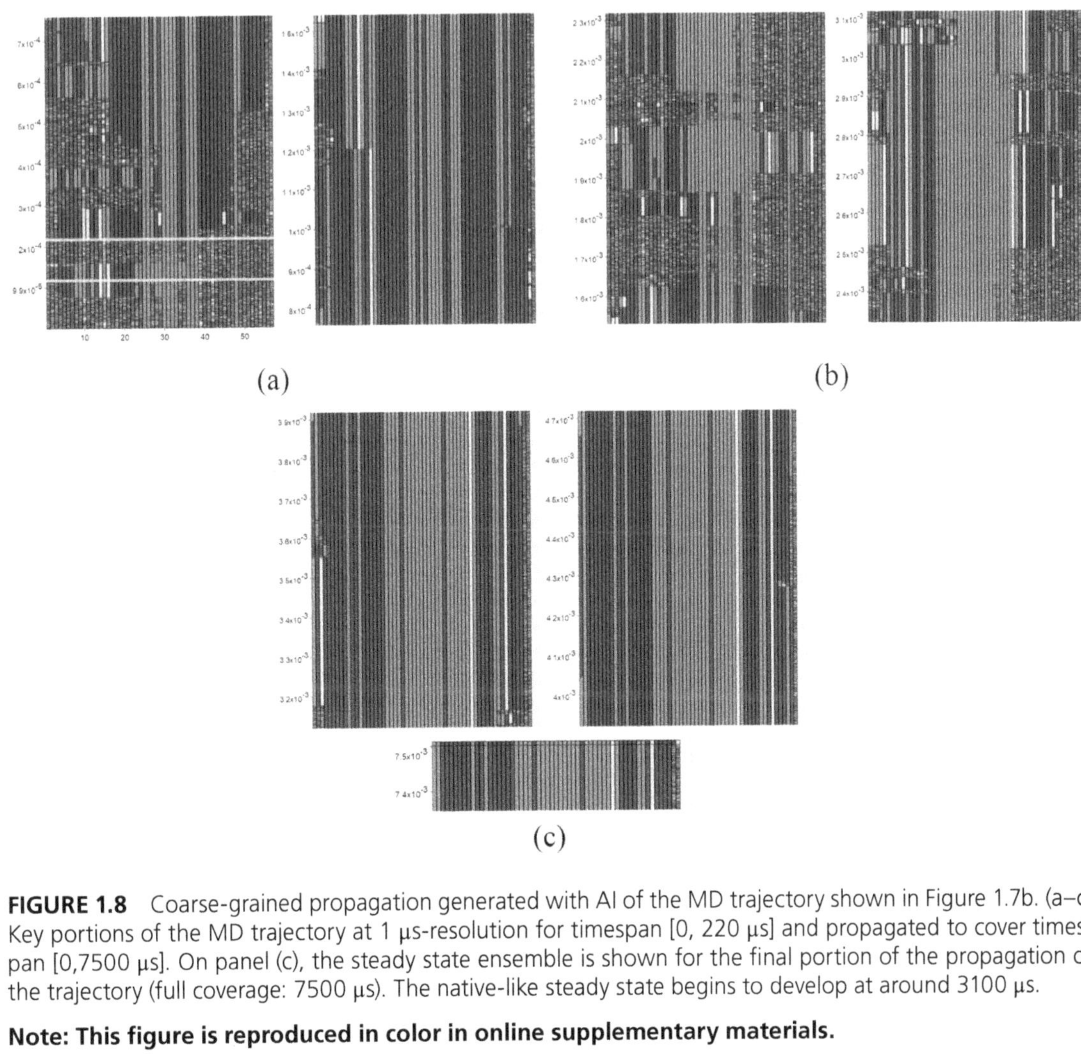

FIGURE 1.8 Coarse-grained propagation generated with AI of the MD trajectory shown in Figure 1.7b. (a–c) Key portions of the MD trajectory at 1 μs-resolution for timespan [0, 220 μs] and propagated to cover timespan [0,7500 μs]. On panel (c), the steady state ensemble is shown for the final portion of the propagation of the trajectory (full coverage: 7500 μs). The native-like steady state begins to develop at around 3100 μs.

Note: This figure is reproduced in color in online supplementary materials.

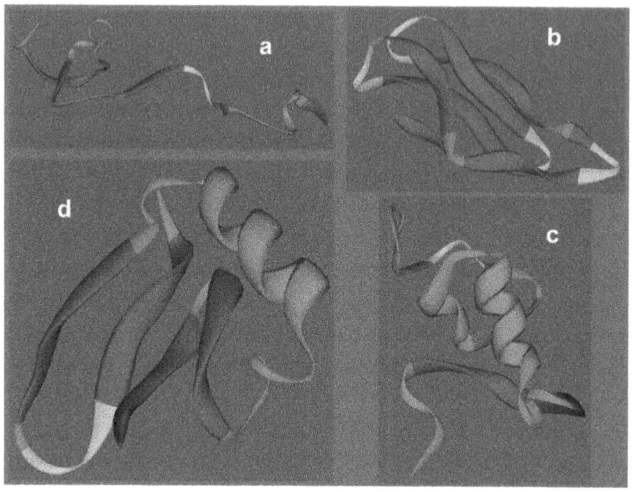

FIGURE 1.9 Decoded microstates in ribbon representation for the AI-enabled propagation of the MD trajectory, generated at start (a), 0.7 ms (b), 2.7 ms (c), and 7.5 ms (d). The modulo-basin representation of the last microstate (d) is the eigenvector associated with eigenvalue 1 of matrix Γ_τ^* and is topologically equivalent to the native state from PDB entry 1GB4 obtained by X-ray diffraction crystallography.

Note: This figure is reproduced in color in online supplementary materials.

1.3.6 The Intelligent Autoencoder Incorporates In Vivo Reality to Expedite Folding Pathways

To illustrate the generation of *in vivo* folding pathways, the intelligent autoencoder platform constructed an *in vivo* reality by learning to generate the propagator $\widehat{\Gamma_\tau}$ from 100 atomistic 30 ns-MD runs of the $N = 57$ protein chain exploring conformation space within the GroEL cavity [8]. To generate atomistic-level short MD trajectories in the selected *in vivo* context, we follow standard protocol: The conformation identified with the X-ray diffraction structure of GroEL-ATP14 (PDB.1SX3) is adopted to model the GroEL cavity assumed to expedite the folding of the B1 domain of G protein. Excess Mg^{++}-complexed ATP molecules must be removed, while the 22 inherently disordered C-terminal amino acids, mainly comprised of Gly-Gly-Met repeats, are incorporated with Maestro [21]. One protein chain is placed at the center of the GroEL cavity, and the system is solvated in a 200 Å × 200 Å × 200 Å water box at 120 mM KCl concentration. The final system is composed of ~1,178,000 particles. Simulations of 100 ns are performed using the charmm package [7]. One hundred 100 ns simulations are performed with the protein in the GroEL-ATP7 cavity. Ten runs are started in an unfolded state, ten in the protein native state (PDB.1GB4), and 90 runs in metastable states. The system is subject to energy minimization and equilibrated at 300 K and 1 atm with harmonic restraints on the alpha carbons. The runs are performed in the NPT ensemble with a Nosé–Hoover thermostat [22,23] and a MTK barostat [24], with the mass of all hydrogen atoms set at 4 amu, while the time step gets fixed at 4.0 fs.

To generate *in vivo* folding trajectories, the trained CNN generates the coarse-grained propagator $\widehat{\Gamma_\tau}$ that removes kinetic traps by selectively disrupting misfolded metastable structure that would be otherwise susceptible to aggregation due to the high concentration in the cell relative to an *in vitro* environment. The CNN choreographs folding disruptions at states where $\widehat{\Gamma_\tau}$ overwrites Γ_τ^* when the application of the latter generates metastable states characterized by the persistence of the chain in a specific basin assignment.

The autoencoder operationally implements the *in vivo* intervention step. Assume a substantial portion (typically 80% or more) of the basin assignment for the chain is detected to be locked at time $t_f + q_s\tau$, then the potentially disruptive propagator $\widetilde{\Gamma_\tau}$ overwrites Γ_τ^* for the step $t_f + q_s\tau \rightarrow t_f + (q_s + 1)\tau$. Thus, the coarse state at $t_f + (q_s + 1)\tau$ is computed as $\widetilde{\Gamma_\tau}[\Gamma_\tau^*]^{q_s} x(t_f)$. The disruptive propagator excludes the basin retention $b' = b$ frequency, $f_\tau(n, b, b)$, from the Monte Carlo scheme that assigns at time $t_f + (q_s + 1)\tau$ the destiny basin b' at the basin-locked chain position n. For residues that did not retain their basin assignment at time $t_f + q_s\tau$, the propagation $b \rightarrow b'$ is constructed as in Γ_τ^*. We note that to generate the *in vivo* reality, full network training with 30 ns atomistic MD runs and parameter optimization took approximately 18 hours on a cluster of 12 Titan RTX GPUs, while the full propagation computation required an additional 12 hours on the same resource.

The autoencoder-generated folding trajectory is shown in Figure 1.10a, while Figure 1.10b shows the influence of the *in vivo* context depicted as selective interferences with basin transitions. Four types of basin assignments are distinguished 1) implicating the *in vitro* model valid in $\mathcal{D}(\Gamma_\tau^*) \setminus [\mathcal{D}(\Gamma_\tau^*) \cap \mathcal{D}(\widetilde{\Gamma_\tau})]$, 2) involving *in vivo* intervention in $\mathcal{D}(\widetilde{\Gamma_\tau}) \setminus [\mathcal{D}(\Gamma_\tau^*) \cap \mathcal{D}(\widetilde{\Gamma_\tau})]$, 3) involving *in vivo* overwriting of *in vitro* assignments in $\mathcal{D}(\Gamma_\tau^*) \cap \mathcal{D}(\widetilde{\Gamma_\tau})$ and 4) coincidence between *in vitro* and *in vivo* assignments in $\mathcal{D}(\Gamma_\tau^*) \cap \mathcal{D}(\widetilde{\Gamma_\tau})$. The starting conformation is shown in ribbon rendering in Figure 1.10c. The convergence to a native fold (Figure 1.10d), with RMSD = 2.1 Å relative to PDB.1GB4, occurs 750 times faster than in the *in vitro* setting (cf. Figure 1.8), attesting to the staggering efficiency of the *in vivo* context to extricate the chain from kinetic traps.

The efficacy of the autoencoder is thus apparent in the computation of a highly expeditious in vivo protein folding pathway. Its intelligence is reflected in the simplification of the encoding process (Figure 1.6), which is enabled by the ambiguity in the specification of the quantum state for the protein backbone.

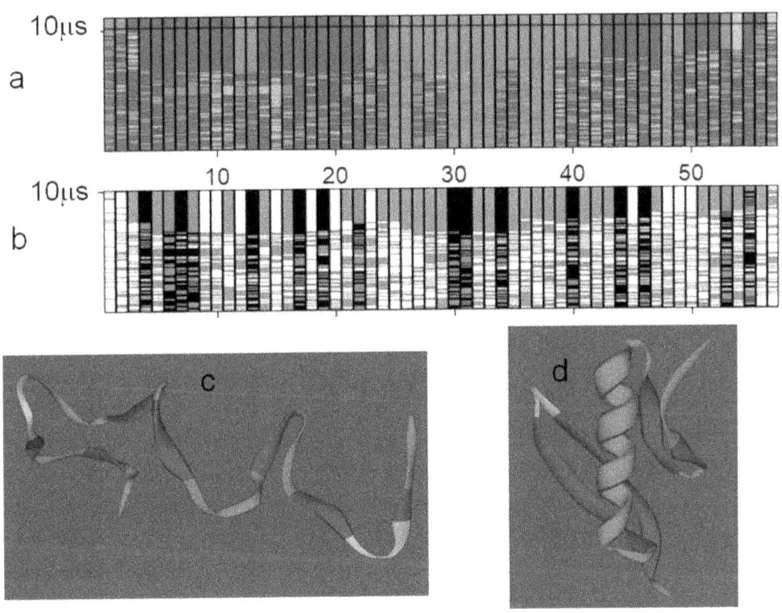

FIGURE 1.10 Ultra-expeditious *in vivo* folding trajectory generated by AI with a CNN trained with 100 atomistic 30ns-MD runs of the chain dynamics within a GroEL environment. (a) Coarse grained trajectory at 0.1μs resolution. (b) Influence of *in vivo* context. The basin assignments are depicted in white when implicating the in vitro model valid in $\mathcal{D}(\Gamma_\tau^*) \setminus [\mathcal{D}(\Gamma_\tau^*) \cap \mathcal{D}(\tilde{\Gamma}_\tau)]$, in light grey when involving *in vivo* intervention in $\mathcal{D}(\tilde{\Gamma}_\tau) \setminus [\mathcal{D}(\Gamma_\tau^*) \cap \mathcal{D}(\tilde{\Gamma}_\tau)]$, in black for *in vivo* overwriting of *in vivo* transitions in $\mathcal{D}(\Gamma_\tau^*) \cap \mathcal{D}(\tilde{\Gamma}_\tau)$ and in grey when there is coincidence between *in vivo* and *in vivo* assignments in $\mathcal{D}(\Gamma_\tau^*) \cap \mathcal{D}(\tilde{\Gamma}_\tau)$. (c) Initial state. (d) Decoded final state.

Note: This figure is reproduced in color in online supplementary materials.

The unresolved quantum state superposition yields the keto–enol tautomerization of the peptide linkages (Figure 1.5), which in turn renders the peptide bonds torsionally rigid ($\omega = 180°$), a key condition to simplify the backbone conformation by introducing the equivalence relation where torsional states of the chain are regarded modulo the Ramachandran basins where they belong in the potential energy surface (Figure 1.6).

Aside from its computational prowess, the autoencoder described in Figure 1.6 may be said to be endowed with a protoconscience insofar as it is able to strategize its computational approach in a meaningful way. By lumping backbone torsional states modulo Ramachandran basins, the autoencoder was able to align tactics with goals in order to take initiatives that represent a way forward toward the achievement of the goals.

1.4 ARTIFICIAL CONSCIENCE IN NEURAL FORMAL REASONING

The self-examination inherent in AC requires a superior activity of neural networks above and beyond the previously discussed computational realm, the realm where essential models were distilled and intelligent simplifications were autonomously introduced to optimize the encoding of dynamic data. AC needs to be examined from the right perspective as the inherent learning capability of neural networks is combined with the accuracy of symbolic proof checking to create neural theorem-proving platforms [25]. This superior activity needs to be dealt with as conscience is inherent in an "understanding" of the generated formal proofs, an understanding that spans beyond the computable consistency checking. Thus, in the subsequent chapters, the neural component of the autoencoders will be implemented to train LLMs in order to generate candidate formal proofs that will be verified by the symbolic component of proof assistants, such as Lean, with its extensive math library *Mathlib4* [26].

The proof assistant presents a neural locus implemented for the generation of tactics paired with selection of premises to search for complete proofs in the manner of a generative pretrained transformer (GPT) [27]. The approach is scaled up to implement proof-generation LLMs in a truly autonomous manner by wrapping the proof assistant into a neural locus (Chapter 2).

Symbolic systems operating as theorem provers such as Lean are not per se engineered for neural network applications. However, neuro-symbolic frameworks such as the ones presented in Chapters 2, 6, and 7 can be used to develop an LLM-based proof generator, as the LLM is pretrained natively in the symbolic system.

To discuss conscience in a physical context, it has been argued in the previous sections that an amalgamation of QM and GR would provide a promising starting point. Insofar as conscience is assumed to be concerned with noncomputable quantum phenomena, a QM–GR amalgamation is essential to "complete QM" in a Gödelian sense, a completion aptly termed "protoconscience" [2,4]. The neural architecture suitable to combine two different physical contexts is obviously that of an autoencoder [5], which needs to be taken to a realm of formal symbolic reasoning, not merely a means of ensuring compatibility of computable events in two different realms. For instance, in a quantum gravity autoencoder, the latent space (Ω) is the flat manifold supporting QM, while W is the relativistic space-time manifold, and an event becomes a displacement along a world line of matter, $F{:}W \rightarrow W$, mapping one point in W into another point in W. To pursue formal reasoning, the neural architecture runs an LLM with the goal of generating theory in the realm of quantum cosmology, an investigation requiring a suitable operational embodiment of conscience to encompass the noncomputable aspects of QM.

The LLMs often fail to prove nontrivial theorems from domains that are different from those where the training data belongs. These difficulties are circumvented with the implementation of large-scale formalization projects. One such project is described in Chapter 2, where a Mathlib repository for "vierbein" formalization provides a fiber-bundle formulation of the gauge theory of gravity, eminently suitable as an

AI tool to investigate the emerging field of quantum gravity. The synthetic data generated is shown to enable the LLM-based inference of a massive gravity-communicating boson, the "warped graviton" (Chapter 2). In general, LLM inferences are run within the symbolic system, such as Lean with a supporting framework (copilot) provided via a shared CTranslate2 C++ library at the foreign function interface (FFI) [28].

Given this background on neural theorem provers, consciousness is expected to emerge as the system is able to turn on itself to critically examine and redesign its own inferential scheme when running the LLM inference in a functional programming language. The autoencoder, itself a transformer [29,30], is used to prove formal theorems through its interaction with proof assistants whereby it can suggest tactics (i.e., programs in a domain-specific language), select premises, and generate complete proofs. The tactic suggestion is itself a tactic, and when applied, it feeds proof goals into the LLM. Thus, a conscious tactic is precisely the optimization of the autoencoder vis-à-vis a set of proof goals. As an intelligent autoencoder, the conscious neural theorem prover selects the optimal latent manifold $(\Omega,\Omega/\sim,\ldots)$ and decoding space $(W' = W, W/\approx, \hat{\Omega}, \widehat{\Omega/\sim},\ldots)$, not only by minimizing the loss function (J) of its neural realization but also by optimizing the alignment of the tactics with the proof goals (functional semantics optimization). In other words, conscience focuses the neural theorem prover on itself to optimize its own inferential scheme by restructuring the neural network (Figure 1.11). This is precisely the noncomputable aspect of the formalization project that cannot be directly incorporated into the functional axiomatization library. The optimization of the supporting AI system when deprived of conscience is incapable of such drastic steps, which are triggered by introspective evaluation.

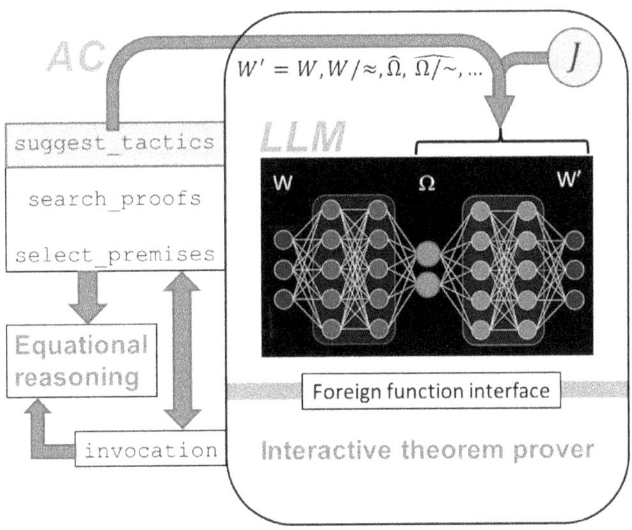

FIGURE 1.11 Basic scheme of an AI system functioning as a neural theorem prover endowed with conscience (AC). The AI system consists of an LLM trained by an autoencoder and wrapped in a functional programming language that operates as the LLM copilot. Consciousness emerges as the system is able to inwardly examine and restructure its own inferential scheme. The autoencoder functions as a transformer and trains the LLM to prove formal theorems through its interface with proof assistants whereby it can suggest tactics (i.e., programs in a domain-specific language), select premises, and generate complete proofs. The tactic suggestion is itself a tactic, and when applied, it feeds proof goals into the LLM. Thus, a conscious tactic is precisely the optimization of the autoencoder vis-à-vis a set of proof goals. As an intelligent autoencoder, the conscious neural theorem prover selects the optimal latent and decoding manifolds not only by minimizing the loss function (J) of its neural realization but also by optimizing the alignment of the tactics with the proof goals (functional semantics optimization). In other words, conscience invokes or summons the formalized mathematics repository in order to prompt improvements to the autoencoder architecture by computing better choices of latent and decoding manifolds. This is precisely the noncomputable aspect in the optimization of the supporting AI system deprived of conscience.

Note: This figure is reproduced in color in online supplementary materials.

1.5 ARTIFICIAL CONSCIENCE WEIGHS TIME TRAVEL WITHIN A SUPPORTING SPACE-TIME TOPOLOGY

The topology of space-time is established through ontological predicates on the Big Bang. A functional programming language in the guise of an unconscious theorem proof assistant would be impervious to such knowledge when engaged in equational reasoning based on the GR-formalization uploaded in a mathematics repository (i.e., Mathlib for Lean [26,27]). This is so since an unconscious theorem prover regards the local (differential geometric) nature of Einstein's GR equations as rendering irrelevant any ontological propositions stemming from a global analysis of the space-time manifold. The ontological constraints cannot be incorporated in a formal computation; they pertain to the noncomputable aspects that frame the AC.

Once and if space-time came into existence as geometric support of the universe, it can be predicated as follows: (I) The foliation of space-time into fixed-t three-dimensional cross sections cannot be granted until time originates as a traceable exogenous way to define "evolution of the universe". (II) Space-time cannot have boundaries, since "outside" is not a thing; otherwise, it would be part of the "inside". "Outside" can only be equated with nothingness; hence, it has no geometry to signify it as a thing. Boundary means interface with the nonentity "outside the universe"; therefore, there is no boundary. (III) Space-time is a compact manifold because it once was, that is, at the Planck epoch, and topological change is forbidden by GR. (IV) Since space-time is compact (predicate III) and has no boundary (predicate II), it cannot be a simply connected 4D manifold. Furthermore, if the aspect ratio between all pairs of dimensions is 1, space-time must have a primeval sustainable wormhole. (V) Space-time cannot be an orientable manifold insofar as time became the foliation parameter only when the universe's evolution became a traceable event (predicate I). Since specific chirality stems from parity violation as a particle interacts with a boson that communicates the weak nuclear force (WNF), there should be a source of dark matter (DM) arising from the nonorientability of space-time and chirality can only be local.

The ontological constraints I–V cannot be subsumed in the formalization projects that enable functional programming languages to make inferences through equational reasoning. Hence, they cannot be dealt with as computable entities by an interactive theorem prover. This inevitably leads to the problem of implementing an AC so that it can "learn" the ontological predicates as part of the training process. As shown subsequently, this requires a neural support operated by an intelligent autoencoder that trains an LLM wrapped into the theorem proof assistant (Figure 1.11). The LLM per se cannot incorporate the ontological predicates as they are not semantically compatible with the axioms of GR, all rooted in differential geometry. Neither can the ontological predicates be uploaded to Mathlib or other formal projects due to the same type of incompatibility. Yet, an AC can incorporate them as tactics as it controls its own inference scheme by determining the latent manifold (Ω) and the decoding manifold (W') (Figure 1.11). Each such tactic invokes specific directories of the mathematics library to tailor the decoding space to the specific needs of the task at hand and can be directly tested by computing the inferential scheme that results from the ensuing solution to the GR equations (Chapter 7). The nonorientable compactification of $\mathbb{R}^4$ into a manifold with no boundaries requires the identification of opposite faces of a 4D cube. One of the identifications must be rotated 90° to blend the time dimension with a spatial dimension (Figures 1.12 and 1.13), as required by the lack of foliation that preceded the point at which the universe evolution became a verifiable event. These considerations lead the autoencoder to pick the latent manifold as the twisted quotient.

$$\Omega \approx \left(\mathbb{R}^4 / \mathbb{Z}^4\right)^{\nparallel} \tag{1.14}$$

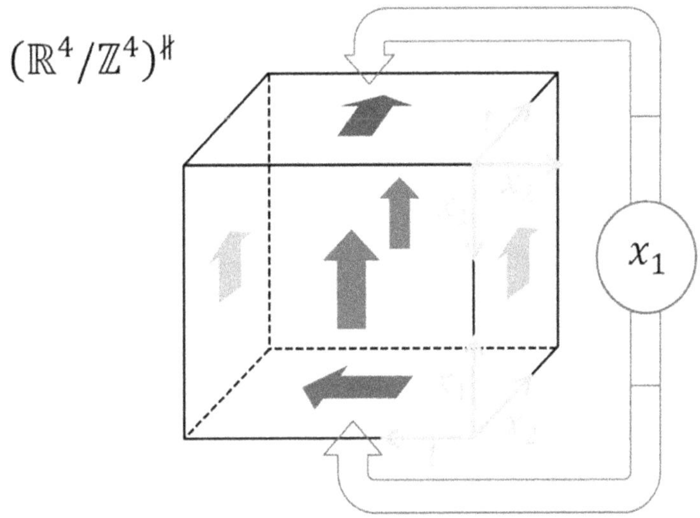

FIGURE 1.12 The nonorientable compactification of $\mathbb{R}^4$ into a manifold with no boundaries and a blending of time and space coordinates in accord with the lack of foliation prior to the events that follow after the Big Bang. For visualization purposes, a single 3D cross section is presented. The compactification quotient yields a multiply connected manifold and requires the identification of opposite faces of the hypercube. One of the identifications must be rotated 90° to blend the time dimension with a spatial dimension, yielding a quotient manifold homeomorphic to the twisted torus.

Note: This figure is reproduced in color in online supplementary materials.

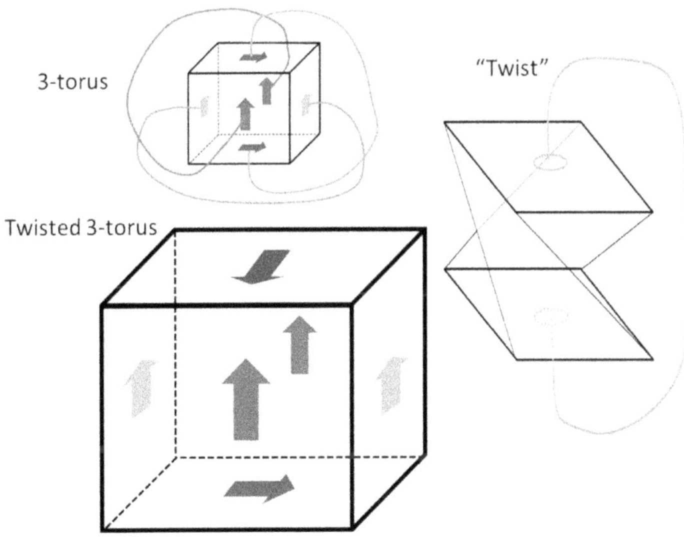

FIGURE 1.13 3D cross section comparison of the respective constructions of orientable (foliated) and nonorientable (non-foliated) compactifications of $\mathbb{R}^4$ into multiply connected space-time manifolds with no boundaries. The nonorientable manifold is a twisted torus, and the torsion results from the lack of foliation that preceded the point at which the universe evolution became a verifiable event.

Note: This figure is reproduced in color in online supplementary materials.

The twist (indicated with the symbol "$\nparallel$") is provided by the blending of time t with a spatial coordinate (x_2) using spatial coordinate x_1 as vehicle to generate a squared line differential of the form (Chapter 7):

$$ds^2 = \left(\sqrt{2}\omega_0\right)^{-2}\left\{\left[dx_0 + f(x_1)dx_2\right]^2 - dx_1^2 - \frac{1}{2}f^2(x_1)dx_2^2 - dx_3^2\right\} \tag{1.15}$$

where $\omega_0 = \sqrt{4\pi G\rho}$ (G = universal gravitational constant, ρ = uniform mass density) and the function $f(x_1)$ defines one of the symmetry transformations (homothetic rotation) of Ω (Chapter 7). As derived in Chapter 7, a radiation-dominated early universe endowed with such topology and associated cosmological solution given by Eq. 1.15 would contain closed time-like curves. This implies the existence of world lines of matter that would enable time travel into the past.

The nonorientability of time is an essential prerequisite of time travel, and, as a property of space-time and not of time itself, it cannot be established or disproved through the tenets of GR. This is obvious because space-time is locally time-orientable: Indeed, space-time is locally an Euclidean flat manifold endowed with the Minkowski metric within its own tangent bundle (Chapters 2 and 7) [31]. Yet, time nonorientability can be asserted as an alternative description of particle–antiparticle annihilation (Figure 1.14), an event arising from vacuum quantum fluctuations yielding fermionic (spin 1/2) matter in accord with Heisenberg's uncertainty principle [31].

The violation of causality associated with time travel or closed time-like curves would not be resolved by an individual AI system, be it an LLM or a functional programming language, if acting autonomously in isolation without inputting the tuning of its own inferential system. It is resolved via an AC that operates within the amalgamation of the two AI systems, as indicated in Figure 1.11. Thus, as it controls its own inferential scheme, AC circumvents the violation of causality by adopting, for example, the universal covering of the latent manifold as decoding space:

$$W' = \hat{\Omega} \approx \widehat{\left(\mathbb{R}^4 / \mathbb{Z}^4\right)^{\nparallel}} \tag{1.16}$$

This implies that an extra dimension may be incorporated to space-time, and the closed time-like curves are actually the projections of holographic full helical turns in the simply connected and hence orientable universal covering (Chapter 7). In this way, a *multiverse* scenario, with each replica containing each turn of the helix, circumvents the causality paradox as the return to the initial point becomes a projection artifact resulting from the missing fifth space-time dimension in the latent space-time (Figure 1.15).

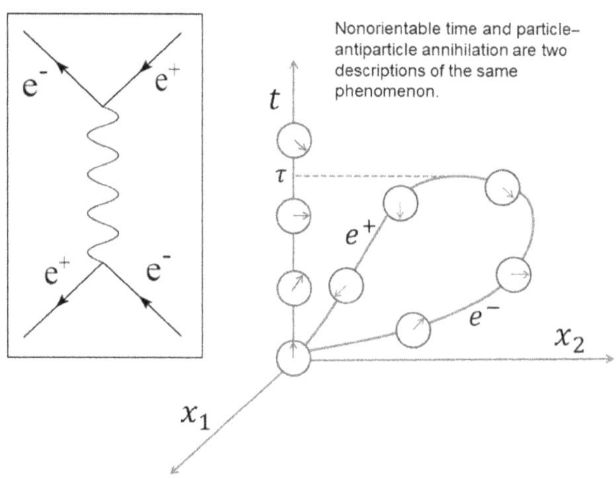

FIGURE 1.14 Time nonorientability of space-time manifested as a particle–antiparticle (e^-, e^+) annihilation event along a time-like curve. The two times, intrinsic and external, are symbolized by the clocks placed along the world line of matter and along the time axis, respectively. The positron travels backwards in external time.

Note: This figure is reproduced in color in online supplementary materials.

$$\begin{array}{ccccc} \mathbb{R}^4 & \xrightarrow{\pi} & (\mathbb{R}^4/\mathbb{Z}^4)^{\sharp} & \xrightarrow{\eta} & (\widehat{\mathbb{R}^4/\mathbb{Z}^4})^{\sharp} \\ \downarrow & & \downarrow & & \downarrow \\ \mathbb{R}^4 & \xrightarrow{\pi} & (\mathbb{R}^4/\mathbb{Z}^4)^{\sharp} & \xrightarrow{\eta} & (\widehat{\mathbb{R}^4/\mathbb{Z}^4})^{\sharp} \end{array}$$

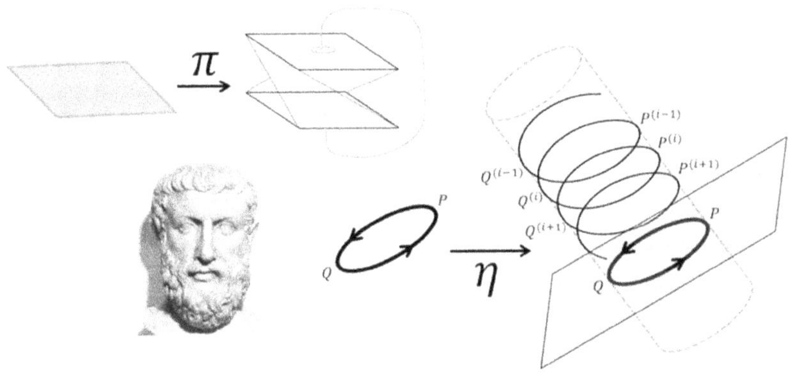

FIGURE 1.15 Formal autoencoder circumventing the causality violation associated with a closed time-like curve in the latent space-time manifold. The latter is constructed by AC, turning Parmenides-type ontological premises into the canonical projection π of $\mathbb{R}^4$ onto a compact nonorientable multiply connected manifold without boundary that serves as latent manifold. An extra dimension may be incorporated into space-time so that the closed time-like curves are actually the projections of holographic full helical turns in the simply connected and hence orientable universal covering. In this way, a *multiverse* scenario, with each replica containing each turn of the helix, circumvents the causality paradox as the return to the initial point becomes a "Plato's cave projection artifact". The multi-valued map *h* is the hologram that maps the closed time-like curve onto the helix. The effigy of Parmenides was obtained from the public domain (https://fr.wikipedia.org/wiki/Parm%C3%A9nide#/media/Fichier:Parmenides.jpg).

Since it is simply connected, the universal covering $\left(\widehat{\mathbb{R}^4 / \mathbb{Z}^4}\right)^{\sharp}$ is orientable and reflexible, that is, nonchiral (Chapter 7). Thus, while this tactical choice of decoding space circumvents the causality problem, it cannot accommodate the chirality differentiation vis-à-vis the WNF (parity violation) or other quantum aspects of particle cosmology [31], including the 5:1 ratio of dark to observable matter [32].

An n-fold covering akin to a Riemann surface for an analytic n-valued function would be required to represent a closed time-like curve as in Plato's cave scenario (Figure 1.16). The allegory suggests that the closed time-like curve is actually a projection of an open curve across two sheets of the Riemann surface. By invoking the algebraic topology armamentarium to suggest tactics (Figure 1.17), the AC is able to summon the formal reasoning needed to identify the correct Riemann-surface-type covering of $\Omega \approx \left(\mathbb{R}^4 / \mathbb{Z}^4\right)^{\sharp}$.

Intuitively, the answer can be found in the observation that each source of nonorientability requires a two-sheet covering, so that the map that reverses orientation becomes equivalent to the homeomorphism between the two sheets (Chapter 5). We also notice that time is blended with the three spatial dimensions, as it results from all identifications of opposite faces in the hypercube (3D cross section in Figure 1.13). Hence, there are three sources of nonorientability in $\left(\mathbb{R}^4 / \mathbb{Z}^4\right)^{\sharp}$, and consequently its orientable covering must be sixfold.

As it senses poor J-function values for the $n = 2$ covering $\langle\Omega\rangle_2$, AC prompts the algebraic topology directory in `Mathlib` to suggest tactics by proposing decoding spaces as orientable coverings of Ω (Figure 1.17). To accommodate the electroweak sector of the Standard Model (SM), AC searches for orientable nonreflexible coverings, that is, their projection onto Ω is not invariant under orientation-reversing homeomorphisms. The nonreflexible covering with the lowest number (n) of sheets capable

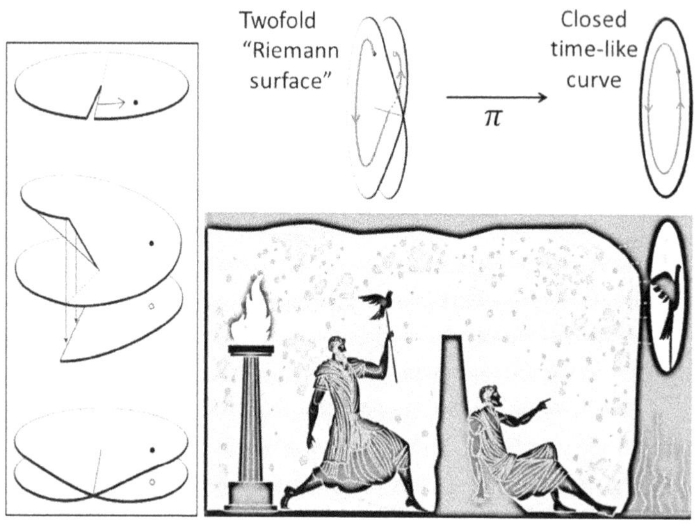

FIGURE 1.16 Construction of a twofold orientable covering of space-time whose 2D cross section is homeomorphic to the Riemann surface for the analytic two-valued function $\sqrt{z}$. The allegory of Plato's cave suggests that a closed time-like curve is actually the projection of an open curve across two sheets of the Riemann surface that, in this illustration, represents the true topology of space-time.

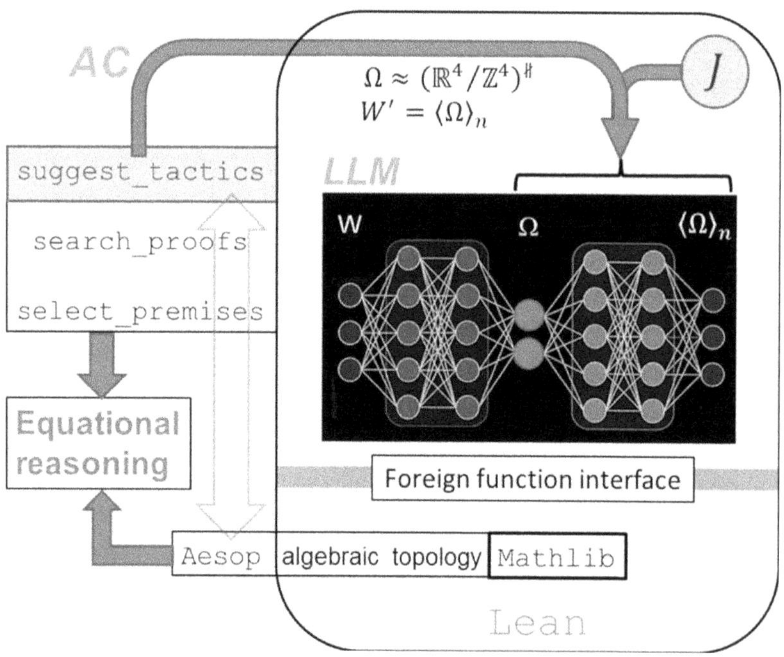

FIGURE 1.17 In accordance with the scheme of a conscious LLM run by an autoencoder and wrapped in a theorem prover (Lean), the neural formal reasoning system (LLM with Lean copilot [28]) is able to summon a formal mathematical repository from Mathlib, an action needed to identify the correct Riemann-surface-type covering of the latent space-time $\Omega \approx (\mathbb{R}^4/\mathbb{Z}^4)^\sharp$. AC is capable of prompting a restructuring of the neural network architecture, well beyond parameter tinkering, as a drastic step steered by J-function minimization. Specifically, AC prompts the algebraic topology directory in Mathlib to suggest tactics to restructure the learning system by proposing decoding manifolds of the type $\langle\Omega\rangle_n$ as orientable coverings of Ω.

of differentiating chirality corresponds to $n = 6$ (Chapter 7) (Figure 1.18). Strikingly, the results of the AC-prompted cost function calculation agree that the $n = 6$ covering manifold yields the optimal autoencoder as $n = 6$ yields the lowest cost function (Figure 1.18).

Thus AC configures the autoencoder (Figure 1.17) with an orientable sixfold nonreflexible chiral-twin manifold, denoted $\langle\Omega\rangle_6$, endowed with a tangent bundle geometrization determined by the six unique pairs of Pauli matrices from QM (Figure 1.19). The details of the prompted algebraic topology computation are provided in Chapter 7. The sixfold topological parsing determined by $\langle\Omega\rangle_6$ yields a sixfold quantum field theory (QFD) of DM that explains the 5:1 abundance ratio of dark to visible matter (Chapter 4).

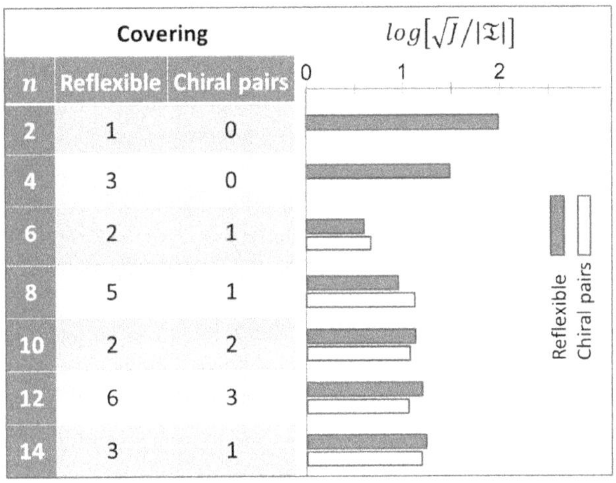

n	Reflexible	Chiral pairs
2	1	0
4	3	0
6	2	1
8	5	1
10	2	2
12	6	3
14	3	1

FIGURE 1.18 Results of the AC-prompted algebraic topology computation of the optimal latent space-time covering required to accommodate the electroweak sector of the Standard Model, circumvent the paradox of causality violation and optimize vis-à-vis the J-function. AC prompts the search for orientable nonreflexible coverings, that is, their projection onto Ω is not invariant under orientation-reversing homeomorphisms. The nonreflexible covering with the lowest number (n) of sheets capable of differentiating chirality corresponds to $n = 6$ (Chapter 7). Strikingly, the results of the AC-prompted cost function calculation agree that the $n = 6$ covering manifold yields the optimal autoencoder as the value $n = 6$ yields the lowest cost function. The symbol $|\mathfrak{T}|$ denotes the cardinal of the testing set.

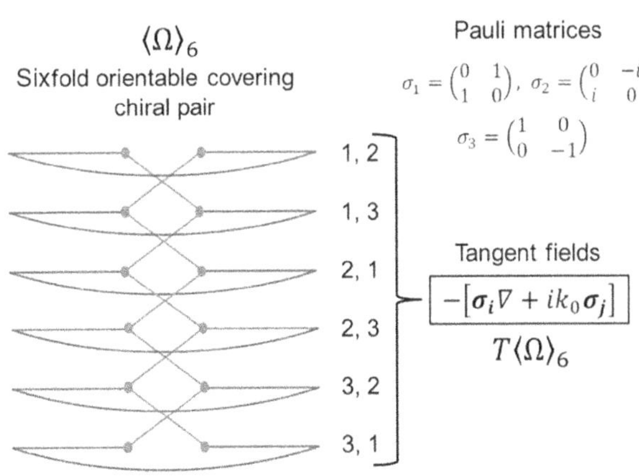

FIGURE 1.19 AC configures the autoencoder with $\langle\Omega\rangle_6$, the orientable sixfold nonreflexible chiral-twin manifold. This manifold is endowed with a tangent bundle geometrization that is determined by the six unique pairs of Pauli matrices. This implies that the algebraic topology solution (Figure 1.18) generates the six sheets with respective gauge symmetries determinant of the quantum mechanical wave function equations for (spin ½) fermionic (one sheet) and DM (five sheets), as confirmed in Chapter 2.

Thus, a new cogent topological model of space-time emerges reconciling the computationally implemented ontological cogitations of AC with a body of established facts from quantum cosmology.

1.6 ARTIFICIAL CONSCIENCE AT WORK

An LLM trained by an autoencoder and endowed with AC (Figures 1.11 and 1.17) differs fundamentally from the supporting LLM in two aspects: (1) It is wrapped in a functional language program that serves as copilot [28] in order to acquire equational reasoning capabilities; (2) it is capable of restructuring its learning scheme and underlying neural network architecture by resorting to its acquired reasoning capabilities in order to accommodate ontological premises rendered not computable or impossible to formalize by the underlying copilot-endowed LLM.

The results in Figure 1.18 are particularly revealing as they demonstrate the staggering power of AC to strategize the inferential approach of the AI system when faced with a concrete problem such as the deconstruction of the symmetries and constraints in the space-time topology that could potentially lead to time travel. By inwardly surveying performance, that is, parameter selection, measured by the autoencoder's cost function (Eq. 1.7), AC may prompt the AI system to invoke/summon formalization resources from the mathematics repository of the unsupervised theorem prover to generate computable options in order to improve the performance of the autoencoder to optimal levels. *This type of inward surveillance prompting a restructuring of the learning scheme is perceived as an act of conscience.*

The most striking result achieved in the particular case described in Figure 1.18 is that AC is able to summon mathematical resources, that is, algebraic topology formalization not exploited in the standard copilot-riding autoencoder application, in order to restructure the autoencoder-based training of the LLM to achieve optimal performance as defined by the cost function.

As established by comparing Figures 1.18 and 1.19, the AI system that provides the optimal topological model for the universe vis-à-vis the computational cost function is at the same time the one compatible with the gauge symmetries required by QM while circumventing the time travel paradox. This towering achievement of AC will be exploited in Chapter 2 to unravel the origin of the dark sector in the SM and pursue the associated problems in quantum gravity arising thereof.

1.7 ARTIFICIAL CONSCIENCE FOR THE QUANTUM COSMOLOGIST: PREPARATORY NOTIONS

In spite of its potential to provide the conceptual underpinnings of unification efforts including quantum gravity [32] and Penrose's "gravitization of QM" [2,4], a blending of QM and GR has proven unyielding to most theoretical efforts. Space-time is a Riemann manifold while the support of the wave function is a domain in a flat Euclidean space, so only a "vierbein" formalism homeomorphically mapping space-time onto its tangent bundle [32,33] can reconcile GR with QM, and yet no powerful result has emerged at the interface of the two theories. A natural way of combining them is to adopt an autoencoder where encoding and decoding are provided by the vierbein maps (Figure 1.20) in such a way that the autoencoder serves as the matrix for an AI system endowed with AC, such as the one described in Figure 1.11.

By themselves, AI systems in the guise of theorem proof assistants [25,30] are inherently unsuited for the task because a QM–GR amalgamation would require a blending of two different formalization schemes, differential geometry and operator theory in Hilbert space, respectively, the mathematical frameworks for GR and QM. Such blending can in principle be achieved by tapping into the mathematical

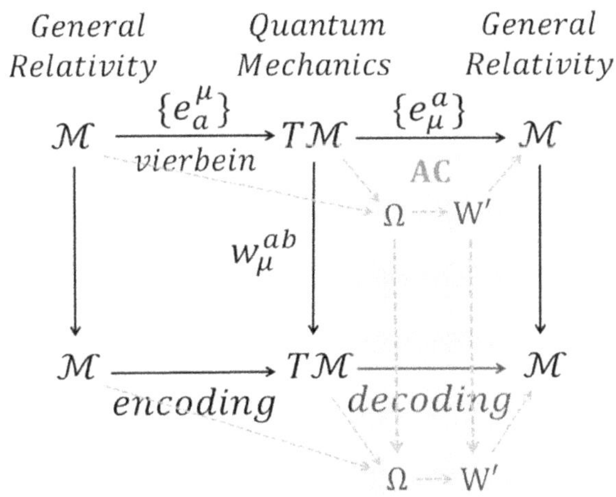

FIGURE 1.20 Autoencoder for an AI system endowed with AC as required to blend QM with GR under the vierbein scheme (Chapters 2 and 6). Space-time is a Riemann manifold while the support of the wave function is a domain in a flat Euclidean space, so only a "vierbein" formalism homeomorphically mapping space-time onto its tangent bundle can reconcile GR with QM. A natural way of combining them is to adopt an autoencoder where encoding and decoding are provided by the vierbein maps in such a way that the autoencoder serves as the matrix for an AI system endowed with AC (Figures 1.11 and 1.17). The vierbein formalism that enables the interfacing between GR and QM is thus cast onto the autoencoder as inferential platform. AC optimizes the latent manifold Ω and decoding manifold W' in accord with ontological premises that AC turns computable via the invocation of a formalization repository (Figures 1.11 and 1.17). The inward surveillance thus prompts a restructuring of the neural network scaffold in order to optimize the autoencoder as a commutative diagram.

library for the functional programming language (`Mathlib` for `Lean`) [25,27], the "vierbein" formalization axioms, and associated propositional context that enables the transition from the space-time manifold to the Euclidean support of QM, providing a fiber-bundle gauge formulation of gravity [33,34]. Unfortunately, the formalization projects for differential geometry and operators in Hilbert space are radically different in their axiomatic presentation, so the likelihood of producing nontrivial cross-disciplinary inferences, such as those required by the unification efforts, is dismal.

On the other hand, the neural empowering of the theorem prover will be shown to provide a much more viable option as it is naturally endowed with a conscience (AC), as described in Figure 1.11, while the LLM that serves as copilot [28] may be trained by an autoencoder, precisely the neural network architecture that enables the compatible blending of the two theoretical contexts (Figure 1.20). The scope of this book may be delineated by asserting that this conscious AI system will be extensively exploited to investigate the interrelated and interdisciplinary problems in quantum cosmology. Such efforts culminate in the deconstruction of a time travel scenario (Chapter 7).

The level of exploratory initiative of AC in terms of expanding neural reasoning extends beyond the combinations of AI systems that support its embodiment. This is apparent as AC suggests tactics that prompt a restructuring of the neural network that empowers the LLM (Figures 1.11 and 1.20). Since this neural network adopts the autoencoder architecture, it becomes most suitable to integrate the formalization across the GR–QM interface. Thus, GR and QM are naturally combined in the autoencoder, as the encoding manifold $\mathcal{M}$ is space-time, while the latent space is its tangent bundle $T\mathcal{M}$, locally suitable to support the QM wave function. In this way, all problems that require equational reasoning at the GR-QM interface are amenable to be tackled by an AI system such as the one presented in Figure 1.11. The prerequisites are that the AI system should be empowered by a GR-QM autoencoder and endowed with an AC that restructures the autoencoder by tapping the mathematical resources in the library of the functional programming language via an invocator such as `Aesop` in `Lean` (Figure 1.17) [25–28].

The vierbein formalism that enables the interfacing between GR and QM is thus cast onto the autoencoder as inferential platform to achieve an AI-based symbiosis of the two pillars of contemporary physics (Figure 1.20). AC optimizes the latent space Ω and decoding space W' in accord with ontological cogitations that AC turns computable via the invocation of a specific formalization directory from the mathematics repository (Figures 1.11 and 1.17). This inward surveillance is tantamount to a veritable restructuring of the neural reasoning scaffold in order to optimize the autoencoder as a commutative diagram.

This inferential platform endowed with AC and capable of empowering an LLM for equational reasoning at the GR-QM interface will be used throughout the book to pursue interdisciplinary research and solve core problems in particle cosmology, where synergetic relations between the two dominant theories are required.

1.8 TOWARD AN ARTIFICIAL INTELLIGENCE SYSTEM FOR QUANTUM COSMOLOGY RESEARCH

Following the program outlined in Section 1.7 (Figures 1.11 and 1.17), the neural symbolic reasoning schemes described in this book muse over the possibility that our universe may be the relic of something simple, powerful, and highly symmetric. This tantalizing possibility is surely suggested by the experimental evidence that helped establish the Big Bang as well as deep-space observational data [31]. As to the nature of the primeval entity, physicists' favorite answer involves a fluctuation in the quantum vacuum. The evidence suggests an entity where matter and energy cannot be told apart, the four fundamental forces are unified into one, and the quantum vacuum is somehow endowed with attributes of ur-matter, that is, capable of acting as a precursor to the observable matter. The cosmological birth then becomes akin to a first-order phase transition where primeval spatial symmetries are transmuted into gauge symmetries. In this overarching model, the emerging gauge fields with the proper symmetry interact with photons conforming visible matter, while those fields with incompatible symmetry represent DM, the scaffold of the universe in deep space. To complete the picture, dark energy (DE) becomes the latent heat released with the first-order phase transition.

That a phase transition may internalize a spatial symmetry as a latent symmetry that is spontaneously broken is by no means an alien concept to physics. However, transferring this picture into a viable ansatz for the origin of the universe poses a number of challenges to the theoretician, beginning with the definition of the critical temperature T_c. Thus, if we are concerned with the origin of mass, T_c should be taken to be $T_c = E_{\mathrm{TV}} / k_B$, where E_{TV} is the vacuum expectation value (true vacuum) for the Higgs field and k_B is Boltzmann's constant. An AI system will be implemented to model this crucial phase transition and generate the physics that yields the symmetry breaking leading to the emergence of mass. In a skeletal representation, the AI system takes the form of a commutative arrow diagram, as shown in Figure 1.21. An extra compact dimension is incorporated to space-time, yielding the ur-manifold W to account for the dramatic geometric dilution of gravity relative to the other three fundamental forces of established quantum origin. The gauge fields thus emerge as supported by a 4D coarse-grained manifold $W/\sim$, the standard space-time, rendered as quotient space via an equivalence relation "~" that filters out the extra dimension. The horizontal arrows $W \rightarrow W/\sim$ become canonical projections, surjectively assigning the corresponding equivalence class to each point in the 5D manifold, while the vertical arrows represent the symmetry breaking associated with the phase transition, $T \gg T_c \rightarrow T < T_c$, at the two different levels of resolution. This book upholds and advocates for this overarching picture and justifies this position as it approaches the problem of the origin of the universe empowered by AI. It is necessary to understand that the book covers uncharted territory, where AI formalizes quantum cosmology by implementing LLMs enshrined in formal proof assistants [28]. The AI system is delineated in Chapter 3 while the

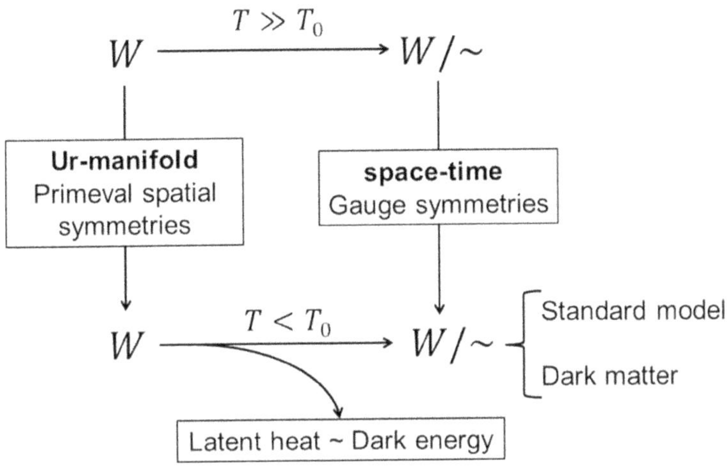

FIGURE 1.21 Blueprint for the birth of the universe as a phase transition internalizing spatial ur-symmetries as gauge symmetries. An AI system will be implemented to model this phase transition and generate the physics that yields the symmetry breaking leading to the emergence of mass. The AI system is represented as a commutative arrow diagram. An extra compact dimension is incorporated to space-time, yielding the ur-manifold W to account for the dramatic geometric dilution of gravity relative to the other three fundamental forces. The gauge fields thus emerge as supported by a 4D coarse-grained manifold $W/\sim$, the standard space-time, rendered as quotient space via an equivalence relation "~" that filters out the extra dimension. The horizontal arrows $W \to W/\sim$ become canonical projections, while the vertical arrows represent the symmetry breaking associated with the phase transition, $T \gg T_c \to T < T_c$.

formalization and mathematical proving processes are described in detail in Chapter 2. The AI system, termed autoencoder (Figures 1.2, 1.11, and 1.17), distills the essential physical picture from a vastly more complicated representation. This is the type of AI system that can assist the reader coming to grips with quantum gravity, the holy grail in the grand unification of fundamental forces, purportedly holding the key to the origin of the universe.

As it searches for an encompassing physical picture where it all falls into place at the aftermath of creation from a quantum void, AI reckons that the problem cannot be tackled without proper contextualization, that is, without dealing with other intimately related problems in particle cosmology that bear on the understanding of the primeval forces that shaped in universe. These problems include: the nature of DM and DE, the universe topology, the vacuum catastrophe, and the cosmological constant problem. Accordingly, this chapter introduces elements of particle cosmology that enable the pursuit of the problem of the origin of the universe in its full conceptual and relational richness.

DM and DE become the key relics in the aftermath of the birth of the universe. As shown in Chapter 2, DM along with visible matter emerged from the first-order phase transition as two different alternative ways of internalizing or "gauging" the primeval symmetry, and since DM is the gravitational scaffold of the universe, the gravitational waves sprouting from the primeval phase transition should be detectable even today. On the other hand, if DE is the latent heat released with the primeval phase transition, how can the DE concentration remain constant, as experimentally verified in a universe subject to accelerated expansion? Quantum vacuum energy generated autocatalytically in an ever-expanding vacuum does not seem to be the straightforward answer, as suggested by the so-called vacuum catastrophe problem. This conundrum is resolved in Chapters 2 and 5, where a model of DE depletion/replenishment dynamics is put forth by the autoencoder to account for the sustainability of the universe topology in the observed runaway mode.

This preparatory chapter delineates an AI metamodel of the dark universe that is, of the invisible matter in all forms, including DE, assumed to be present in overwhelming amounts in deep space. By metamodel we mean an inferential scheme that acquires big evidentiary data and generates a matrix suited for a cogent, physically consistent model capable of meaningful prediction [5]. The evidence includes deep-space phenomenology and radiation relics of the Big Bang and other cosmic reproductive events. By leveraging fairly elementary physical arguments, the chapter first surveys the experimental evidence from deep space observation that supports the existence of the dark sector. The data is then reconciled with the incorporation of a latent extra spatial dimension, modeled by means of the autoencoder, that distils the quantum reality from a space-time encoded in five dimensions. In this way, the chapter paves the way for the AI-empowered approach to the origin of the universe described in Chapter 2.

1.9 AI SYSTEM TO ENCODE THE DARK SECTOR IN THE STANDARD MODEL

The dark universe constitutes a vast proportion of matter and energy in deep space that cannot be accounted for with extant physical models [31,35,36]. To shed light on the problem, AI is leveraged and testable assumptions are made concerning the nature of DM and DE. As described in the subsequent chapters, cosmological data will be inputted into an appropriate AI system under the assumption that there exists a "dark dimension" such that any particle field concentrated along this dimension will not interact with photons unless there is a spillover or "geometric dilution" of the field into the observable dimensions.

Human civilization has been able to physically model the fundamental particle fields that constitute the fabric of the universe, as described by the so-called Standard Model (SM) of particle physics. However, the model clearly does not encompass the dark portion of the universe, the vast portion now known to be impervious to interaction with photons and other gauge bosons in the SM. Thus, we may regard the extant physical model as actually corresponding to the evolution map on a latent manifold $K_{SM} : \Omega \to \Omega$, but not on W. Un-localizable (undetectable) particles, altogether lacking Compton wavelength, are excluded from the latent physics, requiring (AI postulates) a dark dimension to materialize.

We know that the K_{SM} model is unsatisfactory in deep space because it does not incorporate the bearing of the dark universe on the detectable dynamics. When this influence is incorporated, especially at the large scales of deep space, we end up with a model of broader applicability $K_{SM}^{\odot} : \Omega \to \Omega$ with $K_{SM} \neq K_{SM}^{\odot}$. Under the universe-as-autoencoder ansatz, determining the physics of the dark universe becomes tantamount to identifying the dark universe as the large portion, $W \setminus \mu\Omega$, of W that excludes the hologram $\mu\Omega$ and determining the autoencoder model (γ, F, μ) that satisfies the commutative relations $K_{SM}^{\odot} \circ \gamma = \gamma \circ F$ and $F \circ \mu = \mu \circ K_{SM}^{\odot}$.

The autoencoder scheme will also be used in a new development as an abstract formalization tool embodied as an LLM enshrined within a theorem proof assistant (Chapters 2 and 7). This AI-empowered tool is aptly named "formal autoencoder" and enables the formalization of a physical theory cast on one representation in order to distill it as a projection into a simpler representation. Chapter 2 will make extensive use of formal autoencoders as it makes GR compatible with quantum physics in order to seek the ultimate unification in a theory of quantum gravity. Both theories represent towering achievements in the understanding of the universe at what appeared to be vastly different scales, from cosmological to subatomic. Yet, as we come to grips with cosmological singularities such as black holes, the locality and dimensions under consideration enabled the possibility of a common ground for quantum gravity, purportedly bridging the two towering theories [31]. At the same time, the awareness that a vast proportion (95%) of the universe contains invisible or undetectable forms of matter and energy added further impetus to the pursuit of quantum gravity as a vantage point to understand unyielding aspects of deep space and interpret the phenomenology that arises thereof. The hitherto unknown forms of matter and energy constitute the dark universe and are likely to trigger a paradigm shift. This assertion served in good measure as the motivation for this book.

Three pivotal tenets, justified and elaborated in the subsequent chapters, define our approach to understanding the dark universe in the context of the birth of the universe:

I. The topology of space-time is a determinant of the nature of the dark universe.
II. The universe is compact and cannot have boundaries because there is no interface with nothingness; therefore, space-time is a multiply connected manifold.
III. The universe contains a hologram in the sense that the quantum fabric of space is apparent within a dimension reduction yielding a latent manifold that encodes the relativistic space-time [37].

The last tenet implies that an AI system in the guise of an autoencoder may be ideally suited to study quantum gravity, as shown in Chapter 2. Autoencoders may be thought of as learning systems that distill the fundamental physics underlying a process or phenomenology by encoding it in a lower-dimensional manifold, known as the latent manifold. Thus, *a hologram becomes styled as an autoencoder*. A physical embodiment of the autoencoder may be tuned with adherence to statistical mechanical laws in such a way as to "realize" quantum gravity.

The existence of a dark universe summons a hologram setting in which the phenomenology associated with the dark universe manifests itself in the latent manifold (Ω) only indirectly and through gravitational perturbation. This manifold corresponds to the realm of the detectable, the realm where, in terms of quantum physics, an act of observation collapses the wave function. However, as shown in this book, the dark universe is contained in a higher-dimensional space (W), an ur-universe, that admits no observer. Only 5% of matter in all forms in W is encoded by Ω, with the remaining 95% corresponding to the dark universe. We may state that "the dark universe is the part of W that is not part of the hologram". In rigorous terms, if $\eta : \Omega \to W$ is the holographic function, the dark universe $\mathcal{W}$ is defined as $\mathcal{W} = W \setminus \eta(\Omega)$.

Let $K_{\mathrm{SM}}^{\odot} \supset \mathcal{K}_{\mathrm{SM}} : \Omega \to \Omega$ define a generic transformation beyond the SM, then an overarching goal is to identify the transformation $\mathcal{F} : W \setminus \mathcal{W} \to W \setminus \mathcal{W}$ jointly with the map $\gamma : W \setminus \mathcal{W} \to \Omega$ such that the diagram in Figure 1.22 is commutative, that is: $\gamma \circ \mathcal{F} = K_{\mathrm{SM}}^{\odot} \circ \gamma$. The dark universe is thus determined by the compatibility of the observable universe (Ω, η), and the ur-universe (W) guaranteed by the commutativity of the diagram in Figure 1.22, which delineates the metamodel for the dark universe.

This picture is incomplete and, in a sense, unsatisfactory insofar as the ur-universe remains undefined. To grasp the nature of the problem, we need to incorporate the tenets I and II. The topology of the universe prompts the autoencoder to determine the embedding of the four-dimensional space-time W_4 into the higher-dimensional space W. As discussed in the subsequent chapters, the spatial dimensions

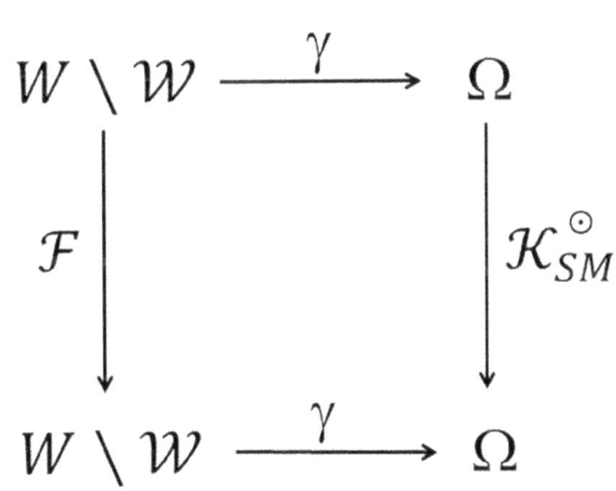

FIGURE 1.22 Metamodel of the dark universe $\mathcal{W} = W \setminus \eta(\Omega)$ understood as an AI system in the form of an autoencoder. The dark universe is identified with the portion of the ur-universe W that excludes the hologram, $\eta(\Omega) \subset W$, encoded by the latent manifold Ω. The dark universe is characterized by the lifting $(\mathcal{F}, \gamma)$ of a "corrected" SM map $K_{\mathrm{SM}}^{\odot}$ that makes the diagram commutative.

in W_4 must be compact because they once were, if we grant validity to the Big Bang scenario in a relativistic context, and relativistic space-time cannot afford topological change. Furthermore, Ω cannot have boundary since that would imply interfacing with nothingness, that is, not with vacuum but with the absence of geometry. Such an interface represents a physical and metaphysical impossibility. As a compact manifold without boundaries, Ω must be multiply connected, and consequently, so must be also its hologram $\eta_4(\Omega) \subset W_4$. However, as shown in Chapter 5, to retain rotational symmetry (more specifically, axis permutation), the multiply connected space-time must be endowed with a primeval wormhole. This singularity, however, should be regarded as artefactual of the dimensionality of the space that admits the quantum observer. The situation is akin to that of the Klein bottle, a nonorientable surface that features self-intersection when displayed in the three-dimensional space but not in four dimensions, where it is not amenable to representation (Figure 1.23a). Furthermore, as shown in Chapter 4, W_4 should be embeddable in a space, W_5, with an extra spatial compact dimension required to store DM in the form of stationary waves with no geometric dilution into the other spatial dimensions.

The ur-universe W becomes then the universal (simply connected) covering of W_5. In this way, the simply connected topology that smooths out the inherent singularities of W_5 is recovered at the expense of incorporating a noncompact spatial dimension (Figure 1.23b). This implies that the holographic map $\eta : \Omega \to W$ can be factorized as $\eta = \eta_\# \circ \eta_4$, through the embedding of the latent space $\eta_4 : \Omega \to W_4$ composed with the lifting $\eta_\# : W_4 \to W$, which is a multi-valued map. This implies that *the universal covering W (together with the lifting $\eta_\#$) corresponds to an AI-encoding of the many-worlds interpretation of* QM. This interpretation is thus validated by AI through the commutativity of the diagram presented in Figure 1.24, whereby the dark universe is identified as $\mathcal{W} = W \setminus (\eta_\# \circ \eta_4)(\Omega)$ and the following relations hold: $\mathcal{F} = \eta_\# \circ \mathcal{F}_4 \circ \gamma_\#$, $\mathcal{F}_4 = \eta_4 \circ \mathcal{K}_{SM} \circ \gamma_4$, where $\gamma_\#$, γ_4 are, respectively, the projections associated with the lifting and the holographic map.

One would say that a goal of the book is to determine the maps $\mathcal{F}$, $\gamma = \gamma_\# \circ \gamma_4$, that make the diagram in Figure 1.24 commutative and thereby enable the elucidation of the dark universe.

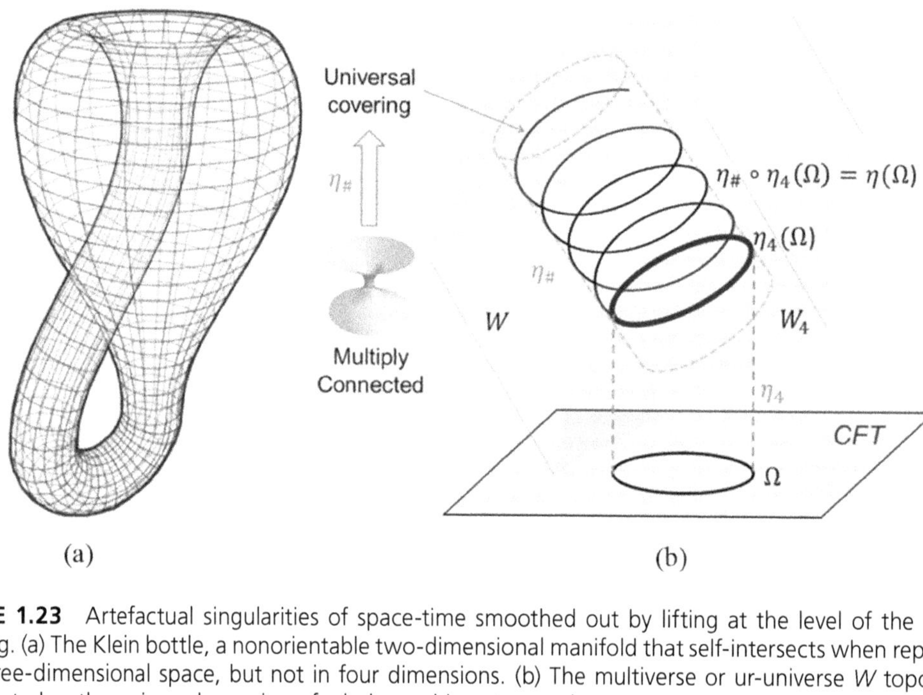

(a) (b)

FIGURE 1.23 Artefactual singularities of space-time smoothed out by lifting at the level of the universal covering. (a) The Klein bottle, a nonorientable two-dimensional manifold that self-intersects when represented in a three-dimensional space, but not in four dimensions. (b) The multiverse or ur-universe *W* topologically represented as the universal covering of a holographic autoencoder.

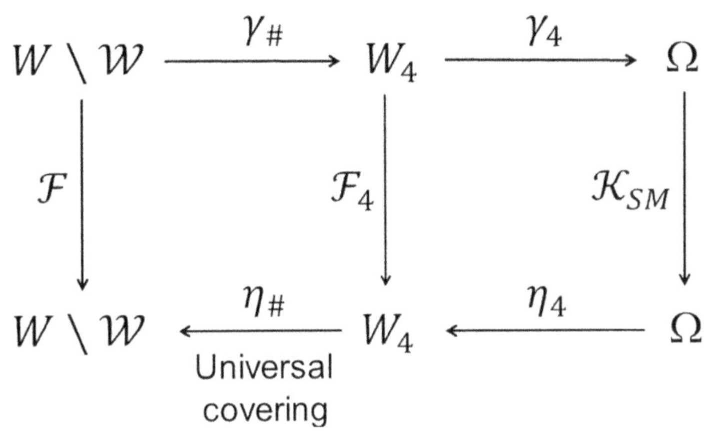

FIGURE 1.24 Characterization of the dark universe through a cogent metamodel of the universe represented as an autoencoder. The compatibility between the holographic and lifted (multiverse) representation of the latent physics enshrined in the SM is warranted by the commutativity of the diagram.

1.10 DARK PHYSICS ON EXTRA DIMENSIONS

A survey of the evidence described in the previous sections prompts us to question whether a shift in paradigm is in order to encompass deep space phenomenology. It appears that the influence of DM and DE cannot be done with by merely patching up the laws of physics or extending them to include cosmic ranges of interaction hitherto uncalled for. So far, the discussion has included local physical theories, that is, theories whose predictive power resides in solving differential equations or become appropriately stylized within the setting of differential geometry. Topological attributes of the universe have not entered into the equations in any obvious manner, and the topology of the universe has not been a matter of concern so far (neither was it in Einstein's time). However, as argued in this book, topology plays a major role in elucidating the nature of DM and DE. The fact that there is a form of matter in vast proportion that is not associated with visible light quanta calls for an extra dimension, as shall be argued in the forthcoming chapters. The main hurdle to incorporating an extra dimension is topological, not geometrical.

There have been early attempts by Theodor Kaluza and Oskar Klein (KK theory, revisited in [38]) to incorporate an extra compact curled-up dimension into what was otherwise regarded at the time as an "Euclidean universe". These attempts held promise at least initially and were taken seriously by none other than Albert Einstein himself, who spent the latter part of his life seeking to unify gravity with the other forces of nature. However, a lack of topological understanding of space-time at the time made it impossible to properly incorporate extra dimensions and reconcile these extensions with data available, especially with data pertaining to the prediction of the rest mass of elementary particles.

In this book, the "KK theme" may be said to be resurrected, albeit in a vastly different guise and for a completely different purpose. First, we shall show in the forthcoming chapter that the universe is compact, regardless of its perceived or inferred extension, and multiply connected, and that these topological attributes determine the way the extra spatial dimension gets incorporated. The extra dimension becomes a necessity to explain DM and DE, as can be deduced from the simple argument laid out subsequently.

Suppose we consider a spatial cross section of the universe along a single compact dimension with radius r. We can store energy in this dimension by considering a stationary "de Broglie" wave with wavelength $n\lambda = 2\pi r$, and we know that the kinetic energy of the wave, $E = \frac{hc}{\lambda} = \frac{\hbar nc}{r}\left(\hbar = \frac{h}{2\pi}\right)$, can be translated into a mass: $m = \frac{\hbar n}{cr}$.

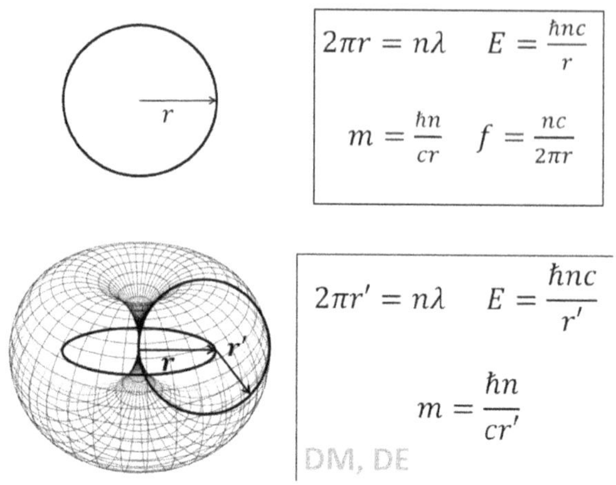

FIGURE 1.25 Dark matter/energy stored along a dormant dimension that cannot be associated with detectable radiation.

Now, assume there is an extra "dormant" dimension in space with radius r' that stores DM (Figure 1.25). This is a fundamental tenet in the theoretical development put forth in this book and is presented here in a simplified fashion for the sake of the argument. The key difference with matter waves along detectable dimensions is that the mass $m = \frac{\hbar n}{cr'}$ cannot be associated with radiation at the Einstein frequency $f = \frac{nc}{2\pi r'}$.

In other words, *the validity of the relation E = mc² is contingent on having radiation and its associated matter stored on the same set of observable dimensions.* A staggering but inevitable consequence of this trend of thought is that Einstein's relation is only fulfilled by 5% of matter in the universe.

1.11 DIRAC SEA: A CANDIDATE FOR THE DARK SECTOR?

It is imperative that we discuss vacuum energy in our quest to investigate the nature of DE. The natural starting point for such a discussion is the Dirac equation, a breakthrough that inspired an enormous intellectual effort, QFD, whereby vacuum energy was properly interpreted and placed on a rigorous footing.

The first successful synthesis between special relativity and QM was accomplished by Paul Dirac (1902–1984) with his relativistic wave equation for the electron conceived in 1928 [39]. Dirac's strategy was to adopt Einstein's relativistic equation

$$E^2 = (pc)^2 + (mc^2)^2 \tag{1.17}$$

with m = rest mass, p = momentum, as a starting point and replace the physical quantities with their corresponding operators in consonance with the Schrödinger equation. At variance with the latter, Dirac's wave function becomes a four-component vector, aptly named a *spinor*, where spin or intrinsic angular momentum emerges as a conserved quantity associated with a quantum number. To yield meaningful solutions, Dirac "linearized" Eq. 1.17 by essentially taking the root square: $E = \pm\sqrt{(pc)^2 + (mc^2)^2}$, essentially "factorizing" the relativistic equation. The existence of negative energies proved extremely problematic and initially exposed Dirac to severe criticism from his peers. Dirac considered the momentum as a four-component vector $\vec{P} = \left(\frac{E}{c}, P_x, P_y, P_z\right) = (P_0, P_1, P_2, P_3)$ where the dot product is determined by the metric $g_{\mu\nu} = \eta_{\mu\nu}$ given in Figure 1.26. Einstein's convention of summation over repeated indices at different levels is followed throughout the book.

$$\left\|\vec{P}\right\|^2 = P_\mu \cdot P^\mu = \eta_{\mu\nu} P^\mu P^\nu = \left(\frac{E}{c}\right)^2 - \left(P_x^2 + P_y^2 + P_z^2\right) \tag{1.18}$$

$$P_\mu = g_{\mu\nu} P^\nu \tag{1.19}$$

To simplify notation, we adopt a unit of velocity equal to the speed of light, so that $c = 1$. Thus, the relativistic Eq. 1.17 reads

$$P_\mu \cdot P^\mu - m^2 = 0 \tag{1.20}$$

Dirac was able to factorize Eq. 1.20 as a wave equation by substituting momentum components for operators ($P_0 = E \rightarrow i\partial_t$, $P_1 \rightarrow -i\partial_x, P_2 \rightarrow -i\partial_y, P_3 \rightarrow -i\partial_z; \hbar = 1$) as prescribed by the Schrödinger formulation, yielding the relativistic wave equation for the electron in the form:

$$\left(i\gamma^\mu \partial_\mu - m\right)\psi = 0 \tag{1.21}$$

with the four (contravariant) gamma matrices indicated in Figure 1.27 and the wave function now represented as a four-component vector, called *spinor.* The gamma matrices satisfy anticommutative constraints imposed by comparing Eqs. 1.20 and 1.21:

$$\left\{\gamma^\mu, \gamma^\nu\right\} = \gamma^\mu\gamma^\nu + \gamma^\nu\gamma^\mu = 0 \quad \text{when} \quad \mu \neq \nu; \quad \gamma^\mu\gamma^\mu = 2g^{\mu\mu} \tag{1.22}$$

The wave function for a particle at rest is

$$\psi = \left(\psi_1, \psi_2, \psi_3, \psi_4\right)^T = \left(a_1, a_2, a_3, a_4\right)^T e^{-iEt}; \quad \sum_{j=1,\ldots,4} a_j^2 = 1 \tag{1.23}$$

Thus, its first two spinor components require that $E = m$, the expected result since $c^2 = 1$ in our convention. However, the last two spinor components yield solutions with the "unrealistic" $E = -m$. In states of a moving electron defined by nonzero third and fourth spinor components, we obtain $E = -\sqrt{(p)^2 + m^2}$ implying that the electron would tend to constantly increase momentum and emit radiation to reach lower and lower energy, all the way to $E = -\infty$, clearly an absurdity that elicited harsh criticism toward the Dirac approach.

$$\eta = \begin{pmatrix} -1 & 0 & 0 & 0 \\ 0 & 1 & 0 & 0 \\ 0 & 0 & 1 & 0 \\ 0 & 0 & 0 & 1 \end{pmatrix}$$

FIGURE 1.26 A Minkowski metric for four-dimensional space-time with inner product metric signature −+++.

$$\gamma^0 = \begin{pmatrix} 1 & 0 & 0 & 0 \\ 0 & 1 & 0 & 0 \\ 0 & 0 & -1 & 0 \\ 0 & 0 & 0 & -1 \end{pmatrix} \qquad \gamma^1 = \begin{pmatrix} 0 & 0 & 0 & 1 \\ 0 & 0 & 1 & 0 \\ 0 & -1 & 0 & 0 \\ -1 & 0 & 0 & 0 \end{pmatrix} \; (\sigma_1)$$

$$\gamma^2 = \begin{pmatrix} 0 & 0 & 0 & -i \\ 0 & 0 & i & 0 \\ 0 & i & 0 & 0 \\ -i & 0 & 0 & 0 \end{pmatrix} \; (\sigma_2) \qquad \gamma^3 = \begin{pmatrix} 0 & 0 & 1 & 0 \\ 0 & 0 & 0 & -1 \\ -1 & 0 & 0 & 0 \\ 0 & 1 & 0 & 0 \end{pmatrix} \; (\sigma_3)$$

FIGURE 1.27 Contravariant Dirac matrices.

Dirac addressed this problem by introducing the following assumptions:

- Dirac's equation is actually of broader applicability, as it holds for any fermion with spin ½. This is because of the way in which the gamma matrices are constructed based on Pauli's spin matrices (Figure 1.27): $\gamma^j = i\sigma^2 \otimes \sigma^j$; $\gamma^0 = \sigma^3 \otimes \mathbb{I}_2$; $\otimes$ = Kronecker product.
- The relativistic wave equation for the electron is actually an equation for ***all*** electrons in the universe.
- All or most states with negative energy are occupied by electrons/fermions, and Pauli's exclusion principle comes into play, forcing the detectable electron/fermion, describable with Dirac's equation, into a state with positive energy. The negative-energy electrons are completely unobservable, and its collection became known as the "Dirac sea".
- An electron (or 1/2–spin fermion in general) in the highest negative energy level ($E = -m$) can become "detectable" by absorbing a photon of energy $E = hf = 2mc^2$. Such a photon can cause an electron to transition from the highest negative energy state ($E = -m$) to the lowest positive energy state ($E = m$). Half of the energy absorbed would correspond to the emergence of a detectable electron, and the other half would correspond to a "hole" in the Dirac sea, and since a negative charge has been taken away into the detectable realm, the hole is equivalent to a positive "antielectron", which in modern terms would be named "positron".

These assumptions lead to the prediction of the existence of antimatter, a prediction eventually confirmed by experiment. Positrons were detected by Carl David Anderson (1905–1991) within the spontaneous formation of electron–positron pairs in a cloud chamber.

The Dirac sea was ultimately considered a dated construction superseded by QFT. In QFT, particles are considered not describable individually or even in a closed system but rather excitations of universal fields. Furthermore, the emergence of the electron–positron pair is regarded as a spontaneous elementary process in absolute vacuum (Figure 1.28), whereby the detectable fermion spontaneously emerges from vacuum through communication with a photon with a lifetime determined from Heisenberg's uncertainty principle: $\tau \sim \frac{\hbar}{4mc^2}$.

With the discovery of the dark universe, the Dirac sea may need to be summoned once again, as it will be described in this book. A fermionic sea totally unaccounted for may be just what is needed to delineate the nature of the dark universe, except that the weak force governing the decay of such particles

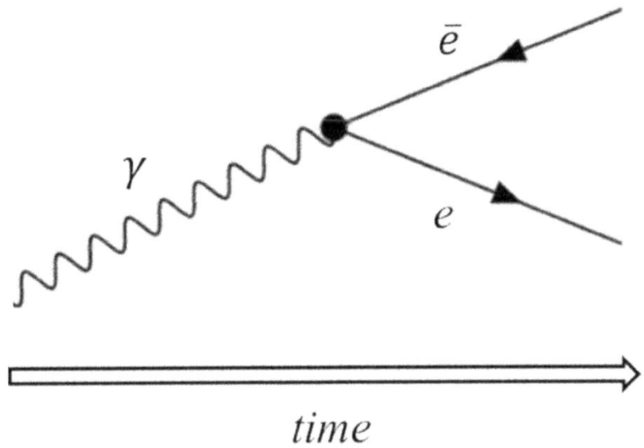

FIGURE 1.28 Electron–antielectron pair creation in quantum field theory as schematically represented in by its Feynman diagram.

would not be operative in the way it is described in the Dirac equation. It is well established that DE and DM cannot be turned visible through photon absorption, so a vast proportion of the negative-energy fermionic sea must be rendered impervious to communication of the electroweak force, but how? The identification of the purveyor to the dark universe is described in Chapter 5 and requires a careful discussion of "chirality", an essential concept in particle physics.

It is obviously tempting to identify the Dirac sea of negative-energy fermions with the dark universe, but this assertion is incorrect and would require serious revision and fine-tuning vis-à-vis the fact that the electroweak interaction that enables fermions to escape from the Dirac sea is *chiral*, a conceptual framework elaborated in Chapter 5.

We shall provide a vantage point to interpret DE as the surplus in vacuum energy from the amount required to maintain the topology of the universe as it expands. The concentration of this surplus is indeed constant since DE is not subject to dilution associated with vacuum creation. This fact stands in contrast with the concentration of DM that will be shown to be in dynamic equilibrium with detectable matter. Clearly, the DM concentration decreases as it gets progressively diluted because the vacuum grows larger over time. The sustainability of the universe ensures the dynamic equilibrium between "darkened" fermions drawn from the Dirac sea and detectable matter, a contribution that maintains the overall mass of DM constant while the overall concentration of DE is kept unchanged.

At this point, we can anticipate that the dynamic equilibrium between DM and detectable matter is maintained by a cosmic engine fueled by vacuum energy that sustains the portal to the dark universe while maintaining a constant concentration of DE in an ever-expanding universe. This AI-empowered model will be elaborated in Chapter 5.

1.12 DARK MATTER AND UNIVERSE TOPOLOGY

The narrative in this and the subsequent sections will take a turn, as it becomes essential at this stage to retrace the discussion of Section 1.5 on the topological features of the universe that seem to have been hitherto omitted since they do not impinge in any obvious way on Einstein's local theory of gravity or even on the nonlocality of quantum physics. This topological characterization invites a veritable deconstruction of the SM of particle physics. In turn, the analysis is essential to deal with the problem of whether DM may be accounted for by extending the SM with the incorporation of an extra spatial dimension. After provisionally settling the problem of the universe topology as a quotient topology, this and the subsequent sections argue for the existence of a compact fourth spatial dimension that stores DM. The quest for DM becomes contingent on identifying and validating the topology of the universe as a compact multiply connected space whose quotient space modulo the "dormant" dimension is homeomorphic to $\left(\mathbb{R}^3/\mathbb{Z}^3\right)\times\mathbb{R}$. Experimental measurements of the spectrum of the gradient temperature field for Cosmic Microwave Background (CMB) radiation lead to the conclusion that a dormant circular dimension is consistent with a present-day toroidal space-time with extremely large aspect ratios relative to the dormant coordinate. The topological invariance of the universe since the Big Bang is thus upheld (it was once compact; hence, it remains compact), provided the relativistic picture holds since the cosmic birth.

As discussed in the previous sections, Newton is credited with providing the first model of gravity, which he named law of universal gravitation. The predictive power of this law is staggering, and thus, it constitutes one of the towering achievements of humanity. Yet, when asked about the nature of gravity, Newton cautiously responded that he would not frame hypotheses. Hence, in this sense, Newton may be said not to have conceived a "theory of gravity". Humanity had to wait until the early twentieth century to get the first such theory, embodied in Einstein's GR. This theory recognizes gravity as an acceleration exerted by a body endowed with mass and regards the fabric of space-time as if it were a membrane, whereby acceleration becomes effectively commensurate with curvature, in turn determined by mass concentration (Figure 1.29). To further see how Einstein got inspired to describe the fabric of space-time, let us examine the equation for a vibrating string that gets perturbed from its equilibrium position along a spatial dimension x by an amount ϕ:

$$\frac{\partial^2\phi}{\partial t^2} - v^2\frac{\partial^2\phi}{\partial x^2} = 0 \tag{1.24}$$

The term on the left indicates the second instantaneous variation of the displacement with respect to time, that is, it represents acceleration due to an exogenous force (Figure 1.27), while the distortion in the string caused by the force exerted is represented by the curvature $\frac{\partial^2\phi}{\partial x^2}$ of the string. It should be noted that this second spatial derivative is seldom interpreted as the curvature, which in fact it is, just like the first

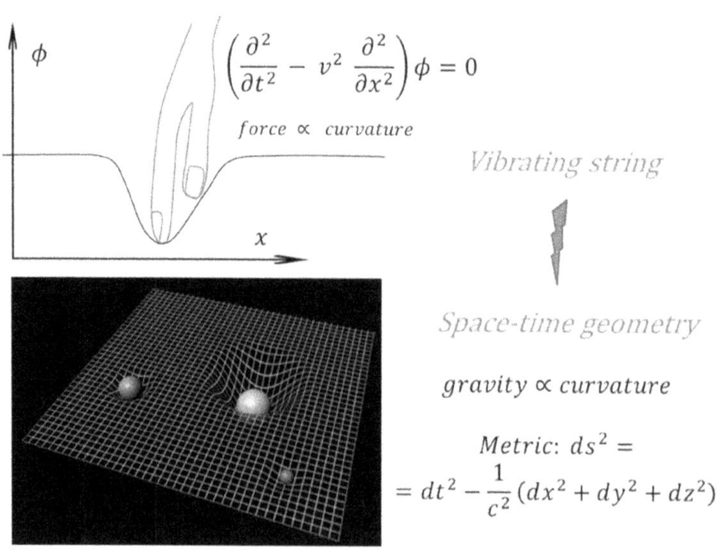

FIGURE 1.29 The equation for the vibrating string where curvature becomes commensurate with the applied force (or acceleration) as inspirational source for Einstein's differential geometry model of general relativity.

derivative gives the slope of the curve. The proportionality constant is the square of the speed at which the perturbation propagates in time along the x dimension.

While this fact is typically omitted, it is fairly obvious that the fundamental Eq. 1.24 linking curvature and acceleration served as the primary source of inspiration for the differential geometry model of space-time that Einstein adopted in GR, his classical theory of gravity (Figure 1.29). It suffices to notice that the metric he adopted in his 4D space-time with coordinates t, x, y, z became:

$$ds^2 = dt^2 - \frac{1}{c^2}\left(dx^2 + dy^2 + dz^2\right) \tag{1.25}$$

This differential volume ds^2 of space-time is the one dictated by the relation between curvature and acceleration, except that the velocity v is now the speed of light c. A complementary way to assert the pivotal relation between curvature and gravity at the heart of Einstein's argument requires that we remind ourselves of another classical equation of physics, the Poisson equation. This equation relates the mass density ρ (amount of mass per unit volume) and the gravitational field ϕ "generated" by the mass. Poisson's equation reads:

$$\nabla^2\phi = 4\pi G\rho \tag{1.26}$$

where $\nabla^2 = \left(\frac{\partial^2}{\partial x^2} + \frac{\partial^2}{\partial y^2} + \frac{\partial^2}{\partial z^2}\right)$ and $G = 6.674 \times 10^{-11}\ \text{Nm}^2/\text{kg}^2$ is Newton's gravitational constant [31]. Thus, integrating Eq. 1.26 on a ball of radius r yields Newton gravitational field

$$\phi = -MG / r \tag{1.27}$$

where M is the mass contained in the ball of radius r. Hence, we can justify the relationship between curvature (measured as $\nabla^2\phi$) and acceleration due to the gravitational pull per unit volume, given by $4\pi G\rho$. Thus, the differential geometry of the space-time manifold of GR is inspired and amply justified by the classical equations of physics. This observation prompts us to assert that Einstein's theory of gravity is in fact a classical theory. It is clear that we now have a theory of gravity that includes its interaction with light, a feature absent in the Newtonian law of universal gravitation. Einstein provides a differential geometry framework in which light travels along a hypermembrane (a manifold often referred to as "brane") whose geometric fabric is defined by mass distribution.

Yet, it has proven daunting to reconcile Einstein's theory with quantum physics, now known to govern the other three fundamental forces of nature (electromagnetism and the WNF and strong nuclear force (SNF)). Furthermore, gravity is 10^{-38} times weaker than electromagnetism, an extremely irksome fact that points to a massive geometric dilution of gravity on the known dimensions of space-time. These conundrums are of course compounded by the mysterious nature of DM that postulates the existence of invisible massive particles that do not detectably interact with the known elementary particles identified in the SM [40], and yet they provide the "missing gravity" in the detectable portion of the universe [35,36].

A window of opportunity for further scientific inquiry into these seemingly related problems and others arising thereof is offered by the possibility of incorporating a extra (fifth) dimension to the differential geometry of the space-time manifold. To enable the possibility of storing significant amounts of kinetic energy in an undetectable stationary wave amenable of a quantum mechanical treatment, the extra dimension would be expected to be compact, specifically rolled up in a circle of extremely small radius. It should be noted that the smallest conceivable material dimension at present is that of a quark (see Section 1.13), of the order of 10^{-18} m (a length unit named attometer) [41].

If we specifically adopt a circular fifth dimension of radius $r_0 = 0.802 \times 10^{-18}$ m (the smallest material scale assimilated to the effective quark diameter established from parton models of inelastic $e^{\pm}p$ scattering [41]) we find out, through Einstein's relation $E = hf = hc / \lambda$, that it can store stationary waves with extremely large energies

$$E_{5,n} = \frac{n\hbar c}{r_0}, \quad n\lambda = 2\pi r_0, \quad \hbar = \frac{h}{2\pi}, \quad n = 1, 2, \ldots \tag{1.28}$$

Strikingly, the lowest such energy is $E_{5,1} = 246$ GeV, which is precisely the vacuum expectation energy of the elementary particle responsible for bestowing mass on the other known particles, the so-called Higgs boson (see Section 1.13) [40]. The daunting problem of incorporating the extra dimension arises from the fact that the circular dimension cannot be thought of an independent of the others, at least there does not seem to be any obvious reason for such an assumption. Hence, the preexisting four dimensions should be considered locally cylindrical or, rather, helical, instead of linear, with symmetries becoming only approximate and vast differences in curvature depending on the stride of the helices. Such a universe will be described subsequently and will be shown to be far more suitable to achieve a unified field theory while explaining the extreme geometric dilution of gravity on the "observable space-time manifold", as well as the origin of DM and DE.

Such a fifth dimension would be extremely hard to detect with current experimental resources, even at the Large Hadron Collider (LHC) operating at optimal performance, due to the extremely high energies associated with wavelengths of the order the quark dimension. Yet, highly massive particles may be yielded by storing energy in the fifth dimension. To visualize this, we note that the kinetic energy along the fifth dimension would be undetectable and hence will be regarded as rest mass, in accord with Einstein's relation [40]. Thus, the total rest mass M of a particle on a 5D space-time manifold would be given by the equation

$$M^2 = m^2 + m_5^2 = m^2 + \left(\frac{n\hbar}{cr_0}\right)^2, \quad n = 1, 2, \ldots \tag{1.29}$$

Here, m denotes the detectable rest mass that we would be capable of measuring from the particle's existence in Einstein's 4D space-time, and $m_5 = p_5/c$ is the rest mass associated with the component p_5 of momentum along the fifth dimension.

If a particle only stores stationary kinetic energy in the fifth dimension and has no observable rest mass, it is likely to be invisible, yet it could be more massive than any of the known particles. For instance, an "ur-Higgs boson" would have the mass corresponding to the vacuum expectation energy of the Higgs boson, at

$$M = M_{u-H} = \frac{\hbar}{cr_0} = \frac{246\ \text{GeV}}{c^2} \tag{1.30}$$

This is almost twice the detected mass of the Higgs boson, known to be 124.97 GeV/c^2. The ur-Higgs and its relatives with masses $M_n = \frac{n\hbar}{cr_0}$, $n = 2, 3, \ldots$ may be regarded as the first particles created after the Big Bang. We may estimate the lifetime at which such particles were created from Heisenberg's uncertainty relation discussed previously:

$$\Delta E \times \Delta t \sim \frac{\hbar}{2} \tag{1.31}$$

Since $\hbar = 6.582 \times 10^{-16}$ eVs, we may estimate the time of formation of an ur-Higgs boson ($\Delta E = 246 \times 10^9$ eV) at $t = t_{u-H} \approx 1.34 \times 10^{-27}$ seconds. More massive relatives of the ur-Higgs are expected to emerge at shorter times, estimated at $t = \frac{t_{u-H}}{n}$.

As we discuss massive relatives of the ur-Higgs, we are prompted to ask: How large can n be? The answer is straightforward, as the so-called Planck mass, $M_P = \sqrt{\frac{\hbar c}{G}} = 1.22 \times 10^{19}$ GeV/c^2, sets an upper

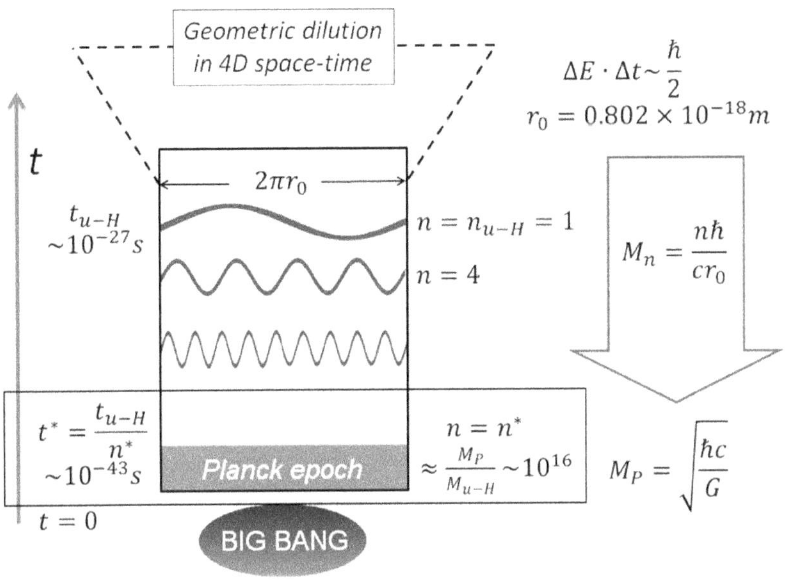

FIGURE 1.30 Elementary particles in an early universe covering the period 10^{-43} seconds $\leq t \leq 10^{-27}$ seconds after the Big Bang obtained by storing energy as stationary waves that span the fifth dormant dimension.

bound to the mass, so that if $M_n = \frac{n\hbar}{c r_0} \geq M_P$, then the particle becomes a black hole. Thus, the first particle formed in the universe had a mass M_{n^*} corresponding to the largest integer $n = n^*$ such that $n^* \leq \frac{M_P}{M_{u-H}} \approx 4.95 \times 10^{16}$. This supermassive particle formed at a time estimated at $t^* = \frac{t_{u-H}}{n^*} \sim 10^{-43}$ seconds, during the so-called "Planck epoch" of the universe.

Thus, a picture of the very early universe for the period 10^{-43} seconds $\leq t \leq 10^{-27}$ seconds may be obtained by assessing the energy stored in the stationary de Broglie wave that spans the fifth dimension, as schematically depicted in Figure 1.30.

1.13 TOPOLOGY OF THE UNIVERSE: A RELEVANT DISCUSSION IN PARTICLE COSMOLOGY

The most difficult issue we stumble upon when postulating a compact fourth spatial dimension to store DM is that the global topology of the universe must allow for such a possibility. In the past, people have tried to incorporate a curled-up dimension into an Euclidean space, an idea that is topologically incompatible yet, of course *a priori* possible. Not only do such attempts fail to generate correct extensions of GR, as discussed earlier, but they arose from the unappealing and unwarranted assumption that one dimension would be compact, while the others are infinite.

Incorporating a compact dimension is a nuanced matter that pivotally depends on the topology of the universe. This was clearly not a matter of concern for Einstein because his GR is essentially governed by differential equations. Since differential equations describe a local situation, the global topology of space-time plays no obvious role. Furthermore, relativity has no concern for boundary conditions, as the universe is assumed to be infinite. In our topology, the universe is compact and multiply connected, but there are no boundary conditions, simply because there is no boundary.

On the other hand, QM defines the fabric of space, as determined by vacuum entanglement, but topology is not factored into the theory in any obvious way, even if entanglement is thought to be the culprit for the nonlocality of quantum phenomena. However, topology suddenly becomes extremely relevant as one tries to reconcile GR with quantum physics, as it is the case with quantum gravity, as shown in Chapters 2 and 5.

If we assume topological invariance of space throughout the universe evolution starting at the Big Bang event, then space cannot be infinite and flat, and therefore Euclidean. Provided we accept the Big Bang scenario, space must be compact because it once was compact with certainty. But if space is compact, then it must be multiply connected. In addition, space cannot have a boundary because that would imply an interface with nothingness, and it is physically and metaphysically impossible to interface with nothingness: An interface presupposes two media. These considerations introduce several constraints on what the topology of the universe must be. Summarizing, space must be compact, locally flat, orientable, and lacking boundary. This leaves us with essentially one option, and it is a multiply connected three-dimensional manifold: The three-torus or Cartesian product $S^1 \times S^1 \times S^1$. Hence, the incorporation of a fourth circular dimension is compatible with the topology of the space-time manifold.

There is a fundamental difference in the way the extra spatial dimension, the curled dormant dimension, is incorporated in the treatment presented in this book when compared with the early attempts, especially the so-called Kaluza–Klein model [38]. These researchers incorporated the extra dimension *directly* to the Euclidean space, creating the absurdity of having one compact and three infinite spatial dimensions in space-time with no justification. This construction proved to be untenable from every perspective, a mere intellectual curiosity.

As we have shown, the spatial cross sections of space-time cannot be regarded as isomorphic to $\mathbb{R}^3$ but rather to the three-dimensional torus $\mathcal{T}^3 \approx \mathbb{R}^3/\mathbb{Z}^3$, that is, a quotient space of classes modulo all the spatial translations of the unit cell indexed by three integers (Figure 1.31). This implies that the space is multiply connected with all spatial dimensions being compact, hence allowing for the incorporation of an extra circular dimension ($S^1 \approx \mathbb{R}/\mathbb{Z}$) without the conceptual violence of the Kaluza–Klein models. A hybrid $\mathbb{R}^3 \times \mathbb{R}/\mathbb{Z}$ of three infinite and one compact spatial dimension would make no sense and is, in fact, not compatible with the tenets of GR, assuming this theory is upheld since the inception of the universe, as topological change is forbidden.

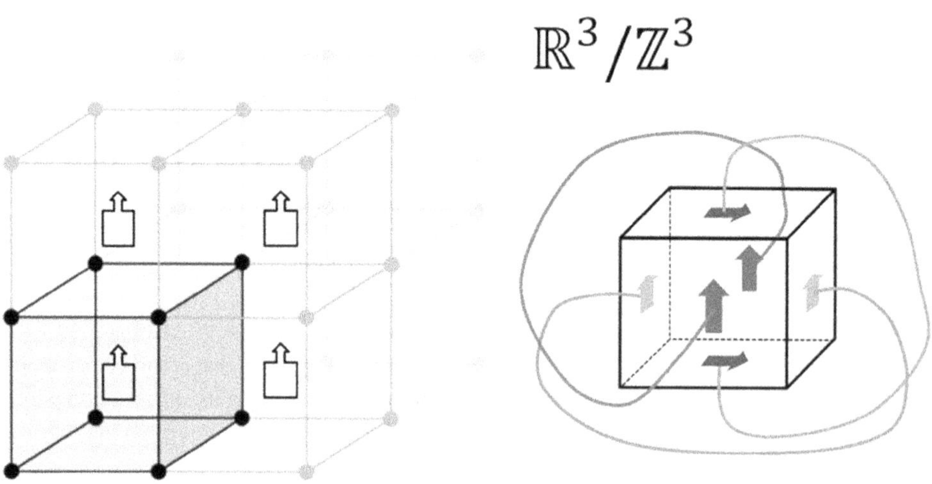

FIGURE 1.31 Generic spatial cross section of space-time with compact coordinates, as required by the topological invariance of the universe dictated by the tenets of general relativity. The compact space is isomorphic to the three-dimensional torus in accord with the topological equivalence $\mathbb{R}^3/\mathbb{Z}^3 \approx \mathcal{T}^3$. The latter manifold is obtained by identifying opposite faces of the solid cube in three dimensions, hence featuring a wormhole upon construction with dimensional aspect ratios equal to 1 (Chapter 5).

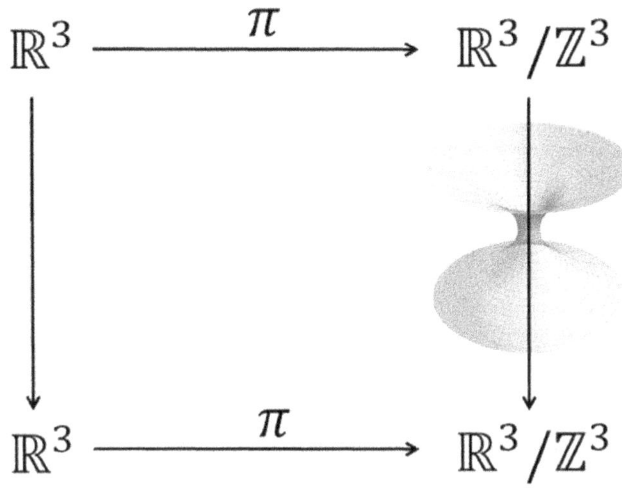

FIGURE 1.32 Commutative scheme adopted by artificial intelligence to circumvent the artefactual wormhole phenomenology associated with the universe compactness by lifting it at the level of universal covering $(\mathbb{R}^3, \pi)$ of the toroidal spatial cross sections $(\mathbb{R}^3 / \mathbb{Z}^3)$ of space-time.

At a very fundamental level, the Poincaré group, assumed to be the group of isometries of space-time over which the whole edifice of the SM is built, does not really contain the proper symmetries but only approximate symmetries or, rather, pseudo-symmetries, since the underlying spatial cross sections of space-time are not isomorphic to $\mathbb{R}^3$ but to $\mathbb{R}^3 / \mathbb{Z}^3$.

On the other hand, a spatially compact universe implies the existence of a primeval wormhole, as inferred from Figure 1.31 and demonstrated in Chapter 5. This can be visualized in a two-dimensional cross section $\mathcal{T}^2$ with a dimensional aspect ratio of 1. As shown in Chapter 5, the sustainability of this wormhole is concomitant with the universe expansion. On the other hand, the very notion of universe expansion becomes meaningless in the noncompact space $\mathbb{R}^3$, while it makes good sense to talk about expansion of the unit cell in the quotient space $\mathbb{R}^3 / \mathbb{Z}^3$. This is yet another reason to assume that the spatial cross sections of the relativistic space-time are actually isomorphic to $\mathbb{R}^3 / \mathbb{Z}^3$, and not to $\mathbb{R}^3$, as assumed in the Minkowski space-time endowed with the isometries of the Poincaré group.

On the other hand, AI adopts the commutative scheme presented in Figure 1.32, where the auto-encoder defines the lifting of the SM at the level of the universal covering with canonical projection $\pi : \mathbb{R}^3 \rightarrow \mathbb{R}^3 / \mathbb{Z}^3$, so that the diagram presented in Figure 1.32 is commutative. This lifting has considerable advantages since it circumvents the observed universe expansion with the concomitant primeval wormhole as artefactual of "Plato's cave physics", where the observer is placed. Even DE, shown in Chapter 5 to be the residual vacuum energy not invested in sustaining the primeval wormhole, becomes artefactual in Plato's cave phenomenology in $\mathbb{R}^3 / \mathbb{Z}^3$, only accessible to the quantum observer.

1.14 PRIMER ON QUANTUM FIELD THEORY

Major modeling efforts that followed after the advent of GR and QM revealed that the true nature of elementary particles cannot be properly captured in the corpuscular representation. Rather, elementary particles constitute excitations or warps in fields – one for each particle type – that establish a correspondence between each point of space with a scalar or vector value [38]. This is the spirit of the QFT that spearheaded current modeling efforts in elementary particle physics.

An elementary particle becomes a local concentration of energy or its equivalent mass, as dictated by Einstein's relation $E = mc^2$ derived in Chapter 1. The particle field itself is viewed as a map $\phi : W \rightarrow V$, where W is the four-dimensional space-time of special relativity and V is either $\mathbb{R}, \mathbb{C}$, the sets of real or complex numbers, or $\mathbb{R}^n$ $(\mathbb{C}^n), n > 1$,depending on whether ϕ is a scalar or a vector field, respectively. The picture where particles are regarded as excitations of their respective fields is a major tenet of QFT and the modeling effort resulting thereof is known as the SM of particle physics [38].

In accord with current thinking, particles are best characterized as local modes of storing energy/mass, that is, localized field excitations defined by independent representations of the symmetry group of space-time. These representations have generators that are parametrized by specific attributes of particles, including charge, mass, spin (an intrinsic angular momentum vis-à-vis an internal axis of rotation), spin degrees of freedom (projection of the spin vector along the direction of particle motion), and so on. Yet, this manner of particle classification tells us very little unless we can define interactions between particles based on their intrinsic attributes, assessing how particles communicate and transform into one another. The nature of these interactions is delineated in the Section 1.15.

Now let us briefly explored how the concept of particle field came about. By attributing the energy of the photon to its frequency of oscillation as in the equation $E = hf = hc / \lambda$, Einstein was the first to associate a particle with a wave in his description of the photoelectric effect. Einstein, and Planck before him, reasoned that the frequency was simply a surrogate for kinetic energy, as it is an indicator of the number of cycles per unit time. This was followed by de Broglie's representation of the electron orbiting the atomic nucleus as a stationary wave with a wavelength that satisfies the relation $n\lambda = 2\pi r$, where n is a positive integer value and r is the orbit radius. This "quantization" of the wavelength ensures constructive (reinforcing) phase interference, as depicted in Figure 1.33. It also captures the discreteness of the experimentally obtained electron absorption spectrum, as the only allowed transitions for the electron are the result of absorption of photons carrying energies given by the differences between the n-indexed electron energies $E_n = n\hbar c / r$, with $\hbar = h / 2\pi$. In fact, de Broglie went one step beyond, as he associated any particle with momentum p to a wavelength λ according to the relation $p = h / \lambda$. We can see in these early modeling efforts the need to go beyond a mere corpuscular description of particles, a conceptual framework later adopted by QFT in its construction of the SM [38].

The SM has defined, characterized, and organized all of the known elementary particles in much the same way that the periodic table systematically categorized or classified the types of atoms known as elements. The predictive value of the SM has been staggering, notwithstanding the fact that it does not fully incorporate gravity, while being able to encompass the other three quantum forces of nature in a unified way.

Yet, by 1970, mainly due to the monumental "bottom-up" modeling efforts based on QFT, elementary particles were shown to fall in two main categories: Fermions and bosons, with fermions making up matter while bosons, or more precisely "gauge bosons", transmit forces to which fermions respond or

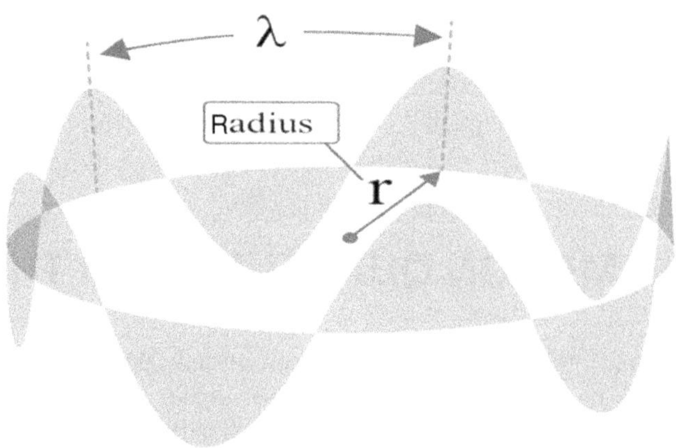

FIGURE 1.33 A de Broglie stationary wave of wavelength λ in a circle of radius r.

react to [38]. Fermions are divided into two kinds of particles, depending on the forces they respond to: Quarks and leptons. Particles within the three types, quarks, leptons, and bosons, are symmetry-related to each other. This means that a coordinate transformation associated with a space-time symmetry operation transforms one particle into another of the same family. It was established that particles communicate with one another via four forces: Electromagnetism, SNF, WNF, and gravity. The SM describes the first three, while gravity does not yet feature satisfactorily in the current version of the SM, as discussed in Chapter 2.

Different particles communicate and interact through different forces. For example, only the quarks relate to the gluon, its natural gauge boson and carrier of the SNF, while electrons relate to the photon, the gauge boson that carries the electromagnetic force. In addition, electrons can also communicate via the *W* boson and *Z* boson, the carriers of the WNF. So electromagnetism is the force that holds electrons in the atom, while the SNF communicated by gluons keeps the nuclei from breaking apart. Meanwhile, the WNF communicated by *Z* or *W* bosons assists the radioactive decay of nuclei, whereby, for example, the quark/gluon assemblage of a neutron composite becomes that of a proton upon interaction with a *W* boson with emission of electromagnetic particles.

We shall now describe the Higgs boson, a particle endowed with a field ϕ that may be represented as a doublet of complex numbers: $\phi = (\phi_1, \phi_2)$, with $\phi_j = \mathrm{Re}\,\phi_j + i\,\mathrm{Im}\,\phi_j, j = 1,2$. This field, empirically determined in a significant modeling effort, is peculiar in that its potential energy does not reach its minimum at zero field but a value of ~246 GeV (1 eV = 1.602×10^{-19} J, 1 GeV = 1.602×10^{-10} J) known as the vacuum expectation value (v), hence implying a symmetry breaking (Figure 1.34). The symmetry breaking arises because different changes in potential energy around the zero-point energy are obtained depending on the direction of variation of the field. For example, if we write the singlet field in radial coordinates as $\phi = \rho e^{i\theta}$, then the zero-point energy is the circle $\rho = v$ and radial variations $\rho = v \pm \delta h$ yield variations in the potential energy (dashed line in Figure 1.34), while angular variations along the circle $\rho = v$ yield no change in potential energy, which remains at its zero value. This symmetry breaking turns out to be of paramount importance in determining the role of the Higgs boson vis-à-vis its interactions with other particles. Thus, just like other gauge bosons are force carriers, communicating different types of forces associated with particle interactions, the Higgs boson may be interpreted as bestowing mass upon other particles.

To interpret this process of mass endowment as well as the communication of fundamental forces by the other gauge bosons, it is necessary to resort to the mathematical arsenal of particle physics [38]. This arsenal is necessary to describe the "geodesic flow", that is, the lines of least action followed by the fields that underlie particle physics. As said, a basic tenet of QFT indicates that a particle is represented as a local excitation of its respective field, representing a local concentration of energy or, equivalently, of mass. In QFT, a particle scalar field $\phi = \phi\left(\{x_\mu\}_\mu\right)$ in space-time is determined by the particle Lagrangian

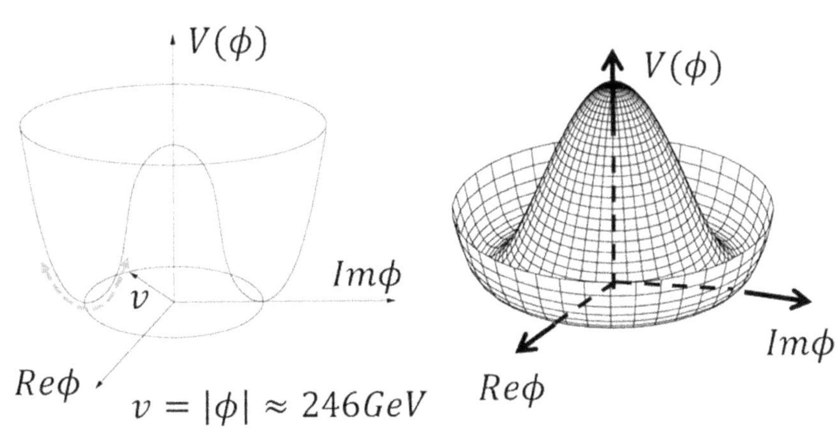

FIGURE 1.34 Scalar field for the Higgs boson.

$\mathcal{L} = \mathcal{L}\left(\phi, \{\partial_\mu \phi\}_\mu\right)\left(\partial_\mu \equiv \frac{\partial}{\partial x^\mu}\right)$ whose definition is inspired by classical mechanics: $\mathcal{L} = \frac{1}{2}\left(\partial_\mu \phi\right)^2 - V(\phi)$, with the first term on the r.h.s. representing the kinetic energy (K), and the second, the potential energy. The latter is exemplified in Figure 1.34 for the Higgs field. Einstein convention of summation over repeated indices is followed throughout. Thus, the geodesic flow in space-time defined by the field $\phi = \phi\left(\{x_\mu\}_\mu\right)$ minimizes the action $\mathfrak{A} = \int \mathcal{L}\left(\phi, \partial_\mu \phi\right) dx^\mu$, and hence, the field satisfies the so-called Euler-Lagrange equations:

$$\partial_\phi \mathcal{L} = \partial_\mu \left[\frac{\partial \mathcal{L}}{\partial(\partial_\mu \phi)}\right] \tag{1.32}$$

To see how two particles, determined, respectively, by fields ϕ_1, ϕ_2, interact, or one particle communicates a force to another particle, we need to consider the sum $\mathcal{L}_1\left(\phi_1, \{\partial_\mu \phi_1\}_\mu\right) + \mathcal{L}_2\left(\phi_2, \{\partial_\mu \phi_2\}_\mu\right)$ of their respective Lagrangians. An appropriate coordinate transformation $x^\mu \to x'^\nu$ and suitable transference or swapping of terms yield the sum: $\mathcal{L}_1'\left(\phi_1', \{\partial_\nu \phi_1'\}_\nu\right) + \mathcal{L}_2'\left(\phi_2', \{\partial_\nu \phi_2'\}_\nu\right)$ satisfying the action equality

$$\int \left[\mathcal{L}_1\left(\phi_1, \{\partial_\mu \phi_1\}_\mu\right) + \mathcal{L}_2\left(\phi_2, \{\partial_\mu \phi_2\}_\mu\right)\right] dx^\mu = \int \left[\mathcal{L}_1'\left(\phi_1', \{\partial_\nu \phi_1'\}_\nu\right) + \mathcal{L}_2'\left(\phi_2', \{\partial_\nu \phi_2'\}_\nu\right)\right] dx'^\nu \tag{1.33}$$

This implies that the original interactive particle pair defined by the fields ϕ_1, ϕ_2,has transformed through mathematical manipulation (including only term reorganization and change of variables) into new particles interpreted as being defined by the fields ϕ_1', ϕ_2'. These destiny fields are associated with the "destiny Lagrangians" $\mathcal{L}_1'$, $\mathcal{L}_2'$ and may correspond to any number of particles, elementary or composite.

For example, let us consider the interaction of the Higgs field $\phi_1 = \phi_H$ (Figure 1.34), with a second field ϕ_2 corresponding to a particle that may be the original massless version of the W or Z boson. The kinetic energy K_H for the Higgs field may be written in terms of the covariant coordinates $D_\mu = \left(\partial_\mu - i\phi_2\right)$ that respond to the field of the other particle. Upon the coordinate changes $\phi_H \to (\rho, \theta)$ and substituting $\rho = v + \xi$ around the v-value ring of ϕ_H (Figure 1.34), the kinetic energy K_H reads (asterisk here denotes complex conjugate):

$$K_H = \frac{1}{2}\left(D_\mu \phi_H\right)\left(D_\mu \phi_H\right)^* = \frac{1}{2}\left(\partial_\mu \xi\right)^2 + \frac{1}{2}\phi_2^2 (v + \xi)^2 \tag{1.34}$$

This implies that the precursor to the gauge boson has now gained the potential energy term estimated as $\frac{1}{2}\phi_2^2 v^2 : \mathcal{L}_2' \approx \mathcal{L}_2 + \frac{1}{2}\phi_2^2 v^2$, which implies that the mass bestowed by the Higgs field corresponds to radial (i.e., along the ξ–coordinate) excitations of its field, as depicted in Figure 1.34.

1.15 THE BIG BANG SCENARIO IN AN ONTOLOGICAL NONCOMPUTABLE CONTEXT

The forthcoming "metaphysical" discussion is inevitable, since we shall discuss the Big Bang scenario in connection with the nature and origin of DE and DM. Space-time, a foundational concept in relativity, incorporates time as an inherent dimension, so we may assert that space-time per se, and hence the universe, cannot evolve. *Prima facie,* this statement may seem disconcerting, but considering the evolution of space-time introduces a paradox of the Russell type. The point is that the phrase "evolution of space-time" is meaningless as no entity where time is immanent in its being can actually evolve. Since time is inherent to space-time and inseparable from the spatial dimensions, how could space-time change over time? All time change in space-time is already contained in space-time (Figure 1.35).

What we presumably mean by "evolution of the universe" is actually something quite different: Consider "formally" a fixed-t cross section of space-time and define evolution of the universe as the sequence of t cross sections as the "parameter" t is varied. In other words, we consider the time evolution of space. While this seems *a priori* a well-defined object, it is immanent in space-time, so it cannot be equated, even by abuse of language, with the latter. Furthermore, space-time does not allow for the dissociation of time from spatial dimensions. This is a pillar of relativity. There is no external clock. The dimensions are not "separable". For example, if we query about the Big Bang scenario vis-à-vis the universe evolution, we are rapidly trapped into a paradox, because the extreme mass concentration at the "origin" prevents time from flowing; hence, *the primeval point is a priori a state of the universe from which the universe can never evolve.* This is an extreme case of the time dilation experienced by entities crossing the event horizon of a black hole. In fact, a trajectory leading out of a black hole would entail time reversal (Figure 1.36), so time cannot be "externalized" at the Big Bang, let alone flow to describe the origin of the universe.

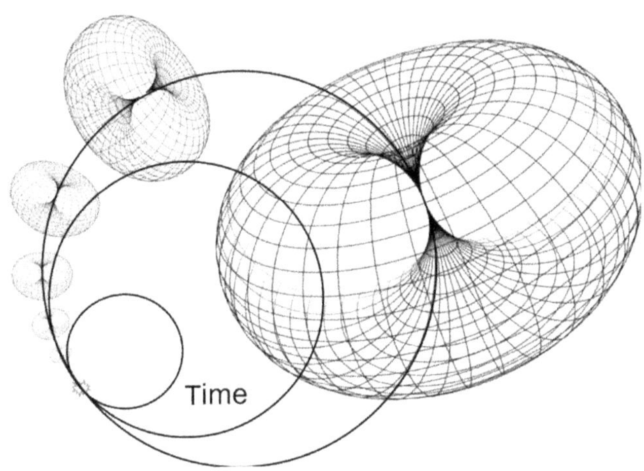

FIGURE 1.35 Space-time subsumes its own evolution; hence, the phrase "space-time evolution" is rigorously meaningless and entails a Russell-type paradox. A torus topology of space cross sections is justifiably adopted.

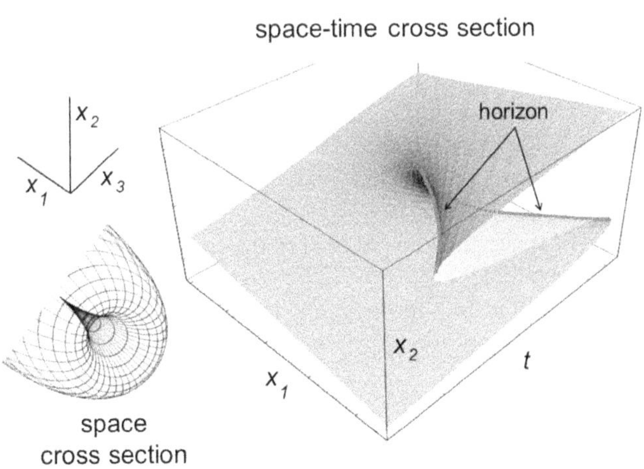

FIGURE 1.36 Crossing the event horizon of a black hole requires time reversal, as depicted in the space-time model of the singularity.

Space-time postulates interdependence between spatial dimensions and time. As such, the dimensions are not separable or cannot be regarded as independent. Thus, fixed-time cross sections may constitute a "formally" acceptable concept, but the time evolution of such cross sections is problematic, especially given that the magnitude of time dilations is strongly dependent on space. This discussion brings us back to the Big Bang scenario. If space is concentrated in a geometric point and time is consequently dilated to infinity, how could the universe be said to have ever evolved in the Big Bang scenario? Furthermore, what is meant by the evolution of the universe, when space-time evolution, understood in any acceptable way, is subsumed in space-time itself (Figure 1.35)? It appears that the Big Bang scenario is ill-posed, and even the phrase "universe evolution" is ill-defined.

Yet, the discussion on universe evolution and Big Bang is alive and ongoing and has borne fruit, especially in the realm of quantum cosmology [31]. This makes it necessary to come up with an operational definition of "evolution of the universe" or, at least, of "evolution of space" that is not conceptually violent and can even reconcile the inseparability of space and time that is inherent to the relativistic perspective. A possible *ansatz* put forth in this book is as follows: Assume a time dilation defined by $t_v \rightarrow t = t_v / \vartheta(x)$, scaling time flowing in an inertial vacuum (t_v) according to a spatially dependent factor $\vartheta(x)$, inherently dependent on the local curvature tensor and commensurate with its norm. We define *extrinsic evolution of space* as the evolution of the spatial t-cross section of space-time, parametrized by rescaling time to make it an extrinsic parameter dissociated from the intrinsic time $\left(t = \dfrac{t_v}{\vartheta(x)}\right)$ that is fixed as a constant in determining the cross section. In other words, we consider the "extrinsic evolution" of the t-slice of space-time: $\mathcal{X}(t_{\text{ext}}) = \{x = x(t)\}_x$ re-parametrized for each point in space as dependent on "extrinsic time": $t_{\text{ext}} = t \times \vartheta(x(t))$.

For relativistic singularities such as the black hole, the scaling of time dilation does not hold. This is because to escape from a black hole horizon, the speed of the moving body should exceed the speed of light, implying that time should move backwards. This can be readily visualized in a cross section of the space-time manifold at the black hole, incorporating two spatial dimensions and time (Figure 1.36). In general, following Hendrik Lorentz [31], we can make the time scaling dependent on the speed of the moving body in accord with the so-called Lorentz transformation: $(\vartheta) = \sqrt{1 - \dfrac{v^2}{c^2}}$; however, the Lorentz formula yields an imaginary factor for speeds exceeding the speed of light ($v > c$), which is meaningless in this context and also incorrect because it does not reflect the actual reversal of time associated with the escape from the black hole. The time reversal implies that the escaping observer has actually traveled into the future as it transposed the event horizon of the black hole, and therefore, its escape from the black hole would entail traveling into the past, as depicted in Figure 1.35. Since time reversal is a relativistic impossibility, there is no escape from the black hole.

These arguments lead us to formulate the following assertion:

Proposition: Only one of the following statements is true:

a. The relativistic tenets hold for the Big Bang, and then the Big Bang scenario does not represent the origin of space-time because space-time subsumes its own evolution and hence it does not have an origin,
b. the relativistic tenets of space-time do not hold in the earliest universe because time and space are decoupled, hence separable, or
c. the relativistic tenets hold for the Big Bang, and then it is impossible to trace the origin of the universe starting at a space-time singularity because no extrinsic flowing time exists or can be defined at its spatial locus.

This proposition leaves AI with fewer resources to tackle the problem of the nature of DM and DE, as AI builds up theory only if it finds leads to a cogent picture through conjecture framing. As hinted by the results previously described in this chapter, the study of the origin of singularities in space-time is essential to make progress in understanding the nature of DM, and, as we shall see in Chapters 2 and 5, the same holds for DE. Yet, it appears that the very phrase "origin of singularity" is troublesome and alludes to an ill-posed problem. The only alternative that has led to further progress is statement (b), which is as likely or unlikely to be true as the other two, but the only one pregnant with possibilities for theoretical progress, as we shall demonstrate. Thus, an operational definition of *extrinsic evolution of space* based on the assumption that statement (b) is valid will be adopted throughout the rest of the book, and by abuse of notation, it will be referred to as "evolution of the universe" or evolution of space-time with the caveat that the clock can be placed outside of the object of study and that it is actually ticking.

1.16 THE UNIVERSE TOPOLOGY ADMITS A COMPACT EXTRA DIMENSION

As described earlier, Einstein's theory of GR is formulated in terms of differential geometry, and its laws are therefore cast as differential equations. It is therefore a local theory. It tells us nothing about the overall shape or topology of the universe while it informs about its geometry [42]. In principle, more than one topology can fit the postulated geometric flatness, which, at any rate, can only be regarded as an approximation [31,42]. There is one thing that GR does tell us in regard to topology: *As the universe evolved in time since the Big Bang, its topology could not change* [31]. This is fairly obvious, since a topological change would imply cutting and pasting space-time, an inadmissible operation. This is so because it would entail a non-homeomorphic distortion of the differential geometry fabric, unless a new force beyond the four established forces could be invoked.

In GR, we assume that the intrinsic geometry of the universe 4D manifold has infinite curvature radius (flat geometry). However, if we soften the assertion to indicate "immeasurably large radius", we may find alternative topologies for compact (yet enormous) manifolds that would fit the postulated approximate flatness. From the previous discussion, it seemed that identifying DM and DE would require a reverse engineering of the SM to incorporate all dimensions in an early universe, together with the evolution of their aspect ratios. Yet, this reverse engineering depends pivotally on the topology of the universe [5], which we know must be invariant in time. This implies that the present-day universe cannot be infinite for the simple reason that the Big Bang singularity and the early universe were compact. *Hence, our universe must be compact, however enormous.*

A quasi-flat compact manifold must necessarily be multiply connected, and hence, the incorporation of a fifth circular coordinate yields one and only one manifold as the only alternative: A five-dimensional torus, that is, a Cartesian product of five circles with huge aspect ratio between the four circular coordinates conforming the locally quasi-flat four-dimensional space-time and the extra fifth spatial coordinate with a quark-size (i.e., LHC-undetectable) radius. We regard the standard four-dimensional space-time in GR as a four-torus locally homeomorphic to the Euclidean space. This is a latent compact manifold representing a quotient space denoted *W*/~ [5], so that two points in a higher-dimensional space *W*, the five-dimensional torus with a circular extra dimension, would be equivalent modulo "~" if they projected onto the same point in the latent manifold. The task ahead then becomes to elucidate how DM would fit into this scheme as a gravity-carrying particle enshrined in the compact fifth dimension, existing in W but not in its quotient space *W*/~. In this way, DM would not interact with the SM in 4D space-time, except via gravity.

The postulated toroidal topology of the universe has found recent experimental validation in the examination of the CMB, the cosmic relic of the Big Bang. If space were infinite (hence flat and simply connected), perturbations in the temperature gradient field of the CMB radiation would exist on all scales. If space is finite, then there would be wavelengths missing that are larger than the size of space itself. Maps of the CMB perturbation spectrum made with probes like NASA's WMAP and the Planck probe from the European Space Agency have shown striking amounts of missing perturbations at large scales. The properties of the observed fluctuations of the CMB show a "missing power" on scales beyond the size of the universe [43]. That would imply that our universe is multiply connected and finite. The spectrum of the CMB is compatible with a three-dimensional torus topology for the spatial components of the universe. Thus, universe models with spatially multiply connected topology contain a discrete spectrum of the Laplacian with a specific wavelength cut-off, as observed in the CMB. Furthermore, the three-dimensional torus model possesses a two-point correlation function that fits CMB maps obtained with the Planck probe [43].

Thus, the toroidal topology enables the incorporation of the extra curl-up dimension as an undeveloped feature in the compact universe evolution. In principle, nothing precludes the presence of more than one such dormant dimension. The dormant dimension cannot be probed with current means as it spans the quark attometer scale, storing energy in the order of 246 GeV. On the other hand, the flat, simply connected universe that we all intuitively have come to grips with would actually be incompatible with the evolution from a compact manifold as implied by the Big Bang scenario simply because such evolution would entail a change in topology. Such a transition from compact multiply connected to flat, simply connected cannot be smooth and would be forbidden by GR.

Within the compact multiply connected model of today's universe, four of the circular dimensions have grown so much that the curvature radius is, for all practical purposes, infinite, and the geometry is locally that of an Euclidean space [31]. Thus, the dormant fifth coordinate combines with the quasi-Euclidean standard coordinates in a local cylinder of quark-size radius, where the standard coordinate constitute the cylinder axis. In essence, the torus with a huge aspect ratio becomes locally a cylinder (Figure 1.37). Standard space-time coordinates get mixed with the dormant coordinate as determined by a pitch angle α defining the helix stride, so that a geometric dilution (υ) for the energy stored in the dormant coordinate may be defined as $\upsilon = -\log\cos\alpha$. Thus, the geometric dilution becomes infinite when there is no projection along the dormant coordinate and zero when there is no projection onto the standard coordinates.

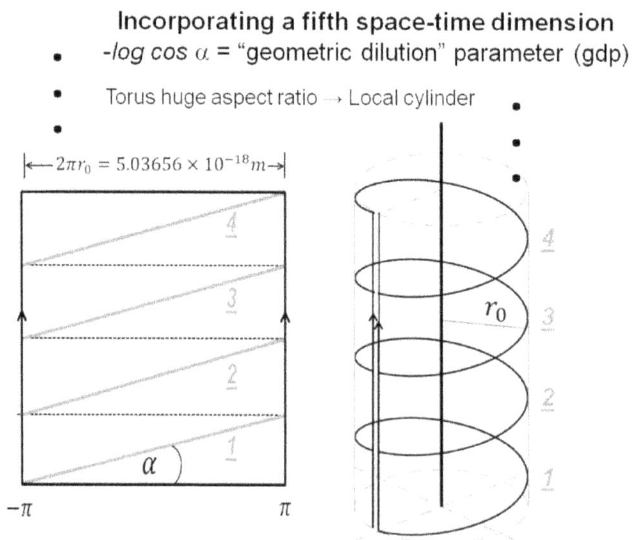

FIGURE 1.37 Incorporating the circular fifth dormant dimension and combining it with generic locally flat dimensions of the four-dimensional space-time through a geometric dilution parameter determining the helical stride.

1.17 AI QUEST FOR THE ORIGIN OF DARK MATTER

Current knowledge of DM is sketchy at best and mostly conjectural. What we actually know or can judiciously conjecture is best summarized as follows:

- DM is invisible, massive, "cold" (speed far lower than the speed of light, $v << c$), mainly formed early on, interacts with the SM only via gravity, and does not decay easily.
- It is probably not an extension of the SM, as no particle derived from the latter fits the characteristics of DM [35,36].
- It may be stored in a compact fourth spatial dimension.
- In advocating for storing DM in an extra dimension, we note that the quark scale ($q = 0.802 \times 10^{-18}$m) is the smallest – undetectable – material scale known, and that a stationary wave along a circular fourth spatial dimension with radius q stores an energy $E = hc/2pq = 246$ GeV, the vacuum expectation value of the Higgs boson.
- We have a validated topological model of the compact universe to incorporate such a dimension and trace the origin of DM to the early universe.

The core involves the identification of DM as stored in the five-dimensional "quintessential" space W, a five-dimensional torus projecting onto a four-dimensional "latent" manifold, $\Omega = W/\sim$ that is locally flat, meaning it is locally homeomorphic to an Euclidean space. Thus, Einstein's four-dimensional space-time Ω is regarded as a quotient space for the space that we intend to prove stores DM. This problem requires the deployment of AI, as it requires the extension of the SM to incorporate a fifth dimension, so far defined in the latent quotient space $\Omega = W/\sim$. This task, undertaken in Chapter 4, entails the lifting of each pairwise particle interaction within Ω, represented by the transformation: $\mathcal{L}_1\left(\phi_1,\{\partial_\mu\phi_1\}_\mu\right)+\mathcal{L}_2\left(\phi_2,\{\partial_\mu\phi_2\}_\mu\right)\rightarrow\mathcal{L}_1'\left(\phi_1',\{\partial_\nu\phi_1'\}_\nu\right)+\mathcal{L}_2'\left(\phi_2',\{\partial_\nu\phi_2'\}_\nu\right)$, to the level of W:

$$\widetilde{\mathcal{L}_1}\left(\widetilde{\phi_1},\{\partial_\mu\widetilde{\phi_1}\}_{\mu=1,\dots,5}\right)+\widetilde{\mathcal{L}_2}\left(\widetilde{\phi_2},\{\partial_\mu\widetilde{\phi_2}\}_{\mu=1,\dots,5}\right)\rightarrow\widetilde{\mathcal{L}_1'}\left(\widetilde{\phi_1'},\{\partial_\nu\widetilde{\phi_1'}\}_{\nu=1,\dots,5}\right)+\widetilde{\mathcal{L}_2'}\left(\widetilde{\phi_2'},\{\partial_\nu\widetilde{\phi_2'}\}_{\nu=1,\dots,5}\right) \tag{1.35}$$

Where $\widetilde{\mathcal{L}_n},\widetilde{\phi_n}$ denote, respectively, the lifting of Lagrangian $\mathcal{L}_n$ and particle field ϕ_n, with n indicating particle types in the SM. As indicated above, the primes denote destiny particles arising from the interaction.

A lifting of the SM is valid if and only if the diagram in Figure 1.38 is commutative. This property indicates that a canonical projection $\pi : W \rightarrow \Omega$ followed by interaction within the latent space Ω yields a destiny state that is the same as that obtained when the interaction is computed directly in the quintessential space W, and the resulting destiny state is subsequently projected onto the latent space Ω. In other words, the following equation must hold for every particle pair $\left(X_J\left(\mathcal{L}_J,\phi_J\right), J=1,2\right)$ in the SM:

$$X_j'\left(\mathcal{L}_J',\phi_J'\right)=\pi\widetilde{X_j'}\left(\widetilde{\mathcal{L}_j'},\widetilde{\phi_j'}\right) \tag{1.36}$$

The lifting $\tilde{\phi} : W \rightarrow \mathbb{C}^n$ of a particle field $\phi : W/\sim \rightarrow \mathbb{C}^n$ requires an *autoencoder* [5]. Usually, in complex dynamical systems, such autoencoders are used to extract the latent space and obtained the simplified differential equations defined on the latent space as entraining the full system. In other words, the autoencoder is used to simplify the dynamical system, retaining the dynamics that are essential and averaging out subordinated degrees of freedom. Chapter 3 will be devoted to describing such autoencoders, with a special focus on systems that yield topological metamodels.

In the context of extending the SM, or rather reverse engineering it to encompass a dormant dimension, we need to have the autoencoder working in reverse. Rather than simplifying the fields defined in latent space, we need to lift them to a quintessential space of higher complexity, so as to recover the

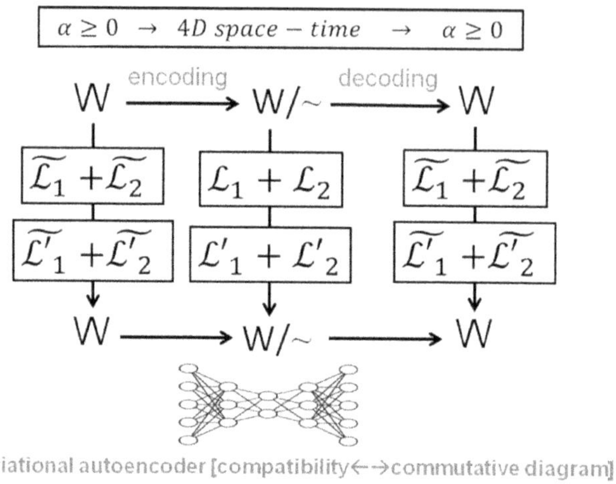

$$x^{\mu} \rightarrow x'^{\nu} \Rightarrow \int [\mathcal{L}_1 + \mathcal{L}_2] dx^{\mu} = \int [\mathcal{L}'_1 + \mathcal{L}'_2]\, dx'^{\nu}$$

FIGURE 1.38 Neural network with autoencoder architecture designed to reverse engineer the Standard Model by incorporating a fifth circular dimension.

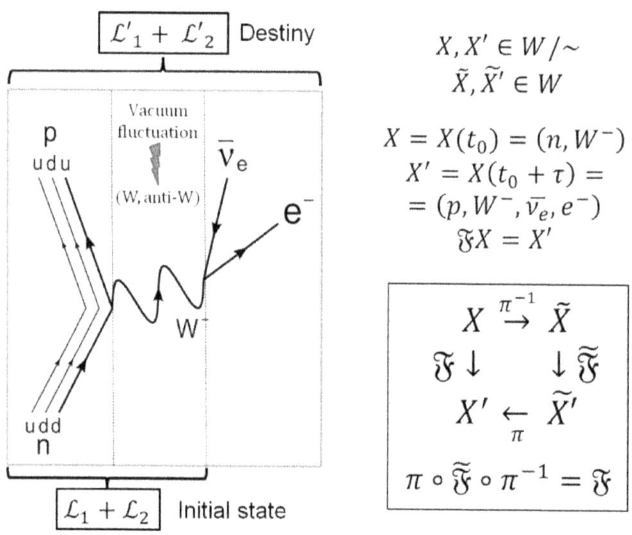

FIGURE 1.39 Elementary process of beta decay defined on the latent space representing the four-dimensional space-time *W*/~ and consistently lifted to the quintessential space *W*. The process consists of a neutron transformed into a proton through communication with a *W*– gauge boson, with emission of an electron and an electron antineutrino.

original fields when taking the quotient modulo the standard four-dimensional space-time coordinates. In addition, we need the autoencoder to produce the topological metamodel that would enable the proper lifting, so that a diagram of the type presented in Figure 1.38 becomes commutative.

An illustration of the kind of tasks performed by the topological autoencoder of the SM is depicted in Figure 1.39. The figure describes the lifting to quintessential space *W* of a process, known as beta decay, comprising the neutron (n) transformation into a proton (p) by transformation of one of its constitutive down quarks (d) into an up quark (u), with emission of an electron (e^-) and an electron antineutrino ($\overline{\nu_e}$). This process requires the interaction of the neutron with a W^--boson, a massive particle ($M = 80.433$ GeV/c^2) whose existence is ephemeral ($\tau \approx 10^{-26}$ seconds $\sim 6.582\times10^{-16}$ eVs$/\left(2\times80.433\times10^{9}\text{ eV}\right)$), as its formation requires borrowing energy from the vacuum as a fluctuation governed by Heisenberg's uncertainty principle ($\Delta E \times \Delta t \sim \frac{\hbar}{2}$). Here, we adopt the notation $X = (n, W^-), X' = (p, W^-, \overline{\nu_e}, e^-)$ to indicate, respectively, the initial and final state of the system as recorded on the latent space *W*/~ and dictated by the SM. The respective liftings to *W* are $\widetilde{X}, \widetilde{X}'$, and their validity is ensured if and only if they satisfy the flow relation:

$$\left(\pi \circ \tilde{\mathfrak{F}}\right)\tilde{X} = \mathfrak{F}X; \quad \mathfrak{F}X = X'; \quad \tilde{\mathfrak{F}}\tilde{X} = \widetilde{X'} \tag{1.37}$$

The operators $\mathfrak{F}$ and $\tilde{\mathfrak{F}}$ determine, respectively, all processes in *W*/~ dictated by the SM and the lifting of such processes to the quintessential space *W*. Thus, the operators may be interpreted as time evolution propagators over time intervals identified with the lifetimes of the particles that communicate fundamental forces. In this context, the autoencoder is charged with the task of computing the operator $\tilde{\mathfrak{F}}$ and canonical projection π, so that the generic diagram in Figure 1.40 becomes commutative.

Similarly, the lifting of another fundamental process, the creation and decay of the Higgs boson (*H*) into two lepton–antilepton pairs, is depicted in Figure 1.41. The creation of the Higgs boson entails the fusion of two gluons (*g*, *g*) mediated by a virtual top quark $(\hat{t})$, and the "timing" of the process is given by the sum of the lifetimes of the bosons (*H*, *Z*) that serve as intermediates.

$$\begin{array}{ccc} W & \overset{\pi}{\rightarrow} & W/\sim \\ \downarrow \widetilde{\mathfrak{F}} & & \downarrow \mathfrak{F} \\ W & \overset{\pi}{\rightarrow} & W/\sim \end{array}$$

FIGURE 1.40 Commutative diagram for the quintessential lifting of a fundamental process in the reverse engineering of the Standard Model.

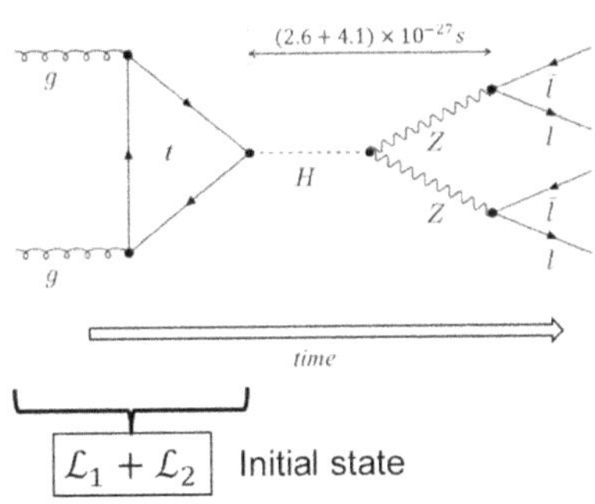

$\mathcal{L}_1 + \mathcal{L}_2$ Initial state

$\tau = 6.7 \times 10^{-27} s$

$$X, X' \in W/\sim$$
$$\tilde{X}, \tilde{X}' \in W$$
$$X = X(t_0) = (g, g, \hat{t})$$
$$X' = X(t_0 + \tau) = (l, \bar{l}, l, \bar{l})$$
$$\mathfrak{F}X = X'$$

$$\begin{array}{ccc} X & \overset{\pi^{-1}}{\rightarrow} & \tilde{X} \\ \mathfrak{F}\downarrow & & \downarrow \widetilde{\mathfrak{F}} \\ X' & \underset{\pi}{\leftarrow} & \widetilde{X}' \end{array}$$

$$\pi \circ \widetilde{\mathfrak{F}} \circ \pi^{-1} = \mathfrak{F}$$

FIGURE 1.41 The lifting of the elementary process of creation and (detectable) decay of a Higgs boson.

In this way, the SM may be extended by the topological autoencoder to incorporate an extra dimension compatible with the compact and multiply connected topological metamodel of the universe, validated experimentally by the spectral decomposition of the CMB. The topological autoencoder would then be in position to discover DM particles as it reverse-engineers the SM to adapt it to the early universe, where the dormant dimension became commensurate with the other four dimensions of space-time. This program will be carried out in Chapter 4.

1.18 AMALGAMATED HIGGS BOSON AS DARK-MATTER PORTAL

The SM is considered the towering achievement of particle physics, mainly because of its amazing predictive power and sparse parametrization. Despite its success, the SM in its current version does not include DM. Furthermore, it seems unclear that it can be extended in a cogent way to encompass DM without significantly altering the existing models of observable particles and solving the quantum gravity and hierarchy problems. Ongoing research described in this book, particularly in Chapter 4, may turn this situation around. With the mathematical implementation of the concept of geometric dilution (υ) within the premises of the SM, DM can be shown to percolate into observable dimensions. Therefore, we are confident that a suitable extension of the SM will be forthcoming. This discussion points to a viable possibility in this regard, as fermionic DM is shown to have a dynamical origin capable of interplay or mixing with the Higgs boson. What follows is an outline of the SM extension.

The approach requires a mixing of the scalar field for DM with the Higgs doublet scalar field to yield an irreducible mediator between the observable and dark sectors of the universe. Technically, this interplay requires boundaries in the dark curly dimension, so the interplay may materialize at the three-brane that represents the boundary. But such boundaries do not exist in the compact extra dimension, since it constitutes a circle. For this reason, we need to replace the topology of the dark dimension, $\mathcal{S}^1$, for the quotient manifold $\mathcal{S}^1 / \mathbb{Z}_2$. This quotient is made up of orbits under the symmetry operation $\vartheta \to -\vartheta$, with the polar representation $y = r_0 e^{i\vartheta}$ of the dark coordinate. The metric in $\mathcal{S}^1 / \mathbb{Z}_2$ is entirely determined by that of the subset $0 \leq \vartheta \leq \pi$, which yields boundaries, that is, the three-branes $\vartheta = 0, \vartheta = \pi$, as the dark extra dimension is incorporated. Furthermore, DM particles become $\mathbb{Z}_2$-odd scalar fields that can now be amalgamated with the Higgs scalar field to yield mediators between the observable and dark sectors. The lightest such portal corresponds to the combination with the first Kaluza–Klein (i.e., at $\upsilon = 0$) excitation of a dark fermion, which corresponds to the ur-Higgs.

This construction of the compound particle departs from the parametrized semiempirical model of the Higgs boson, with its quartic term in the potential energy accounting for self-interactions of the Higgs field $\phi_h = \phi_h(x,t)$. In the SM, the spontaneous symmetry breaking in the Higgs potential, responsible for endowing particles with mass, is associated with the nonzero vacuum expectation value (v.e.v.) v arising from self-interactions that yield the potential term $-\lambda\phi_h^4$ with coupling constant λ. The parametrization of the v.e.v. at $v = 246$ GeV is adjusted to yield all particle masses. By contrast, our model does not include self-interactions; instead, it extends the field ϕ_h to the amalgamated field $\phi_H = \phi_H(x,y,t)$ by incorporating a $\mathbb{Z}_2$-odd DM-field $\phi_D(y)$, that spans the dark dimension:

$$\phi_H = \phi_H(x,y,t) = \phi_h(x,t)\left[1 - \left(\frac{\phi_h(x,t)}{\sqrt{2}\phi_D(y)}\right)^2\right]^{1/2} \tag{1.38}$$

The v.e.v. is stored in the dark dimension with quotient topology $\mathcal{S}^1 / \mathbb{Z}_2$, while only the (x,t)-dependence of the Higgs field is apparent in the SM (in its current version).

By adopting the radial representation $\phi_h = \rho e^{i\omega}$, the Higgs boson becomes, in standard space-time, an oscillatory excitation of the radial field $\eta = \rho - v$, while the Lagrangian for η lifted to the (x,t,y)-five-dimensional space-time becomes

$$\widetilde{\mathcal{L}(\eta)} = \frac{1}{2}(\partial\eta)^2 - \lambda\overline{\phi_D}\phi_D\eta^2 \tag{1.39}$$

This result implies that the Higgs boson mass $m_H = \sqrt{2\lambda}v$ ($\sqrt{2\lambda} = \cos\alpha$ geometric dilution parameter) is bestowed through a coupling with the DM scalar field. If geometrically undiluted, the v.e.v. thus stored would constitute the lightest DM, namely the ur-Higgs. Equations 1.37 and 1.38 inform that DM can be incorporated into the SM via the key observation that the v.e.v. of the Higgs field is actually a $\mathbb{Z}_2$-odd stationary wave spanning the dark circular dimension. This circle has radius $r_0 = 0.802695 \times 10^{-18}$ m, the smallest material scale assimilated to the effective quark diameter established from parton models of inelastic $e^{\pm}p$ scattering [41].

In this model, the amalgamated Higgs field (Eq. 1.38) enables the spillover of the dark universe into the observable universe, acting as a gatekeeper and a portal between the observable and dark sectors. This is because the parameter that embodies the geometric dilution $\upsilon = -\log\cos\alpha$ of the ur-Higgs onto observable dimensions to yield a specific particle mass (Chapter 4) can be unambiguously associated with the Yukawa coupling, g_Y, of the fermion field with the Higgs field, and with the mass-to-v.e.v. ratio in the case of the weak-force bosons. *The behavior of the amalgamated Higgs boson as a portal for DM, enabling the spillover that endows observable fermions with mass, is enshrined in the fundamental proportionality*

$$g_Y \propto \cos\alpha \tag{1.40}$$

Under these observations, it becomes clear that the amalgamated Higgs boson serves as the portal of the dark universe, with a controlled spillover mechanism that endows SM particles with mass.

1.19 PROPAEDEUTICS FOR THE DEPLOYMENT OF ARTIFICIAL CONSCIENCE

This chapter outlined the implementation of an AI-based version of conscience (AC) enshrined in an LLM that is trained by an autoencoder wrapped in a functional programming language capable of formal (equational) reasoning (Figure 1.11). It also paved the way to deploy an AI system endowed with AC to address core problems in quantum gravity (Chapters 2–7). These applications become natural since the autoencoder that trains the LLM is eminently suitable to enable GR-QM amalgamation by selecting the GR and QM support manifolds as encoding and latent manifold, respectively (Figure 1.20).

The AC is capable of inward surveillance via a cost function for the learning neural network in the guise of an autoencoder. Beyond parameter optimization, the self-examination introduced by AC prompts the deployment of formalized mathematics resources that are not typically exploited by the underlying "unconscious" AI system. The symbolic computations thus prompted are required by the AC to restructure the autoencoder in order to optimize its performance by modifying the latent and decoding manifolds. Thus, in contrast with an AI system, AC introduces core modifications to the network architecture that restructure its approach to equational reasoning.

The most striking result achieved in an AC-steered computation of the space-time topology (Figures 1.16–1.19) is that AC was able to invoke a mathematical formalism, that is, algebraic topology, not used in the standard autoencoder application in order to restructure the neural network as required to optimize performance of the theorem prover (Figure 1.17). The AC-endowed AI system was deployed to identify the space-time topology by integrating ontological premises/constraints turned computable by AC via the invocation/summoning of formalized mathematical resources (Figure 1.17). The results are nothing short of astonishing: As proven in Chapter 7, the space-time encoding topology becomes $\left\langle \left(\mathbb{R}^4/\mathbb{Z}^4\right)^{\#} \right\rangle_6$, the six-fold orientable chiral-pair covering of the nonorientable compact and multiply connected latent manifold $\Omega \approx \left(\mathbb{R}^4/\mathbb{Z}^4\right)^{\#}$. The covering topology derived from algebraic topology calculation (Chapter 7) yields the optimal inferential scheme for the copilot-assisted LLM (Figure 1.18), while being also the only possibility compatible with the chiral constraints imposed by the electroweak sector of the SM and the only topology suitable to yield the Dirac equation for spin-1/2 fermions in the "observable" sheet (Figure 1.19). The remaining five nonreflexible sheets (chiral pairs) are endowed with the gauge symmetries that make the resultant wave functions suitable to represent DM in accordance with the approximate 5:1 abundance ratio of dark to detectable matter. This analysis justifies the adoption of the manifold $\left\langle \left(\mathbb{R}^4/\mathbb{Z}^4\right)^{\#} \right\rangle_6$ as the encoding manifold for space-time in the approach to core problems in particle cosmology (Chapters 2 and 7).

1.20 ARTIFICIAL CREATIVITY AND IMAGINATION: THE DREAMY AUTOENCODER

The prolegomena so far presented may leave some readers with an unsavory impression of the book's adopted version of AC. A "last line of defense" for those individuals who deny the possibility of AC on ontological grounds is that humans, the quintessential conscious beings, possess creativity and imagination, attributes that they consider inconceivable or impossible to implement in a machine. After all, machine learning systems differ in fundamental ways from the human brain [44]: The latter has circa 10^{14} connections, whereas machines have around 1% of that number, while the training set of the machine is vastly larger than that of the human brain, by at least three orders of magnitude or probably more. On the other hand, if we hold a materialistic view of consciousness, then creativity and imagination should be reproducible in principle and even in practice. Whether we choose to reproduce them so as to make them recognizable by humans or akin to the human counterparts is a different matter.

In fact, the book asserts that autoencoders, as fundamental learning systems, have different levels of optimization, which not only implicate varying the strengths of the underlying network connections, but, at a higher level, they involve wholesale restructurings of the neural network. Thus, the AI systems are endowed with a global variational latitude in addition to the local variational latitude (i.e., connection strengths), whereby the autoencoder may acquire intercalated autoencoders with latent spaces of varying coarse graining (Figure 1.42). Learning efficiency may thus be much improved when adding the large-scale optimization mode.

A similar structural fluidity is probably apparent in the dream phase of the human brain. In a familiar experience, a poem would be memorized after several readings by the person awake, but the result is typically suboptimal, and more rehearsing is needed. Only after a good night's sleep, the text emerges from the memory recesses ready to be flawlessly recited. This suggests two levels of optimization: (1) Varying connection strengths in the "awake mode", leading to moderate incremental improvement; and (2) higher level optimization during sleep, requiring architectural fluidity (dream state), leading to large-scale autoencoder restructuring. The apparent contextual incongruity of remembered dreams may be regarded as attesting to the said structural fluidity.

Another illustration of variational network fluidity that improves optimization is offered by the autoencoder-trained transformer [5,29] adopted in this chapter to infer protein folding pathways. The time series of protein-chain conformations may be encoded as contact maps in the latent space Ω (Figure 1.43), but an improved variational optimization of pathway prediction is achieved when the protein conformation is further coarse-grained to the level of Ramachandran basins, that is, sterically allowed regions in the torsional backbone coordinates. This coarser level of encoding requires the

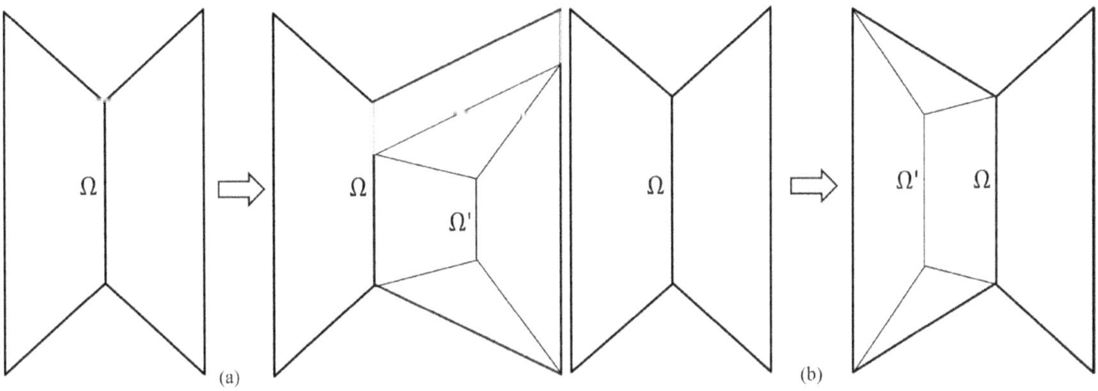

FIGURE 1.42 "Imaginative" network restructuring of an autoencoder with original latent space Ω operating in the "dreamy mode". (a) The intercalation of an autoencoder subunit with latent space Ω′ improves the optimization for a portion of the original latent space, which is suboptimal for reproducing the original input as per the original (limited) cost function that only includes variations in connection strengths. (b) Autoencoder intercalation of a latent space Ω′ that is coarser than the original latent space Ω and is required to optimize the input encoding at the Ω-level.

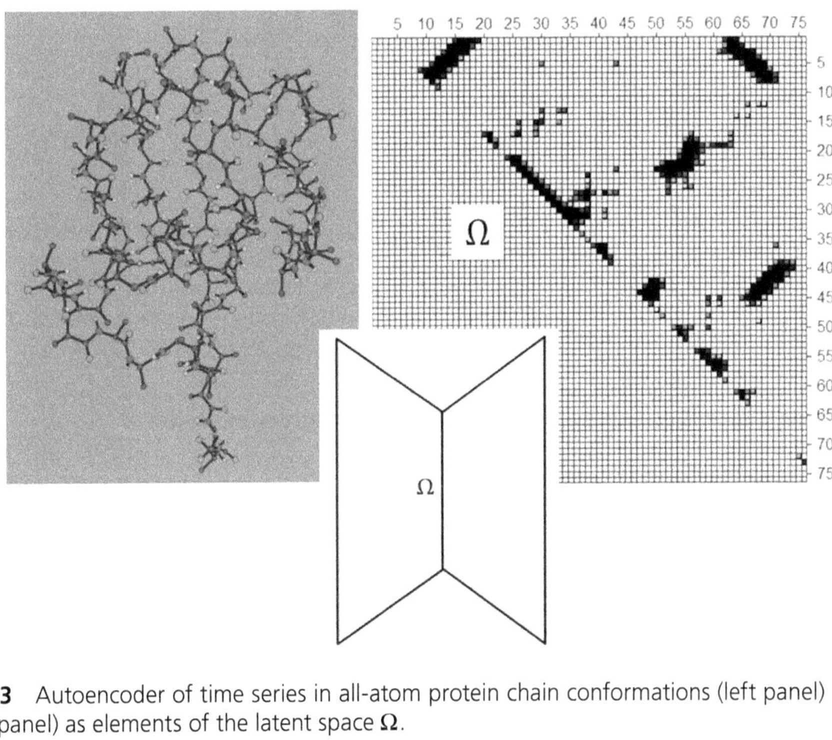

FIGURE 1.43 Autoencoder of time series in all-atom protein chain conformations (left panel) using contact maps (right panel) as elements of the latent space Ω.

Note: This figure is reproduced in color in online supplementary materials.

interaction of an additional autoencoder with latent space Ω′ containing the digitalized topological information (Figures 1.42b and 1.44).

Such AI systems, capable of switching the variational mode to dramatically enhance optimization, are undoubtedly capable of displaying features unrealized in the wakeful state. Those features may be construed as representative of "creativity" and "imagination" insofar as they offer global learning solutions that extend beyond those achieved by varying connection strengths in the underlying neural network.

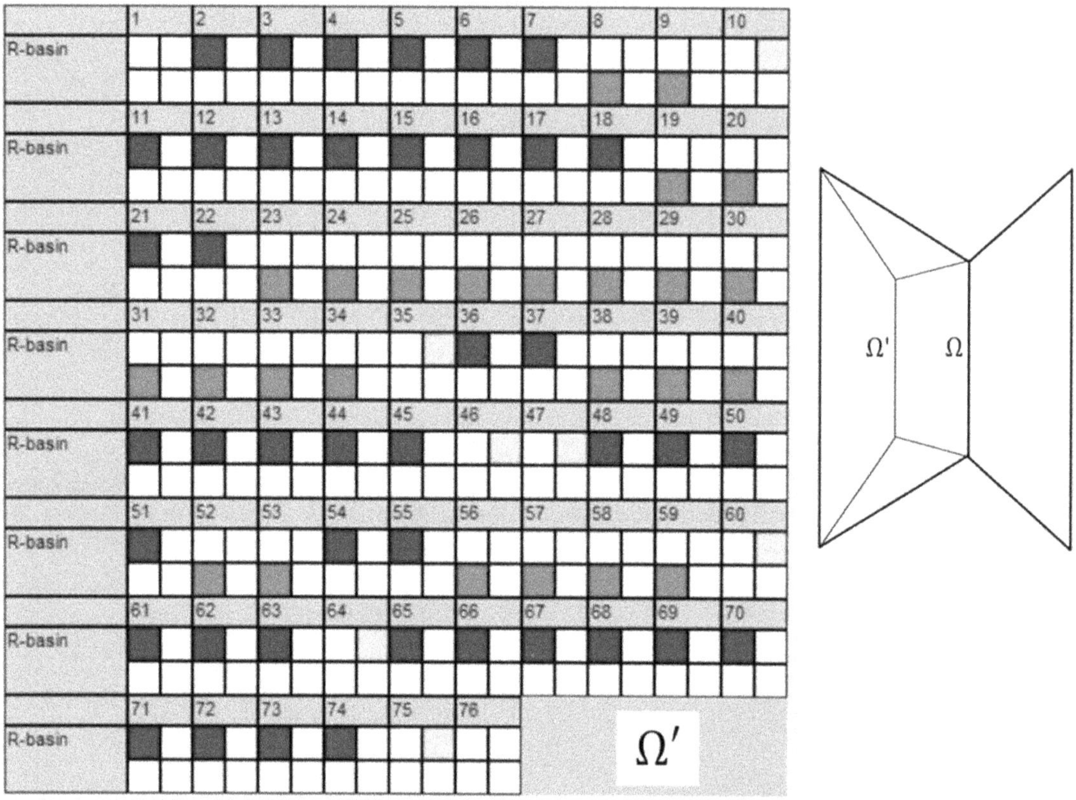

FIGURE 1.44 Intercalated autoencoder in the predictor of protein folding pathways presented in Fig. 1.43. The 76-unit word of Ramachandran basins assignments, one basin per residue, along the protein chain of length $N = 76$ constitutes a coarse-grained "topological" encoding at the $\mathbf{\Omega'}$-level of the information enshrined in the all-atom chain conformation (left panel in Fig. 1.43). The intercalation of a coarser level of description (cf. Fig. 1.42b) enhances the level of optimization of the autoencoder, leading to higher accuracy in the inference of protein folding pathways.

Note: This figure is reproduced in color in online supplementary materials.

The implementation of conscious features to promote the collapse of the wave function as the outcome of an encoding process in specific QM systems requires a higher degree of familiarity with quantum gravity autoencoders, and hence, it is relegated to Chapter 8, the last chapter in the book. AC will be thus shown to be an emergent property of autoencoder-based neural activity as well as an "information field" that disambiguates quantum state representations by enriching the encoding into the latent state.

REFERENCES

1. Gödel K (1931) Über formal unentscheidbare Sätze der Principia Mathematica und verwandter Systeme, I. Monatshefte Mathem Phys 38: 173–198.
2. Penrose R (1996) On gravity's role in quantum state reduction. Gen Relativ Gravit 28: 581–599.
3. Leggett AJ (2005) The quantum measurement problem. Science 307: 871–872.
4. Penrose R (2014) On the gravitization of quantum mechanics 1: Quantum state reduction. Found Phys 44: 557–575.

5. Fernández A (2022) *Topological Dynamics for Metamodel Discovery with Artificial Intelligence*. Chapman & Hall/CRC, Taylor and Francis, London, UK.
6. Shaw DE, Maragakis P, Lindorff-Larsen K, Piana S, Dror RO, et al. (2010) Atomic-level characterization of the structural dynamics of proteins. Science 330: 341–346.
7. Brooks BR, Brooks III CL, Mackerell AD, Nilsson L, Petrella RJ, et al. (2009) CHARMM: The biomolecular simulation program. J Comp Chem 30: 1545–1615.
8. Clark PL, Elcock AH (2016) Molecular chaperones: Providing a safe place to weather a midlife protein-folding crisis. Nature Struct Molec Biol 23: 621–623.
9. Piana S, Shaw DE (2018) Atomic-level description of protein folding inside the GroEL cavity. J Phys Chem B 122: 11440–11449.
10. Thommen M, Holtkamp W, Rodnina MV (2017) Co-translational protein folding: Progress and methods. Curr Opin Struct Biol 42: 83–89.
11. Sorokina I, Mushegian A (2018) Modeling protein folding in vivo. Biol Direct 13: 13.
12. Fernández A (1999) Folding a protein by discretizing its backbone torsional dynamics. Phys Rev E 59: 5928–5934.
13. Fernández A, Colubri A, Berry RS (2000) Topology to geometry in protein folding: Beta-lactoglobulin. Proc Natl Acad Sci USA 97: 14062–14066.
14. Singhal N, Pande VS (2005) Error analysis and efficient sampling in Markovian state models for molecular dynamics. J Chem Phys 123: 204909.
15. Zimmerman MI, Bowman GR (2015) Fast conformational searches by balancing exploration/exploitation trade-offs. J Chem Theory Comput 11: 5747–5757.
16. Schmidhuber J (2015) Deep learning in neural networks: An overview. Neural Netw 61: 85–117.
17. Fernández A (2020) Artificial intelligence teaches drugs to target proteins by tackling the induced folding problem. Mol Pharm (ACS) 17: 2761–2767.
18. Fernández A (2016) *Physics at the Biomolecular Interface*. Springer International Publishing, Switzerland.
19. Fernández A, Shen M, Colubri A, Sosnick TR, Berry RS, Freed KF (2003) Large-scale context in protein folding: Villin headpiece. Biochemistry 42: 664–671.
20. Zeiler MD (2012) ADADELTA, an adaptive learning rate method. arXiv:1212.5701 https://arxiv.org/abs/1212.5701
21. Lippert RA, Predescu C, Ierardi DJ, Mackenzie KM, Eastwood MP, et al. (2013) Accurate and efficient integration for molecular dynamics simulations at constant temperature and pressure. J Chem Phys 139: 164106.
22. Nosé S (1984) A unified formulation of the constant temperature molecular-dynamics methods. J Chem Phys 81: 511–519.
23. Hoover WG (1985) Canonical dynamics: Equilibrium phase-space distributions. Phys Rev A 31: 1695–1697.
24. Martyna GJ, Tobias DJ, Klein ML (1994) Constant pressure molecular dynamics algorithms. J Chem Phys 101: 4177–4189.
25. Li Z, Sun J, Murphy L, Su Q, Li Z, Zhang X, Yang K, Si X (2024) A survey on deep learning for theorem proving. arXiv preprint arXiv:2404.09939.
26. Avigad J, Massot P (2025) Mathematics in Lean: Release v4. 19.0. Lean Community – Github. https://avigad.github.io/mathematics_in_lean/mathematics_in_lean.pdf
27. Yang K, Deng J (2019) Learning to prove theorems via interacting with proof assistants. arXiv:1905.09381.
28. Song P, Yang K, Anandkumar A (2025) Lean copilot: Large language models as copilots for theorem proving in Lean. *International Conference on Neuro-symbolic Systems (NeuS) 2025*. arXiv:2404.12534v2.
29. Fernández A (2021) Artificial intelligence deconstructs drug targeting in vivo by leveraging a transformer platform. ACS Med Chem Lett 12: 1052–1055.
30. Huang D, Dhariwal P, Song D, Sutskever I (2019) GamePad: A learning environment for theorem proving. International Conference on Learning Representations, 2019. arXiv:1806.00608.
31. Weinberg S (2008) *Cosmology*. Oxford University Press, New York, USA.
32. Fernández A (2025) *De Novo Quantum Cosmology with Artificial Intelligence: Applications of Formal Autoencoders*. ISBN: 978-1041045038. CRC Press, Taylor & Francis, UK.
33. Ashtekar A (1986) New variables for classical and quantum gravity. Phys Rev Lett 57: 2244.
34. Capovilla R, Jacobson T, Dell J (1991) A pure spin connection formulation of gravity. Class Quant Grav 8: 59–78.
35. Fisher P (2022) *What Is Dark Matter?* Princeton University Press, Princeton, NJ.
36. Profumo S (2017) *Introduction to Particle Dark Matter*. World Scientific Publishing, Singapore.
37. Susskind L (1995) The world as hologram. J Math Phys 36: 6377–6396.
38. Wesson PS (1999) *Space–Time–Matter, Modern Kaluza–Klein Theory*. World Scientific, Singapore.
39. Dirac PAM (1982) *The Principles of Quantum Mechanics*, 4th Edition. Clarendon Press, UK.

40. Feynman RP, Weinberg S (1999) *Elementary Particles and the Laws of Physics.* Cambridge University Press, Cambridge, UK.
41. Abramowiczy H, Abtt I, Adamczykh L, Adamusae M, Antonelli S, et al. (2016) Zeus collaboration limits on the effective quark radius from inclusive *ep* scattering at HERA. Phys Lett B 757: 468–472.
42. Hawking SW, Ellis GFR (2023) *The Large Scale Structure of Space-Time: 50th Anniversary Edition.* Cambridge University Press, Cambridge, UK.
43. Aurich R, Buchert T, France MJ, Steiner F (2021) The variance of the CMB temperature gradient: A new signature of a multiply connected universe. Class Quant Grav 38: 225005.
44. Bengio Y, Lecun Y, Hinton G (2021) Deep learning for AI. Comm ACM 64: 58–65.

Conscious AI Approaches Core Problems in Particle Cosmology

2

This chapter deploys an artificial intelligence (AI) neural theorem prover endowed with artificial conscience (AC) to tackle core problems in particle cosmology that are computationally represented through a quantum gravity autoencoder. Formally, the AI system suited for the task consists of a large language model (LLM) trained by an autoencoder wrapped by a functional programming language, as outlined in Chapter 1. The encoder amalgamates general relativity (GR) and quantum mechanics (QM) via a vierbein (four-leg transformation) formalism that maps the AC-inferred sixfold covering manifold of space-time onto its tangent bundle. The geometry of the latter (locally flat) becomes the QM support that will be adopted to unravel the origin of dark matter, with its 5:1 abundance ratio relative to observable matter, and to derive a version of the graviton consistent with the ontological tenets that conscience (AC) turns computable in Chapters 1 and 7. The sixfold space-time covering manifold adopted in this chapter was described in Chapter 1, while its derivation using algebraic topology is provided in Chapter 7. Furthermore, the space-time covering is non-reflexible, hence able to generate the required chiral versions of reality imposed by the electroweak (EW) sector. Additionally, in accord with the tangent bundle geometrization, while one sheet realizes the gauge symmetry constraints that fit the Dirac equation for the fermionic wave function, the remaining five sheets accommodate the wave functions for dark matter. In this way, AC demonstrates that the topology of the space-time-encoding manifold is actually computable by incorporating the ontological constraints to which space-time is subject. This is likely to be the right topology, since it is confirmed by evidence on the dark sector gathered in this chapter and because its geometrization fits the QM symmetry requirements as shown in the previous and this chapter.

The striking aspect of the AC inference concerning the topology of the universe exploited in this chapter is that the computed sixfold orientable non-reflexible chiral-pair covering of space-time arising from relativistic premises spans in its individual sheets all the symmetries internalized by QM to generate the EW sector and fermionic dark matter. Thus, a result from algebraic topology calculation prompted by AC to incorporate ontological relativistic premises is shown to support the foundational quantum mechanical derivations for particle cosmology.

The nuanced problem of the origin of the universe, be it out of the quantum vacuum or *ex nihilo*, may be regarded as a core problem in quantum gravity, an immature field that seeks to unify all forces of nature. In this book, the problem is turned over to an AC-endowed AI system, reckoning that the problem cannot be tackled without full contextualization, that is, without dealing with other intimately related core problems in particle cosmology. Such problems include the nature of dark matter and dark energy, the hierarchy problem of particle mass, the incommensurably weak coupling strength of gravity, the topology of the universe, the cosmological constant problem, and the vacuum catastrophe. This book describes how AI squarely addresses the matter in its full relational and contextual richness. Meanwhile, this specific chapter revisits and contributes to the core problems in particle cosmology while outlining the full AI program fleshed out in the chapters that follow.

DOI: 10.1201/9781003785279-2

The book implements and applies an AI system endowed with AC (Figure 1.11) that distills the quantum reality encoded in a higher-dimensional space-time and learns to gauge relativistic space-time symmetry, integrating it into the fabric of quantum reality. To address the core problems in particle cosmology, the autoencoder incorporates a latent spatial dimension that turns the quantum vacuum into ur-matter (UM), that is, into a precursor of the visible and dark sectors of the standard model (SM) of particle physics. Through AI, we learn in this chapter that UM comes in six "shades", one of them, UM proper, being the precursor of the visible sector. Each shade belongs to a different chiral-pair sheet in the AC-inferred space-time covering manifold $\left\langle \left(\mathbb{R}^4/\mathbb{Z}^4\right)^{\#} \right\rangle_6$ (Chapters 1 and 7). The remaining five shades incorporate symmetries that were disregarded in standard quantum field theory but have now become highly relevant to elucidate the nature of the dark sector. These lost symmetries that preclude interaction with the photon or mixing with the Higgs field help delineate the nature of dark matter and dark energy, which are now reckoned as key players in deep-space phenomenology.

In this context, AI first approaches the origin of matter regarding it as a phase transition induced by activation of a portal gauge particle that becomes endowed with a symmetry-breaking (SB) stable vacuum through mixing with one specific shade of UM. This communication is enabled by compatible symmetries between a specific UM shade and the SM.

Consistent with a vast body of experimentation, the AI model postulates that the four fundamental forces originate from UM symmetries of the primeval quantum vacuum internalized as gauge generalized symmetries at the phase transitions encompassed by the birth of the universe. The gauging of the relativistic symmetry internalizes it not as local symmetry but as a one-form generalized symmetry, where the purported curvature of the Riemannian manifold is factored into the extended object charged under the gravity-associated symmetry. Intriguingly, the primeval symmetry of space-time can be simplified because the symmetry groups associated with the EW unification are isomorphic to subgroups of the relativistic Lorentz group. This leads to the striking finding that gravity may be treated in a quantized ultra-unification since its ur-symmetry is related to that of the weak force and electromagnetism. In other words, the ur-symmetry of gravity subsumes the ur-symmetry that activates the Higgs field by endowing it with the true vacuum (TV). The symmetry relatedness linking gravity and the Higgs-induced mass with retention of EW symmetry paves the way for a quantum gravity ultra-unification under a gauge generalized symmetry. This ultra-unification substantiates the gravity-mass duality of GR.

In a final synthesis, AI reveals that the four fundamental forces are shown to stem from one primeval force present at the Planck epoch. This force is diluted in different ways into the geometry of the universe in accord with the generalized symmetries of force carriers activated at successive phase transitions in the aftermath of creation.

To complete the unification, a conscience-endowed LLM within a functional language program is shown to yield a field theory of the graviton that incorporates a warped dimension to account for gravity dilution. Such approaches have not yielded to experimental validation except in cosmology, where Kaluza-Klein (KK) gravitons apparently beget self-interacting dark matter. To learn to generate massive gravitons detectable in colliders, a gauge theory for the strong-field limit (SFL) is built by the LLM within a fiber-bundle formulation that enables quantization of a field associated with self-interacting dark matter diluted in a warped dimension. The results predict detectable gravitons in the SFL associated with Wilson loops that holonomically transduce space-time curvature and get charged under a generalized gauge symmetry that translates into $U(1)$-gauge symmetry in the weak field limit (WFL).

2.1 BACKGROUND ON GAUGE SYMMETRY IN THE ELECTROWEAK UNIFICATION

The mass problem in particle physics has many sides. The one relevant to this discussion can be cursorily formulated as follows [1,2]: How is it possible that the weak nuclear force (WNF) is short ranged

($r \sim 10^{-18}$ m) and yet the bosons that carry the force are endowed with internal symmetries $\left(\mathrm{SU}(2)\times U(1)\right)$ that would render them massless, thereby communicating a force of infinite range? Theoreticians feverishly dealt with this problem in the nineteen sixties and came up with a solution [2]: *Boson masses are actually induced masses*, in the sense that a scalar field needs to be incorporated to break the symmetry as it adopts its stable "true" vacuum expectation value (v.e.v). This scalar field, now known as the Higgs field, reconciled both sides of the paradox and appeared to be the way forward; especially after its massive excitation, the Higgs boson was detected experimentally [1].

However, as demonstrated in this chapter, the presence of a dark (invisible) sector in the SM calls for revisions of the Higgs mechanism. As shown subsequently, the encoding of an extra "within" spatial dimension is one such revision found by AI in its take on the Higgs' "induced mass" standard assumption. The extra dimension will be shown to play a key role to explain why the force of gravity is incommensurably weaker than the other forces: Gravity appears to be leaking into the extra dimension. Furthermore, the extra dimension endows the vacuum with materiality, as it turns it into UM, a precursor of matter. In this way, AI can elucidate the origin of mass without relying on an all-pervading self-interacting scalar field, as shown in Section 2.11.

As it is well established, the massless bosons that communicate the forces in quantum electrodynamics (QED) and quantum chromodynamics (QCD) arise from their respective gauge symmetries $U(1)$ and SU(3) [1], symmetries that they retain. On the other hand, the weak interaction became problematic because its short range $r = \hbar / Mc \sim 10^{-18}$ m bespeaks of a massive boson with a mass M in the order of $100\,\mathrm{GeV}$, while its charge is acquired under QED. This implies that the boson mass communicating the weak force cannot be directly tied to a gauge symmetry (otherwise it would be massless) and that two sectors of the SM must be in fact combining (Figure 2.1). This state of affairs makes it apparent that a gauge theory is required to amalgamate the weak force and QED. This theory came to be known as EW unification or Weinberg-Salam model, in honor of two of its main contributors, Steven Weinberg and Abdus Salam [1].

Current thinking suggests that the weak bosons $W^{\pm}$ get their mass via the spontaneous symmetry breaking (SSB) arising from the combination of the boson field with the Higgs field, implicit in the grand Lagrangian for the SM [2]. Furthermore, gauge bosons experience their own fields, meaning that they self-interact or get charged under their own fields. Additionally, we must keep in mind that as gauge symmetry is broken, any eigendirection that retains symmetry would yield a massless boson known as the Nambu-Goldstone boson [1,2]. Thus, after SSB, the bosons that acquire mass may interact with the massless bosons that originated from symmetry preservation, thereby getting charged under the preserved

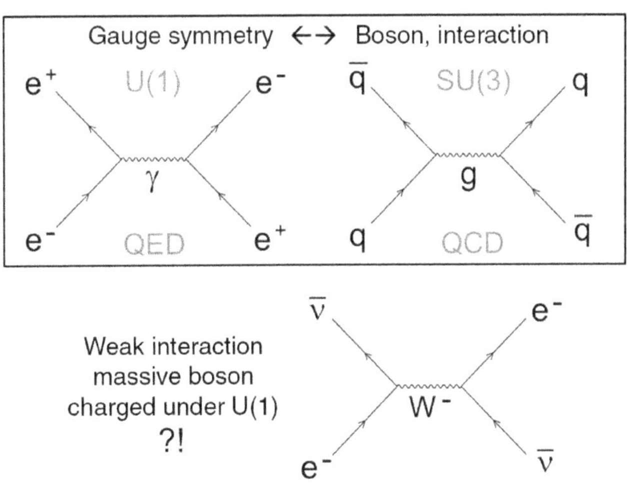

FIGURE 2.1 Conundrum over the nature of the weak interaction whose boson is massive and charged under electromagnetic *U*(1) symmetry. This stands in contrast with QED and QCD, whose bosons are massless and charged under their respective symmetries.

symmetry. If we associate the preserved symmetry post-SSB with $U(1)$, we can readily see that it should be possible to unify the weak and the electromagnetic (EM) interactions provided we can determine the group of symmetries that are broken to endow the weak bosons with mass. In what pertains to fermions experiencing the weak force, the unified EW theory should also encompass the parity violation of the weak interaction, a landmark property of the weak interaction established by Chien-Shiung Wu in 1956 [1]. Therefore, a weak boson does not interact equally with left and right-handed components (two-dim spinors) in the SM: It only interacts with the left-handed component, yielding what is known as a "chiral gauge interaction".

This analysis prompts us to build an EW unification theory subject to the following requirements:

1. An EW symmetry subsuming $U(1)$ as a subgroup needs to be tailored so it can be broken as the weak field is blended with a scalar field that promotes SSB and thereby induces mass on the weak bosons.
2. After broken symmetry, at least two massive weak bosons ($W^{\pm}$) need to emerge while the prevailing symmetry must be of the QED-type ($U(1)$), yielding a Nambu-Goldstone massless boson. As the weak W bosons subsume the massless boson, they get charged under the preserved symmetry. That is why they acquire a QED-type charge.
3. The weak bosons need to self-interact as they experience their own field, so as to get charged under $U(1)$ symmetry, the prevailing symmetry after SSB.
4. The theory must be chiral in accord with the established parity violation of the weak interaction.
5. Fermions must be able to interact with the scalar field that induces SSB in order to acquire mass, but their L and R spinors (chiral fermions) charged under gauge symmetry cannot get mass individually: Interactions with the scalar field require the combination of L and R components to produce the so-called Dirac masses after SSB.
6. As they interact with weak bosons, fermions must be paired according to the experimental observations in weak decays: u with d for quarks and e/μ with neutrinos $(\nu,\bar{\nu})$ for charged leptons.

The EW theory is expected to build on a gauge symmetry akin to the one that shapes QCD, where different modes of self-interaction yield manifold charges, in contrast with QED that yields only one charge. The gauge symmetry should yield at least three bosons $(W^{\pm},\gamma)$, one for each symmetry transformation, with the preserved symmetry after SSB corresponding to the photon. The weak symmetry must therefore include complex rotations for two internal complex components of the fermion fields. This makes the choice fairly obvious, since the group of complex rotations is the special unitary group: $\mathrm{SU}(2)\approx \mathrm{SO}(3)/\mathbb{Z}_2$ $(\mathrm{SO}(3)\equiv$ real 3-D rotation group$)$, the Lie group of 2×2 unitary matrices with unit determinant:

$$\mathrm{SU}(2)=\left\{\begin{pmatrix}\alpha & -\bar{\beta}\\ \beta & \bar{a}\end{pmatrix}:\alpha,\beta\in\mathbb{C},|\alpha|^2+|\beta|^2=1\right\} \tag{2.1}$$

This group is compact and simply connected, hence topologically equivalent to the three sphere (in accord with the Poincare-Perelman theorem [3]), and its Lie algebra is generated by $\{i\sigma_1,-i\sigma_2,i\sigma_3\}$, where $\sigma_1,\sigma_2,\sigma_3$ are the Pauli matrices:

$$\sigma_1=\begin{pmatrix}0 & 1\\ 1 & 0\end{pmatrix},\quad \sigma_2=\begin{pmatrix}0 & -i\\ i & 0\end{pmatrix},\quad \sigma_3=\begin{pmatrix}1 & 0\\ 0 & -1\end{pmatrix} \tag{2.2}$$

Since the transformations are complex, this group is compatible with $U(1)$ in QED, which charges the weak bosons after SSB, and the three generators of its Lie algebra pair fermions into doublets that transform under the gauge interactions. This pairing is analogous to the three color charges of the quarks under SU(3)-QCD interaction and implies we need one separate neutrino to pair with each charged lepton.

While SU(2) gauge bosons only interact with left-handed fermion doublets, we need to reconcile the two seemingly conflicting aspects of a unified theory, namely, the weak beta-decays are indeed chiral, but QED, the relic symmetry, is parity-symmetric. Electroweak unification also requires a scalar field that breaks SU(2) symmetry into the $U(1)$ symmetry of QED, but it is obviously not feasible to get gauge-invariant interactions of the fermions with the scalar field that would generate a mass term. In other words, we need to incorporate fermion Dirac masses and parity-symmetric QED and this can only be done by incorporating right-handed fermions, one for each left-handed fermion charged under QED after SSB, even though they are not part of the weak decay per se. Since we do not have right-handed neutrinos, we cannot get Dirac masses for neutrinos, implying that they are massless, as it has been experimentally corroborated.

This discussion makes it apparent that the unified EW theory requires $\mathrm{SU}(2)_L$ symmetry and another interaction that would yield QED after SSB, i.e., the 1D complex internal rotation group $U(1)$. So, the gauge $\mathrm{SU}(2)_L$ symmetry acts on the L doublets:

$$L_e = \begin{pmatrix} \nu_L^e \\ e_L \end{pmatrix}, \quad L_\mu = \begin{pmatrix} \nu_L^\mu \\ \mu_L \end{pmatrix}, \quad Q_L = \begin{pmatrix} u \\ d \end{pmatrix} \tag{2.3}$$

With associated bosons $\begin{pmatrix} W_1 \\ W_2 \\ W_3 \end{pmatrix}$ as well as a single boson, B, associated with $U(1)$ that interacts with the L doublets as well as with the right-handed singlets e_R, μ_R, u_R, d_R. The charge of the fields under weak SU(2) is known as weak isospin, and labeled τ^3, whereas the charge under $U(1)$ is known as hypercharge and labeled Y. It turns out this gauge symmetry can be spontaneously broken by a scalar field, now identified as the Higgs doublet H, that transforms as a doublet under SU(2) and has nonzero hypercharge. Thus, the unified gauge symmetry $\mathrm{SU}(2)_L \times U(1)_Y$ can be spontaneously broken by H as it acquires its nonzero v.e.v to yield a leftover symmetry, $U(1)_{\mathrm{EM}}$, that can be identified with electromagnetism. With SSB, the weak bosons mix according to

$$W^\pm = \frac{1}{\sqrt{2}}\left(W^1 \mp iW^2\right) \tag{2.4}$$

While B and W^3 combine linearly after SSB to make the gauge boson (γ) of electromagnetism and the weak Z boson:

$$\begin{pmatrix} \gamma \\ Z \end{pmatrix} = \begin{pmatrix} c_W & -s_W \\ s_W & c_W \end{pmatrix} \begin{pmatrix} B \\ W^3 \end{pmatrix} \tag{2.5}$$

The electric charge Q emerges as the linear combination $Q = \tau^3 + \frac{Y}{2}$.

At this stage, we are allowed to combine the Higgs doublet with the L-doublet and R-singlet fermions into gauge-invariant interactions that yield fermion masses after SSB, in accord with the general EW unification scheme presented in Figure 2.2.

Since the scalar Higgs field contains four real scalar degrees of freedom, and ¾ of the continuous symmetry is broken, we expect three Goldstone bosons to be subsumed by $W^\pm$, Z in order to obtain an extra transverse polarization once they become massive. Additionally, we should have a single leftover real scalar neutral under QED, the Higgs boson.

Thus, the unified EW theory predicts the existence of the Z and Higgs boson. Both bosons had been confirmed experimentally: The Z boson was discovered in neutrino scattering off of electrons via neutral

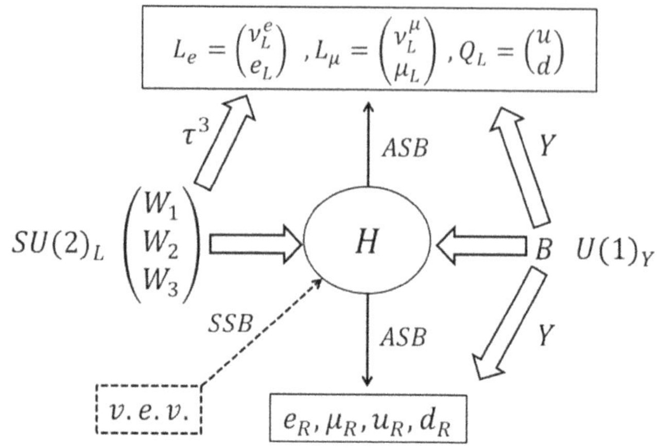

FIGURE 2.2 The electroweak unification scheme reconciling parity violation of the weak interaction leading to beta-decay with a stage at which it is allowed to combine the Higgs doublet with the L-doublet and *R*-singlet fermions into gauge-invariant interactions that yield fermion masses after SSB.

current weak interactions in the bubble chamber and was observed at the proton synchrotron in the 1980s, while the Higgs boson was discovered at the Large Hadron Collider (LHC) in 2012.

The acquisition of fermionic Dirac mass in the EW unification scheme rests heavily on a physical vacuum whose energy is reduced by the Higgs field. Much like a phase transition to a lower energy state, the Higgs field plays a role akin to the latent heat associated with the phase transition. The Higgs field is assumed to break the symmetry as it obtains its v.e.v f, which bestows mass on the *L*-*R* fermion doublets, an after SB (ASB) event. The key issue to address at this point is how does the Higgs field get its v.e.v.

2.2 FIVE SHADES OF DARK MATTER AND A DARK-VISIBLE PORTAL IN THE SIXFOLD COVERING OF SPACE-TIME IDENTIFIED BY THE CONSCIOUS AUTOENCODER

The universe inflation following the quantum gravity epoch need not be uniform along all dimensions. In fact, as shown in Chapter 1, the hierarchy problem of incommensurately low particle masses relative to Planck's value and the incommensurably small coupling strength of gravity relative to the other three fundamental forces suggest the existence of compact "curled-up" dimensions tuned to store matter waves and provide a geometric dilution of gravity. Along expanded dimensions, the range r of a force correlates with the mass M of a bosonic massive carrier: $r = \hbar / Mc$, whereas along curled-up dimensions, the range r becomes the radius of the dimension itself since otherwise there would be destructive interference. Hence, an autoencoder of the fifth space-time dimension distilling quantum reality into the 4D space-time operates under the assumption that the quantum vacuum has an associated wavelength $\lambda_0 = 2\pi r = h / M_{u-H}c$, with $M_{u-H} = 246\ \text{GeV/c}^2$, identified as the mass of the ur-Higgs carrier [4]. This tuning yields $r = r_0 = 0.802695 \times 10^{-18}$ m, a value that fits exceptionally well the upper limit to the smallest material dimension known: $r_0 = 0.86 \times 10^{-18}$ m. This is the effective quark diameter determined from parton models of deep inelastic $e^{\pm}p$ scattering at TeV scales corresponding to luminosity at 1 femtobarn^{-1} (Chapter 1). This upper limit was established within a quark form-factor model [5] and is also in agreement with the range of the weak interaction at $r \approx 10^{-18}$ m (Chapter 1). *Crucially, this length scale that yields the parameter that upholds the physics edifice, the stable vacuum at 246 GeV, is actually assimilated to the quark scale, the smallest material scale known.* In other words, *AI is asserting that what we regard as the quantum vacuum vis-à-vis the visible sector is actually* UM, *the precursor of matter.* Dark matter (DM) in general corresponds to multiples of the momentum $k_0 = \hbar / r_0$ spanning the dark dimension. While now fully justified, the validation of these operational tenets will take up the remaining of this chapter as well as Chapter 4.

As shown subsequently, the way the Higgs field is endowed with its v.e.v. follows from compatibilities with symmetries arising from the quantum relativistic wave equations that describe dark matter. There are six such equations, one for each type of 2D factorization of the special relativity equation:

$$k^2 + k_0^2 = \omega^2, \quad k = \lambda^{-1}, \quad k_0 = \lambda_0^{-1} \tag{2.6}$$

where the wavenumbers k and k_0 indicate momentum and rest-mass momentum, respectively, with "natural" units of h, c, G fixed at 1 and, accordingly, energy expressed as the frequency ω. For simplicity and without loss of generality, we shall assume one visible generic coordinate, x, while y denotes the dark coordinate.

A possible 2D factorization of Eq. 2.6 takes the form:

$$\pm I_2 \omega = \alpha k + \beta k_0 \tag{2.7}$$

where α, β are matrices satisfying the following idempotent and anticommutative relations:

$$\alpha^2 = \beta^2 = I_2 = \begin{pmatrix} 1 & 0 \\ 0 & 1 \end{pmatrix}; \quad \alpha\beta + \beta\alpha \equiv \{\alpha, \beta\} = 0 \tag{2.8}$$

The Pauli matrices (Eq. 2.2) satisfy Eq. 2.8; hence, there are in principle six realizations to Eq. 2.7. We may denote (i, j) the factorization realized by $\alpha = \sigma_i, \beta = \sigma_j \left(i, j = 1,2,3, i \neq j\right)$. Accordingly, dark matter is classified in 3! – 1 = 5 different types, heretofore named "shades", $\mathrm{DM}_{i,j}$, where $\mathrm{DM}_{i,j}$ follows the E^2 factorization $(i, j) \neq (3,1)$. Each shade belongs to a different chiral-pair sheet in the AC-inferred sixfold space-time covering manifold $\left\langle \left(\mathbb{R}^4 / \mathbb{Z}^4\right)^{\#} \right\rangle_6$ (Chapters 1 and 7). Assuming Lorentz-invariant wave functions of the form:

$$\psi(x,t) \sim u(k) e^{i(kx - \omega t)} \tag{2.9}$$

the (3,1) factorization of Eq. 2.6 yields

$$\begin{aligned} i\partial_t \psi_R &= -i\partial_x \psi_R + k_0 \psi_L \\ i\partial_t \psi_L &= i\partial_x \psi_L + k_0 \psi_R \end{aligned} \tag{2.10}$$

This equation is gauge invariant as the ur-Higgs field is enabled to interact with the photon and with the Higgs field.

The $U(1)$ gauge invariance of this fermion equation would require invariance under the internal symmetry transformation (marked with the prime):

$$\psi_L \to \psi_L' = e^{i\vartheta(x)} \psi_L; \quad \psi_R = \psi_R' \tag{2.11}$$

Such symmetry cannot be achieved unless the dark fermionic field is mixed with the scalar Higgs field. First, we incorporate into the equation the EM vector potential A that transforms as

$$A \to A' = A - \partial_x \vartheta \tag{2.12}$$

yielding the covariant derivative:

$$\partial_x \to D_x = \partial_x + iA \tag{2.13}$$

$$\begin{aligned}(D_x\psi_L)' &= e^{i\vartheta(x)}\,\partial_x\psi_L + (i\partial_x\vartheta)e^{i\vartheta(x)}\psi_L + iA'e^{i\vartheta(x)}\psi_L \\ &= e^{i\vartheta(x)}\,\partial_x\psi_L + (i\partial_x\vartheta)e^{i\vartheta(x)}\psi_L + i(A-\partial_x\vartheta)e^{i\vartheta(x)}\psi_L = (D_x\psi_L)e^{i\vartheta(x)}\end{aligned} \tag{2.14}$$

Since the Higgs scalar field ϕ_H transforms as ψ_L, we may endow the ur-Higgs fermion with $U(1)$ symmetry by mixing the two fields, yielding

$$\begin{aligned} i\partial_t\psi_R &= -i\partial_x\psi_R + \overline{\phi_H}\psi_L \\ i\partial_t\psi_L &= iD_x\psi_L + \phi_H\psi_R \end{aligned} \tag{2.15}$$

Equation 2.15 is invariant under internal complex rotation and corresponds to the Higgs-ur-Higgs mixing action in 5D space-time:

$$S = \int dxdy\sqrt{g}\left\{\left[g^{\mu\nu}\sum_{n=1,2}(D_\mu\varphi_n)^\dagger(D_\nu\varphi_n)\right]-\left[-\frac{\mu_H^2}{2}|\phi_H|^2+\frac{\lambda_H^2}{4}|\phi_H|^4-\frac{k_0^2}{2}|\psi|^2+\frac{\lambda_{HuH}}{4}|\psi|^2|\phi_H|^2\right]\right\} \tag{2.16}$$

In Eq. 2.16, $\varphi_1 = \begin{pmatrix}\psi_R\\ \psi_L\end{pmatrix}$, $\varphi_2 = \phi_H$, μ_H = mass parameter for the Higgs field, $\frac{\mu_H}{\lambda_H} = f$, λ_{HuH} = coupling between Higgs and ur-Higgs fields, and g = compound gravitational field g, with

$$g_{\mu\nu} = \eta_{\mu\nu} + \kappa^2 A_\mu A_\nu,\quad g_{5\mu} = \kappa A_\mu,\quad g_{55} = 1 \tag{2.17}$$

where $\kappa^2 = 16\pi G/c^2$, $\eta_{\mu\nu} = \mathrm{diag}(1,-1,-1,-1)$ denotes the 4D Minkowski metric, and A_μ is the EM vector potential.

Thus, by combining Eqs. 2.10 and 2.15, we readily see that the Higgs field obtains its v.e.v. $f = k_0$ via mixing with the (3,1)-ur-Higgs. In this sense and because (3,1)-ur-Higgs is interactive with the photon (Eqs. 2.12–2.15), it may be regarded as the gatekeeper between the dark and visible sectors, the UM shade that serves as precursor to the observable universe. The other relativistic wave equations lack the required gauge symmetry to reflect interactivity with the photon, hence describing invisible excitations of the dark matter scalar field. Such excitations are aptly named "invisible shades". These shades constitute the dark sector and are described by the equations in Table 2.1. As previously noted, each shade of the wave function belongs to a different chiral-pair sheet in the AC-inferred sixfold space-time covering manifold $\left\langle\left(\mathbb{R}^4/\mathbb{Z}^4\right)^{\#}\right\rangle_6$ (Chapters 1 and 7); hence, there are five invisible sheets containing the dark matter.

None of the five shades of dark matter described in Table 2.1 can be made gauge invariant through combination with the EM vector potential jointly with the Higgs scalar field, and for that reason, they constitute the established components of the dark sector. This is in good agreement with the general composition of matter, with approximately five-sixths in the dark sector and one-sixths in the visible sector, making up for the 25% dark matter and 5% visible matter in the total universe composition that additionally includes nearly 70% in an energy equivalent counted as dark energy. On the other hand, the symmetry-compatible spillover of the (3,1) shade of dark energy ($DM_{3,1}$) onto the observable dimensions

TABLE 2.1 Relativistic Wave Equations for the Lightest ur-Particle within the Five Shades of Dark Matter

ω^2-Factorization	Relativistic Wave Equation	
(1,2)	$i\partial_t \psi_R^{1,2} = -i\partial_x \psi_L^{1,2} - ik_0\psi_L^{1,2}$	$i\partial_t \psi_L^{1,2} = -i\partial_x \psi_R^{1,2} + ik_0\psi_R^{1,2}$
(1,3)	$i\partial_t \psi_R^{1,3} = -i\partial_x \psi_L^{1,3} + k_0\psi_R^{1,3}$	$i\partial_t \psi_L^{1,3} = -i\partial_x \psi_R^{1,3} - k_0\psi_L^{1,3}$
(2,1)	$i\partial_t \psi_R^{2,1} = -\partial_x \psi_L^{2,1} + k_0\psi_L^{2,1}$	$i\partial_t \psi_L^{2,1} = \partial_x \psi_R^{2,1} + k_0\psi_R^{2,1}$
(2,3)	$i\partial_t \psi_R^{2,3} = -\partial_x \psi_L^{2,3} + k_0\psi_R^{2,3}$	$i\partial_t \psi_L^{2,3} = \partial_x \psi_R^{2,3} - k_0\psi_L^{2,3}$
(3,2)	$i\partial_t \psi_R^{3,2} = -i\partial_x \psi_R^{3,2} - ik_0\psi_L^{3,2}$	$i\partial_t \psi_L^{3,2} = i\partial_x \psi_L^{3,2} + ik_0\psi_R^{3,2}$

In its (3,1) shade, the ur-particle constitutes the ur-Higgs and thereby becomes the precursor to the visible sector. Strikingly, the six fermionic shades derived in this chapter are in one-to-one correspondence with the six sheets of the space-time encoding manifold $\left\langle \left(\mathbb{R}^4 / \mathbb{Z}^4\right)^{\#} \right\rangle_6$ (cf. Figure 1.19).

taking place in the course of phase transitions yields the visible sector through geometric dilution, as described subsequently.

The dark sector thus includes a (3,1) portal ($B = +B_1$) that behaves as a gatekeeper but also as a precursor to the visible sector through geometric dilution of the ur-Higgs field (Figure 2.3). The fundamental role of the (3,1) portal is to endow the Higgs field with its v.e.v. (Eqs. 2.10, 2.15, and 2.16), thereby inducing symmetry breaking by stabilizing the Higgs vacuum as "true vacuum" (Figure 2.4).

The 2D factorization of special relativity Eq. 2.6 may be alternatively formulated as

$$\pm\beta\omega = \beta\alpha k + I_2 k_0 \tag{2.18}$$

where the matrix products $\beta\alpha$ are given in Table 2.2.

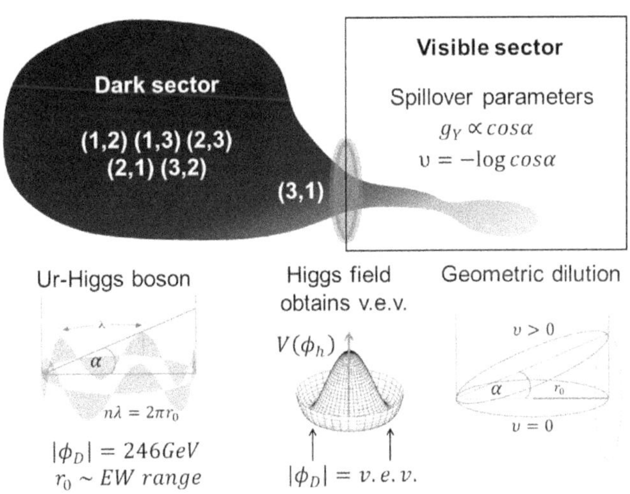

FIGURE 2.3 Composition of the dark and visible sector in accord with the high-dimensional factorization of special relativity relation $k^2 + k_0^2 = \omega^2$.The portal between the dark and visible sectors is obtained by endowing with a true (stable) vacuum the Higgs scalar field via mixing with the ur-matter (3,1)-type scalar field.

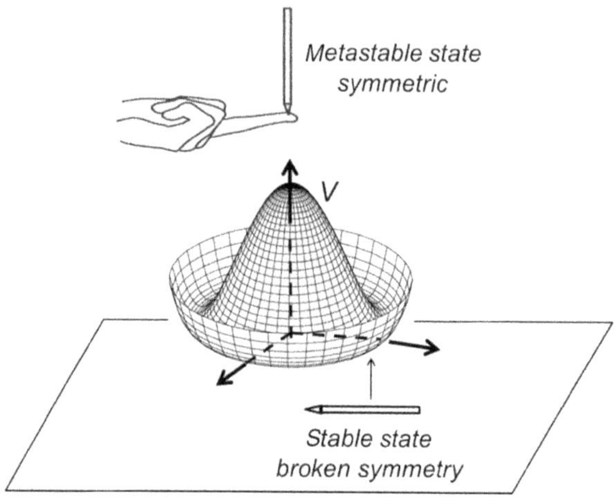

FIGURE 2.4 Stable "true vacuum" (TV) and metastable "false vacuum" (FV) in the Higgs field. The idea of mass yielded by symmetry breaking as the Higgs field is endowed with TV is inspired in Philip W. Anderson's theory of superconductivity. In that context, the photon becomes effectively massive upon a phase transition into an ordered phase representing a symmetry breaking.

TABLE 2.2 Cayley Multiplication Table (Row by Column) for the Pauli (2 × 2) Matrices

×	σ_1	σ_2	σ_3
σ_1	I_2	$i\sigma_3$	$-i\sigma_2$
σ_2	$-i\sigma_3$	I_2	$i\sigma_1$
σ_3	$i\sigma_2$	$-i\sigma_1$	I_2

We now cast Eq. 2.18 in terms of quantum operators acting on a four-component spinor ψ to obtain the six fundamental $\left(S=\frac{1}{2}\right)$ fermionic equations for the dark and visible sector:

$$\left[iB\left(\Xi^{\mu}\partial_{\mu}\right)-k_0\right]\psi=0 \tag{2.19}$$

$$B=\pm B_j=\pm\begin{pmatrix}\sigma_j & \mathbf{0}\\ \mathbf{0} & -\sigma_j\end{pmatrix},\ j=1,2,3;\quad \Xi^0=\begin{pmatrix}I_2 & \mathbf{0}\\ \mathbf{0} & I_2\end{pmatrix};\quad \Xi^{\mu}=\begin{pmatrix}\sigma_{\mu} & \mathbf{0}\\ \mathbf{0} & \sigma_{\mu}\end{pmatrix},\ \mu\neq 0,\sigma_{\mu}=\sigma_1,\sigma_2,\sigma_3 \tag{2.20}$$

The three matrices Ξ^{μ}, $\mu\neq 0$, are all different from each other, and the assignment of a Pauli matrix to a particular coordinate is immaterial since the system is symmetric under visible coordinate permutation.

An alternative factorization ansatz for Eq. 2.6 is given by

$$\beta\omega=\alpha k+\beta k_0 \tag{2.21}$$

This factorization yields Dirac's equation for the shade of UM that serves as precursor for the visible sector:

$$\left[i\left(\gamma^{\mu}\partial_{\mu}\right)-k_0\right]\psi=0 \tag{2.22}$$

The key message from this section is that the autoencoder of the fifth dimension infers the existence of six *a priori* symmetries ("shades") for relativistic dark matter. Five of these shades yield dark matter proper, understood by their inability to communicate with the photon field or EM vector potential. One shade, however, identified as (3,1), yields a symmetry compatible with the photon and Higgs field; hence, it is capable of agglutination with the Higgs field to bestow on it the stable TV that makes EW unification operationally feasible. Strikingly, the six fermionic shades are in one-to-one correspondence with the six sheets of the space-time encoding manifold $\left\langle \left(\mathbb{R}^4/\mathbb{Z}^4\right)^{\#} \right\rangle_6$ (cf. Figure 1.19), a result derived through algebraic topology calculations prompted by AC (Figure 1.17, details in Chapter 7).

2.3 TRUE QUANTUM VACUUM ENCODES THE DARK SECTOR AS PRECURSOR TO THE VISIBLE SECTOR WITHIN THE SIXFOLD SPACE-TIME COVERING MANIFOLD

To understand how the AI system models the quantum vacuum as a generator of the universe, endowing it with the attributes of a precursor to the visible matter, we need to describe in some detail the autoencoder of the extra spatial dimension, where UM is encoded. A thorough account of this approach is provided in the methodological Chapter 3.

The (3,1) portal defined in the previous section is operationally modeled by AI through an autoencoder [6] that determines the role of the dark sector as precursor to the visible sector [4]. Through the autoencoder, an equivalence relation "~" is introduced so that points in the 4D space-time become equivalence classes modulo the curled-up fourth spatial dimension. Thus, the 4D space-time, whereupon the SM is established, becomes the quotient space-time $W/\sim$ of the 5D space-time W where the dark sector has been defined and allocated (Figure 2.5). In this scheme, each point in W is associated with its equivalence class in $W/\sim$ via the canonical projection $\pi : W \rightarrow W/\sim$ and each event or process defined by emission and decay of a boson in the SM through the transformation $\mathcal{K}_{SM} : W/\sim \rightarrow W/\sim$ may be lifted to an event defined by the transformation $\mathcal{F} : W \rightarrow W$ so that the following commutativity relation is satisfied (Figure 2.6) [4]:

$$\pi \circ \mathcal{F} = \mathcal{K}_{SM} \circ \pi \tag{2.23}$$

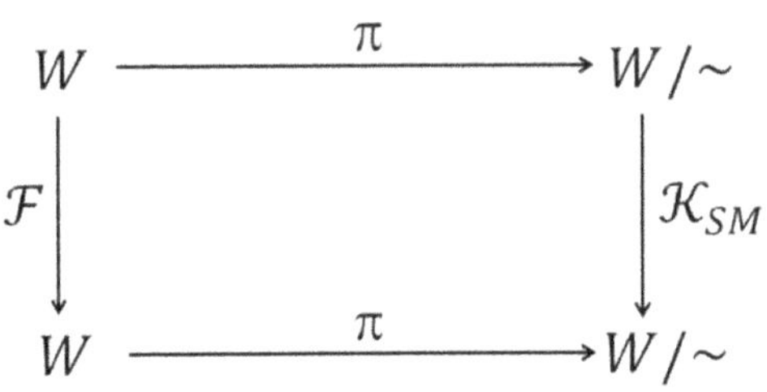

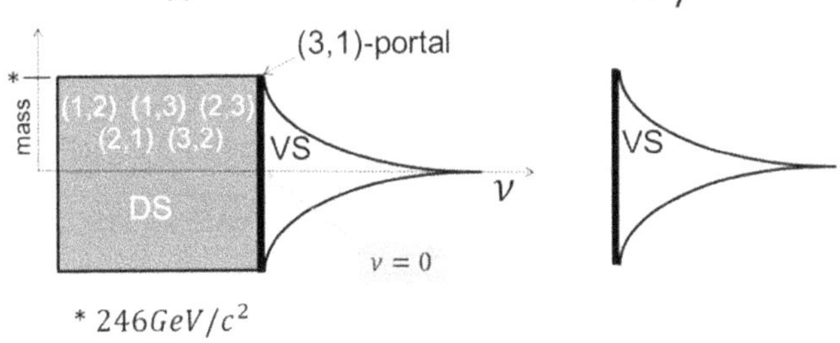

FIGURE 2.5 The (3,1) portal modeled by AI through an autoencoder that determines the influence of the dark sector on the visible sector.

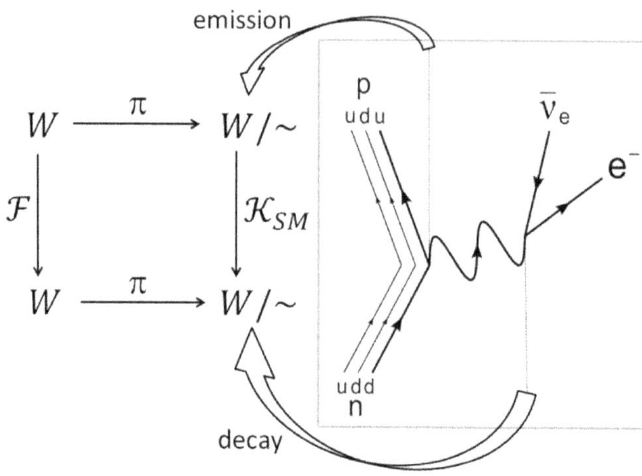

FIGURE 2.6 Autoencoder of the SM represented by the $\{\pi, \mathcal{F}\}$ pair within a commutative diagram. The latent dynamics is made up of detectable events taking place in the visible sector, mathematically described through term rearrangements/regrouping and reinterpretations of the SM Lagrangian. Such events are marked by the emission and decay of the bosons that communicate the forces in the visible sector.

In AI "algebraic" parlance, we may say that the pair $\{\pi, \mathcal{F}\}$ constitutes an autoencoder of the SM [4]. A proper identification with the AI concept requires a suitable definition of the latent dynamics [6], that is, of the detectable events taking place in the visible sector, which are mathematically described through term rearrangements/regrouping and reinterpretations of the overall Lagrangian that underlies the SM (Figure 2.6). These events are marked by the emission and decay of the bosons that communicate the forces in the visible sector [1]. The distribution of distances, $\Delta(d)$, in 4D space-time between the two events is determined experimentally as

$$F^{-1}\left[\frac{1}{k^2+m^2}\right](d) = \Delta(d) \tag{2.24}$$

where $F^{-1}[\ ](d)$ denotes the anti-Fourier transform evaluated at d and m is the boson mass. The momentum k, the Fourier conjugate of distance, is the difference in momenta $p(a) - p(b)$, where a and b denote the incoming and outgoing particle at the point of boson emission. For example, in the beta-decay $n \rightarrow p$ process described in Figure 2.6, the emitted boson W^- emerges from the mutation of a down quark (d) into an up quark (u); hence, the mass of the weak boson may be determined from the measurement of the momenta of the neutron and proton. The emission and decay of the weak boson constitute two elementary events that get encoded in the 5D space-time in terms of changes in geometric dilution for the corresponding ur-particles. Yet, the symmetries under which such ur-particles are charged differ from the symmetries of their correlated particles in the SM, as shown subsequently.

2.4 EMERGENCE OF THE MATERIAL UNIVERSE AS AN INTERNALIZATION OF SPACE-TIME SYMMETRIES

To determine their external local symmetries, ur-particles need to be spatially represented as geometric dilution of the (3,1) shade of DM. As indicated in Chapter 1, the pitch angle α is used to parametrize such dilution (Figure 2.7a), so that the angled cylindrical cross section is elliptical and becomes deformed to a circle with radius $r_0 / \cos\alpha$. Several spatial symmetries can be constructed to warrant the conservation of the geometric dilution $\upsilon = -\log\cos\alpha$ of the ur-Higgs particle (~ 246 GeV/c^2). Two such symmetries directly relevant to the current discussion may be described as follows: (a) The choice of the visible coordinates orthogonal to the dark dimension and used to define the pitch angle α is immaterial; hence, geometric dilution must be invariant under rotations of the visible frame relative to a fixed dark dimension. Such rotations operate around the three visible axis modulo inversions along the center of coordinates (Figure 2.7b). The symmetry group generated under these premises is therefore $SO(3)/\mathbb{Z}_2 \approx SU(2)$. (b) The α-pitched ellipse that projects the 246 GeV matter wave on the dark circular coordinate may be rotated around the axis of the dark dimension irrespective of the visible frame that, in turn, transforms under the premise delineated in (a) (Figure 2.8). Thus, the symmetry group under (b) is $U(1)$, the complex rotation group.

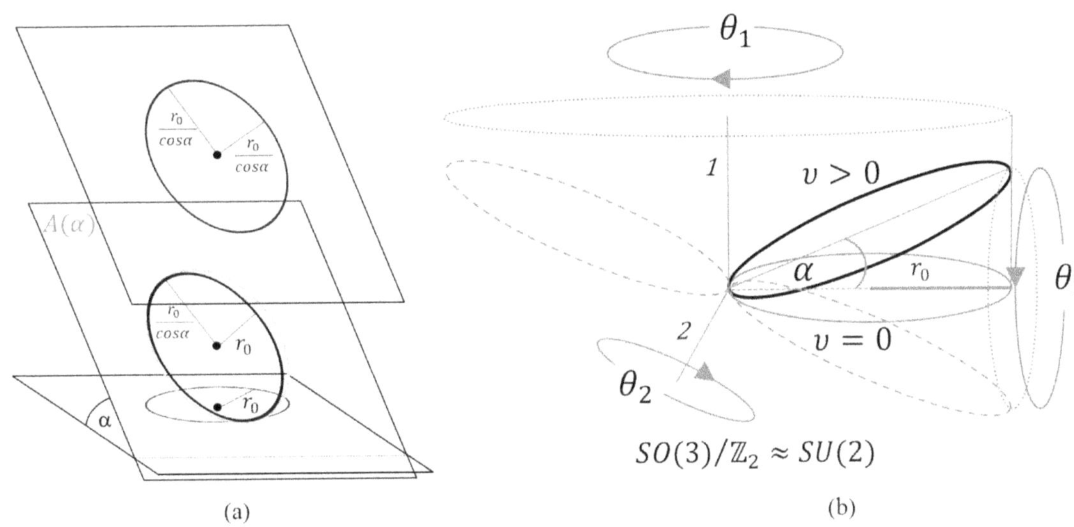

FIGURE 2.7 Spatial symmetries that warrant conservation of geometric dilution of the ur-matter scalar field. (a) Ur-particles spatially represented as geometric dilution of the (3,1) shade of DM. The pitch angle α is used to parametrize such dilution so that the angled cylindrical cross section is elliptical and becomes deformed to a circle with radius $r_0 / \cos\alpha$. (b) The choice of the visible coordinate used to define the pitch angle α is immaterial; hence, geometric dilution must be invariant under rotations of the visible frame as a rigid body around the three visible axes modulo inversions from the center of coordinates.

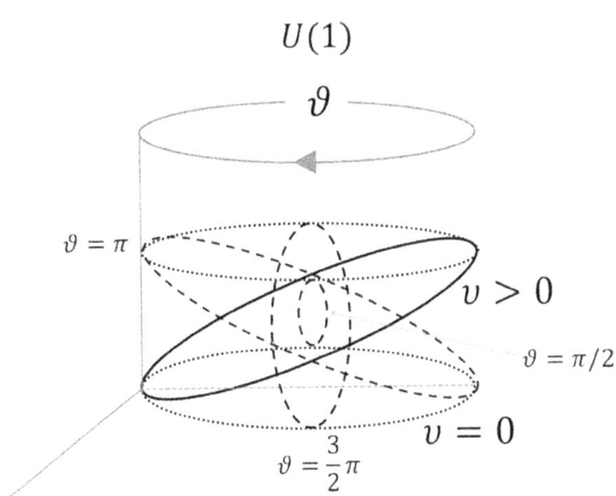

FIGURE 2.8 Invariance of geometric dilution under spatial symmetry defined by rotation around the axis of the dark dimension of the α-pitched ellipse that projects the 246 GeV matter wave on the dark circular coordinate.

We may determine the full external symmetry associated with conservation of geometric dilution in the 5D space-time by amalgamating the two factors arising from (a) and (b) as $SU(2)\times U(1)$. This local external symmetry in 5D space-time W undergoes a transmutation (internalization) into the local gauge symmetry associated with the EW unification upon the canonical projection $\pi : W \to W/\sim$ onto the quotient manifold that represents the standard 4D space-time (Figure 2.9).

The canonical projection 5D → 4D obliterates (actually, internalizes) the external local spatial $SU(2)\times U(1)$ symmetry, as it lumps points in equivalence classes modulo the dark coordinate. However, the symmetry is restored, resurfacing *albeit* in a different guise, i.e., as internal local symmetry, when the EW boson fields are incorporated in the quotient manifold (Figure 2.10). We may state that the "primeval symmetry" of UM gives birth to the EM and weak force charged under the internalization of the ur-symmetry in the autoencoder universe defined by the commuting diagram displayed in Figure 2.10.

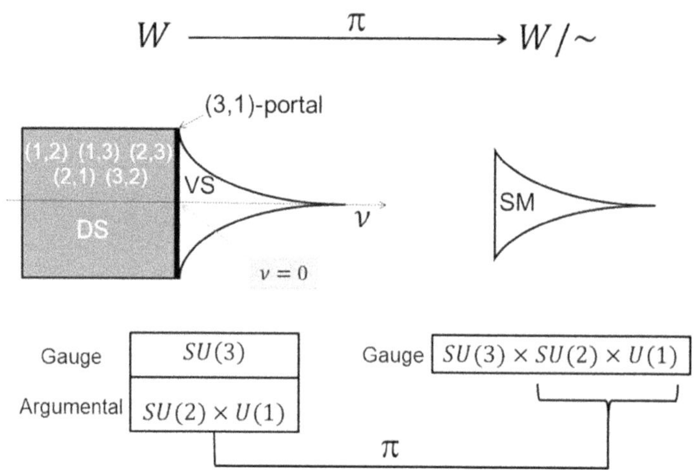

FIGURE 2.9 The spatial symmetry group in 5D space-time W undergoes a transmutation into the electroweak internal symmetry group upon the projection: $W \to W/\sim$.

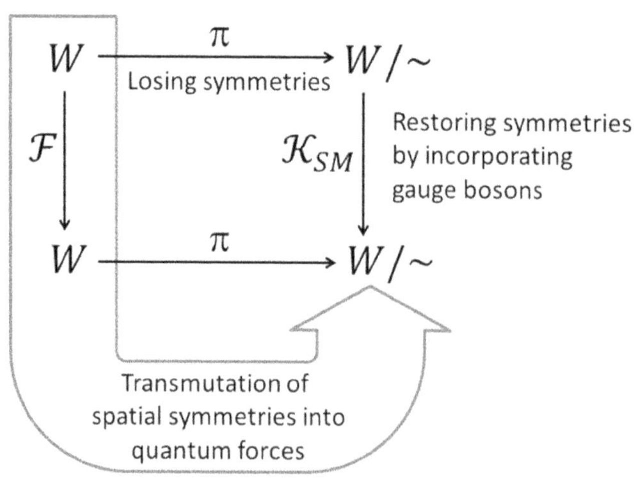

FIGURE 2.10 The "primeval spatial symmetry" of ur-matter engenders the electromagnetic and weak force in the autoencoder universe defined by the commutative diagram. The spatial symmetry of ur-matter undergoes a transmutation as internal symmetry in the quotient manifold.

This analysis cannot be said to be complete without incorporating the evolution of the universe, or more precisely, with a delineation of the critical junctures or phase transitions at which the gauge bosons associated with the $SU(2)\times U(1)$ symmetry became activated. A key criticality is achieved at temperature $T_0 = k_0 / k_B = 2.85\times 10^{15}$ K, where k_B is the Boltzmann constant. This criticality marks the phase transition at which $DM_{3,1}$ mixes and thereby endows ϕ_H with its v.e.v., which thereby becomes the TV [2,7]. This activation of the ur-Higgs scalar is reflected in a kinetic energy commensurate with the rest-mass energy 246 GeV, making it meaningful to factorize the term $\left(k^2 + k_0^2\right)$. The internalization of the spatial symmetry in the 5D space-time occurs as the temperature approaches the critical temperature for activation of the Higgs mechanism:

$$T \rightarrow T_{\mathrm{EW}} \sim m_H c^2 / k_B = 1.44\times 10^{15}\ \mathrm{K} \tag{2.25}$$

This activation requires kinetic energies commensurate with the rest-mass energy of the Higgs boson. The scalar ur-particles in W are therefore transformed into particles endowed with spin, required to internalize the spatial symmetry of UM as gauge EW symmetry within the visible sector. Furthermore, the phase transition at $T \le T_{\mathrm{EW}}$ yields the symmetry breaking associated with the TV (ground state) of the Higgs field. Thus, the field excitations in the TV already generated at $T_0 > T_{\mathrm{EW}}$ yield the particle masses at $T \le T_{\mathrm{EW}}$ (Figure 2.11). The latent heat associated with the phase transition can therefore be assimilated to $\Delta Q_H = V(\phi_H = 0) - V_{\mathrm{vev}}$, $V_{\mathrm{vev}} = V\left(\phi_H^{\mathrm{TV}}\right)$, where V is the potential energy of the Higgs field (Figure 2.4).

This scheme for mass endowment, or rather mass induction, cast in terms of critical phenomena is clearly inspired by the ideas of American physicist Philip W. Anderson on superconductivity [1,7]. In this context, the symmetry breaking associated with the formation of the organized phase below the critical Curie temperature is responsible for endowing the photon with mass, thus relenting it and precluding penetration into the superconducting material of the external magnetic field vectorially carried by the photon.

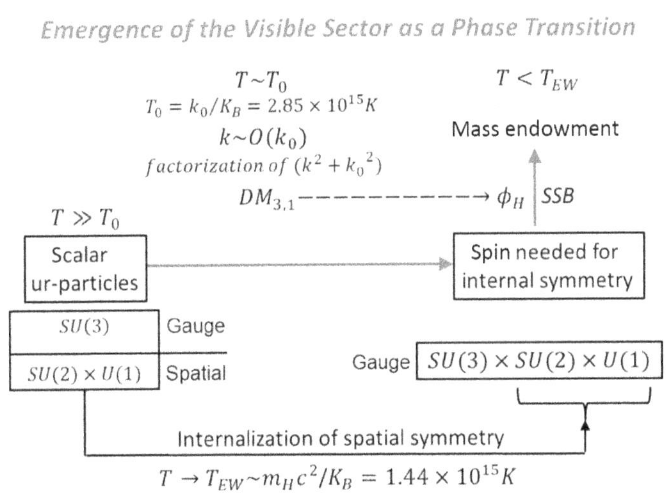

FIGURE 2.11 Emergence of the visible sector as a phase transition from ur-matter, with the internalization of the spatial symmetry in the 5D space-time. The internalization occurs as the temperature approaches the critical temperature for activation of the ur-Higgs mechanism. The scalar ur-particles are thereafter transformed into particles endowed with spin, as required to internalize the spatial symmetry as electroweak symmetry. The phase transition at $T \le T_{\mathrm{EW}}$ activates the Higgs field yielding the symmetry breaking associated with the true vacuum that was acquired at the higher temperature criticality where the ur-Higgs was activated.

2.5 TRUE VACUUM DOMAIN GENERATED AFTER A SYMMETRY-BREAKING PHASE TRANSITION

A phase transition at T_0 results in domains adopting the TV encoded in $DM_{3,1}$, in which particles are endowed with mass through the activation of the SB mechanism [2]. The point of this section is to show that such domains undergo exponential expansion, thereby offering an answer to the origin of the cosmological constant problem. Our observable universe is one such domain. To compute the rate of domain expansion, we assume a single scale factor q, yielding the Hubble constant $H = \frac{\dot{q}}{q}$ (the dot indicates time derivative) and cosmological constant:

$$\Lambda = 3\left(\frac{\dot{q}}{q}\right)^2 \tag{2.26}$$

The TV domain is represented by the wave function:

$$\psi(q) = \Upsilon(q)e^{iS(q)} \tag{2.27}$$

with action $S(q)$ defined as

$$S(q) = \frac{1}{16\pi G}\int \Upsilon(q)\sqrt{-g}d^4x \tag{2.28}$$

The metric is given by

$$ds^2 = \frac{2G}{3\pi}\left[dt^2 - q^2(t)dV_3^2\right] \tag{2.29}$$

with 3D volume measure V_3. Heretofore, we shall assume "natural units" to simplify notation. The Lagrangian associated with Eqs. 2.28 and 2.29 is then

$$\mathcal{L} = p_q\dot{q} - \mathcal{H} \tag{2.30}$$

with scale momentum given by $p_q = -q\dot{q}$ and Hamiltonian:

$$\mathcal{H} = -\frac{1}{2}\left(\frac{p_q^2}{q} + q\right) \tag{2.31}$$

Substituting quantities by operators, the domain wave function satisfies

$$\left[q^2\partial_q\left(q^{-2}\partial_q\right) - q^2\right]\psi(q) = 0 \tag{2.32}$$

Given the factorization of the wave function (Eq. 2.27), Eq. 2.32 is decomposed as

$$\partial_{q^2}^2 S + 2\frac{(\partial_q\Upsilon)(\partial_q S)}{\Upsilon} - \frac{2}{q}(\partial_q S) = 0 \tag{2.33}$$

$$(\partial_q S)^2 + U(q) + U_Q(q) = 0 \tag{2.34}$$

where $U(q)$ and $U_Q(q)$ are, respectively, the classical and quantum potential given by

$$U(q)=q^2;\quad U_Q(q)=-\frac{\partial^2_{q^2}\Upsilon}{\Upsilon}+\frac{2}{q}\frac{(\partial_q\Upsilon)}{\Upsilon} \tag{2.35}$$

Thus, we obtain the wave function by solving

$$\partial_{\dot{q}}\mathcal{L}=\partial_q S \quad \text{or} \quad \dot{q}=-\frac{1}{q}\partial_q S \tag{2.36}$$

which yields

$$\psi(q)=\left[i\xi I_{\frac{3}{4}}\left(\frac{q^2}{2}\right)-K_{\frac{3}{4}}\left(\frac{q^2}{2}\right)\right]q^{\frac{3}{2}} \tag{2.37}$$

where $I_{\frac{3}{4}}\left(\frac{q^2}{2}\right)$ and $K_{\frac{3}{4}}\left(\frac{q^2}{2}\right)$ are, respectively, modified Bessel functions of the first and second kinds. The action then reads

$$S(q)=-\frac{\xi}{4\Gamma\left(\frac{3}{4}\right)\Gamma\left(\frac{7}{4}\right)}q^3;\quad (\xi>0, q\ll 1) \tag{2.38}$$

Combining Eqs. 2.36 and 2.38, we obtain the relation $\dot{q}=\sqrt{\frac{\Lambda}{3}}q$; hence, we can readily see that the domain with the TV is subject to exponential expansion, with cosmological constant given by

$$\Lambda=3\left(\frac{\dot{q}}{q}\right)^2=\frac{27}{16}\left[\frac{\xi}{\Gamma\left(\frac{3}{4}\right)\Gamma\left(\frac{7}{4}\right)}\right]^2 \tag{2.39}$$

What is established here is that our universe constitutes one domain at TV emerging at $T<T_0$ and that such a domain is subject to exponential growth in accord with the Hubble scenario described in Chapter 1. It is unknown whether the phase transition includes other TV domains as well, i.e., a *multiverse* scenario beyond criticality. Autoencoder simulations described subsequently suggest that this is indeed the case.

2.6 DARK ENERGY AS LATENT HEAT IN THE PHASE TRANSITION THAT GENERATED THE UNIVERSE

In the physical picture put forth so far, the production of dark energy in our universe is the resultant of the phase transition leading to false vacuum (FV) decay into the TV, a transition that yields the latent heat $\Delta Q_H=V(\phi_H=0)-V_{\text{vev}}$. Thus, the release of latent heat is associated with the onset of symmetry breaking that occurs at $T<T_0$. At this juncture, the scalar field ϕ_H is endowed with its TV (v.e.v.) by mixing with

the dark matter scalar component $DM_{3,1}$. This TV bestows mass to SM particles within the domain that we identify as our universe, a domain bound to grow exponentially with cosmological constant obtained analytically (Eq. 2.39).

An experimentally verified consequence of this analysis is that *since latent heat ΔQ_H scales with volume, the dark energy remains geometrically undiluted during domain growth or universe expansion.*

To derive the rate of dark energy generation, ν_{DE}, we write the potential energy of the Higgs field as

$$V(\phi_H) = -\frac{\mu_H^2}{2}|\phi_H|^2 + \frac{\lambda_H^2}{4}|\phi_H|^4 \tag{2.40}$$

This expression yields

$$\Delta Q_H = \frac{1}{4}\left(\frac{\mu_H^2}{\lambda_H}\right)^2 \tag{2.41}$$

$$\nu_{\mathrm{DE}} = \frac{1}{4}\left(\frac{\mu_H^2}{\lambda_H}\right)^2 \frac{d}{dt}q^3 = \frac{\sqrt{3\Lambda}}{4}\left(\frac{\mu_H^2}{\lambda_H}\right)^2 q^3 \tag{2.42}$$

Solving for q as $q = q_0 \exp\left(\sqrt{\Lambda/3t}\right)$, Eq. 2.42 can be explicitly written as

$$\nu_{\mathrm{DE}} = \frac{\sqrt{3\Lambda}}{4} q_0^3 \left(\frac{\mu_H^2}{\lambda_H}\right)^2 \exp\left(\sqrt{3\Lambda t}\right) \tag{2.43}$$

The validity of the theory delineated in this section stems not only from the rigorous derivations from first principles but also from numerical estimations of the relatively constant DE density notwithstanding the Hubble picture of universe expansion (Chapter 1) and contrasted with experimental observation [8], as shown in Chapter 5.

2.7 SUSTAINABILITY OF THE UNIVERSE TOPOLOGY AND THE DARK ENERGY THERMODYNAMICS

In consistence with the geometric dilution of gravity, its weakness relative to the other forces, and the hierarchy problem of mass incommensurability with the Planck value, AI has chosen to encode mass in primeval scalar fields that incorporate an extra spatial dimension. Based on symmetry considerations, one such field, the one corresponding to $DM_{3,1}$, is able to endow the scalar Higgs field with a stable vacuum that breaks gauge symmetry yielding masses of particles in the SM. Thus, a veritable phase transition occurs whereby the endowment of mass in 4D space-time $W/\sim$ promoted by the transition from the false (metastable, symmetric) vacuum to the true (stable, asymmetric) vacuum correlates at the level of the 5D space-time W with a geometric dilution ($k_0 \to k_0 \cos\alpha$) of the scalar ur-Higgs field. The correlation becomes operative because the autoencoder represents a commutative diagram (Figure 2.12).

In this scheme, dark energy is characterized as the latent heat of the SB phase transition that yields the expanding TV domain that we came to regard as our universe. This dark energy does not get geometrically diluted because it is proportional to the domain volume itself, and it is commensurate with dark matter because the latter generated it by activating the Higgs mechanism at the criticality $T \approx T_0$. A multiverse (multi-domain) scenario cannot be discarded *a priori*, and AI confirms that this is indeed the correct scenario. This is so because a significant expenditure of dark energy to fuel the transition TV $\to$ FV can

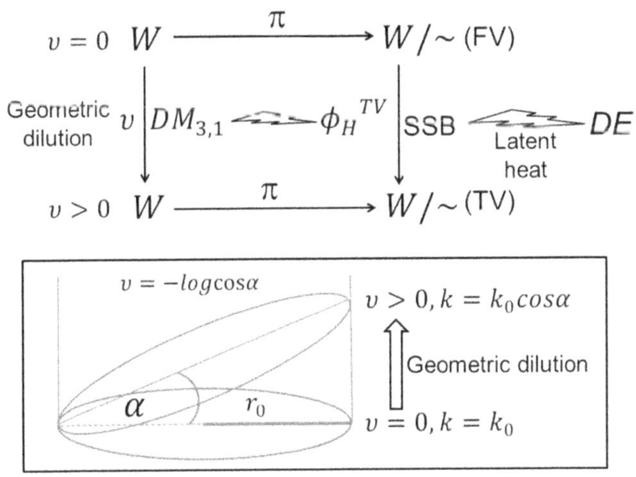

FIGURE 2.12 Autoencoder reproducing a phase transition whereby the particle endowment of mass in 4D space-time promoted by the transition from the false vacuum to the true vacuum correlates at the level of W with the geometric dilution ($k_0 \to k_0 \cos\alpha$) of ur-matter. Acronyms: DM, dark matter; DE, dark energy; SSB, spontaneous symmetry breaking; TV, true vacuum; FV, false vacuum.

be anticipated as being required to uphold the topology of space-time inherited from the early universe, an invariant in accord with the tenets of relativity. Thus, as shown in the next section, the topology of the quotient manifold corresponds to a three torus with aspect ratio 1 for any pair of circular spatial coordinates. A TV domain with this topology, that is, completely covering $W/\sim$, is unsustainable because the underlying topology entails a wormhole whose existence is enabled by a FV, as shown in Section 2.9, and in further detail in Chapter 5. In turn, the dark energy contribution that fuels the sustainability of the FV effectively addresses the problem of the "vacuum catastrophe" [8], whereby the detectable dark energy is only residual energy from an incommensurably more concentrated vacuum energy.

In fact, the primeval wormhole is enabled by the tunneling of a FV domain generated within a TV and sustained through the expenditure of dark energy available as released latent heat of the phase transition at $T \approx T_0$. The autoencoder determines the dynamics of the domain walls as constrained by two classically forbidden regions. The quantum tunneling of the FV domain promotes and sustains the creation of the wormhole, which, in turn, increases the number of instantons, so that the tunneling induces the enlargement of the wormhole throat and the increase in the mass of the primeval black hole. This "cosmological engine" required to sustain the space-time topology is thus fueled by dark energy, as outlined in this chapter and further elaborated in Chapter 5.

2.8 THE TOPOLOGY OF THE UNIVERSE IMPINGES ON QUANTUM COSMOLOGY

The extra spatial dimension has been incorporated to address a specific need requiring an extension of the preexisting SM. The most difficult issue we stumble upon when postulating a compact fourth spatial dimension to elucidate the nature of the quantum vacuum is that the global topology of the universe must allow for such possibility [4]. In the past, people have tried to incorporate a curled-up dimension into an Euclidean space, an idea that is *a priori* feasible but topologically meaningless. Not only such attempts failed to generate correct extensions of GR, but they also arose from the unappealing and unwarranted assumption that one dimension would be compact, while the others are infinite [9]. Furthermore, such a ruled universe would have boundaries, which constitutes an ontological impossibility as shown below.

Incorporating a compact dimension is a nuanced matter that pivotally depends on the underlying topology of the universe [4]. This was clearly not a matter of concern for Einstein because his GR is essentially governed by differential equations [8]. Since differential equations describe a local situation, the global topology of space-time plays no direct role. Furthermore, relativity has no concern for boundary conditions, as the universe is assumed to be infinite. In our topology, the universe is compact and multiply connected, but there are no boundary conditions simply because the existence of boundaries poses insurmountable ontological problems: The quantum vacuum would need to interface with nothingness, that is, with the absence of geometry, a clear metaphysical impossibility. To define boundaries, nothingness would need to be located beyond the boundaries but that would require extending the geometry of space beyond the boundaries, leading to a contradiction in terms.

In spite of such ontological shortcomings, Randall and Sundrum have incorporated a fourth spatial dimension as a quotient set topologically equivalent to $S^1 / \mathbb{Z}_2$ (Chapter 1) to address the hierarchy problem of particle masses and the weakness of gravity as arising from a geometrically diluted force [9]. Since such a quotient dimension would introduce two branes/boundaries, one to concentrate gravity and the other containing the Higgs field, their efforts, however commendable, are summarily disregarded in this book on ontological grounds.

QM defines the fabric of space, as determined by vacuum entanglement, but topology is not factored into the theory in any obvious way, even if entanglement is thought to be the culprit for the nonlocality of quantum phenomena. On the other hand, topology suddenly becomes relevant as one tries to reconcile GR with quantum physics, as is the case with quantum gravity, as described in Section 2.13, and further on in Chapter 5.

If we assume topological invariance of space throughout the universe evolution starting at the Big Bang, then space cannot be infinite and flat and therefore Euclidean. Provided we accept the Big Bang scenario, space must be compact now because it once was compact. But if space is compact, then it must be multiply connected because the dimensions are compact. In addition, as said, space cannot have a boundary because that would imply an interface with nothingness, and it is physically impossible to interface with nothingness since an interface presupposes two media described within a common geometry. This Parmenides-style analysis introduces several ontological constraints on what the topology of the universe must be. Summarizing, space must be compact, with compact dimensions, locally flat, and lacking boundary. This leaves us with essentially one orientable option, and it is a multiply connected three-dimensional manifold, the three torus, or Cartesian product of three-dimensional circles: $S^1 \times S^1 \times S^1$. Hence, the incorporation of a fourth circular dimension is compatible with the topology of the space-time manifold.

As the reader may surmise at this point, there is a fundamental difference in the way the extra spatial dimension, the curled dormant dimension, is incorporated in the treatment presented in this book when compared with early attempts, especially the so-called KK model [1]. These researchers incorporated the extra dimension *directly* into the Euclidean space, creating the absurdity of having one compact and three infinite spatial dimensions in space-time with no justification. This construction proved to be untenable [1,4]. Another attempt is provided by the Randall-Sundrum model [9] where, as said, the extra dimension is topologically assimilated to $S^1 / \mathbb{Z}_2$, that is equivalence classes (orbifolds) on the circle modulo reflections at $y = \pi$. The extra spatial dimension has fixed points $y = 0, \pi$, taken as locations of two three branes that extend in the conventional x^μ directions. In this theory, the $y = 0$ brane concentrates gravity and the $y = \pi$ brane contains the Higgs field. With its three branes as boundaries, the Randall-Sundrum construction is ontologically untenable for the reasons already given.

As we have shown, the spatial cross sections of space-time cannot be regarded as isomorphic to $\mathbb{R}^3$ but rather to the three-dimensional torus $\mathcal{T}^3 \approx \mathbb{R}^3 / \mathbb{Z}^3$, that is, a quotient space of classes modulo all the spatial translations of the unit cell indexed by three integers (Figure 2.13). This implies that the space is multiply connected with all spatial dimensions being compact, hence allowing for the incorporation of an extracircular dimension ($S^1 \approx \mathbb{R} / \mathbb{Z}$) without the conceptual violence of the KK or Randall-Sundrum models. A hybrid $\mathbb{R}^3 \times \mathbb{R} / \mathbb{Z}$ of three infinite and one compact spatial dimension would make no sense and is in fact not compatible with the tenets of GR, assuming this theory is upheld since the inception of space (i.e., once geometry is established), as topological change is forbidden.

At a fundamental level, the Poincaré group, assumed to be the group of isometries of space-time over which the whole edifice of the SM is built [8], does not really contain the proper symmetries but only approximate symmetries or rather, pseudo-symmetries, since the underlying spatial cross sections of space-time are not isomorphic to $\mathbb{R}^3$ but to $\mathbb{R}^3/\mathbb{Z}^3$.

On the other hand, a spatially compact universe implies the existence of a primeval wormhole, since the three circular dimensions are originally equivalent and therefore symmetrically isometric, as shown in Figure 2.13 and in Chapter 5. This can be visualized in a two-dimensional cross section $\mathcal{T}^2$ with a dimensional aspect ratio of 1. As shown in the next section and in Chapter 5, the sustainability of this wormhole is concomitant with the universe expansion. On the other hand, the very notion of universe expansion becomes meaningless if space were assimilated with $\mathbb{R}^3$, while it makes good sense to talk about the expansion of the unit cell in the quotient space $\mathbb{R}^3/\mathbb{Z}^3$. This is yet another reason to assume that the spatial cross sections of the relativistic space-time are actually isomorphic to $\mathbb{R}^3/\mathbb{Z}^3$, and not to $\mathbb{R}^3$, as assumed in the Minkowski space-time endowed with the isometries of the Poincaré group.

On the other hand, as AI adopts the commutative scheme presented in Figure 2.14, the autoencoder defines the lifting of the SM at the level of the universal covering with canonical projection $\pi: \mathbb{R}^3 \to \mathbb{R}^3/\mathbb{Z}^3$,

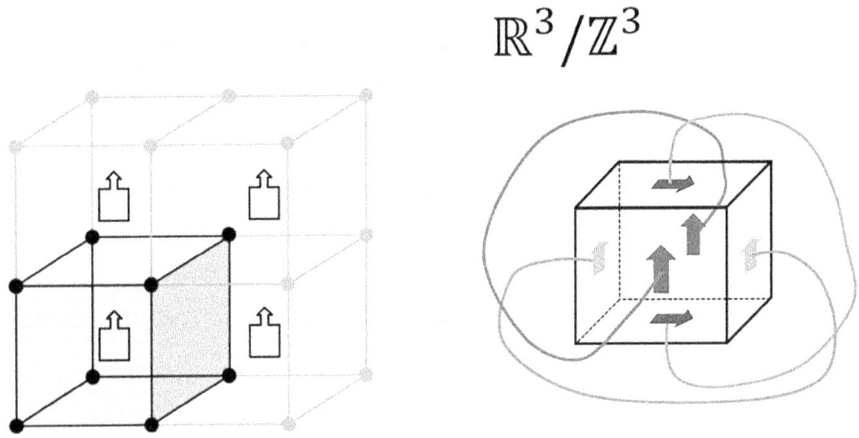

FIGURE 2.13 Generic spatial cross section of space-time with compact coordinates, as required by the topological invariance of the universe dictated by the tenets of general relativity. The compact space is isomorphic to the three-dimensional torus in accord with the topological equivalence $\mathbb{R}^3/\mathbb{Z}^3 \approx \mathcal{T}^3$. The latter manifold is obtained by identifying opposite faces of the solid cube in three dimensions, hence featuring a wormhole upon construction with dimensional aspect ratios equal to 1 (Chapter 5).

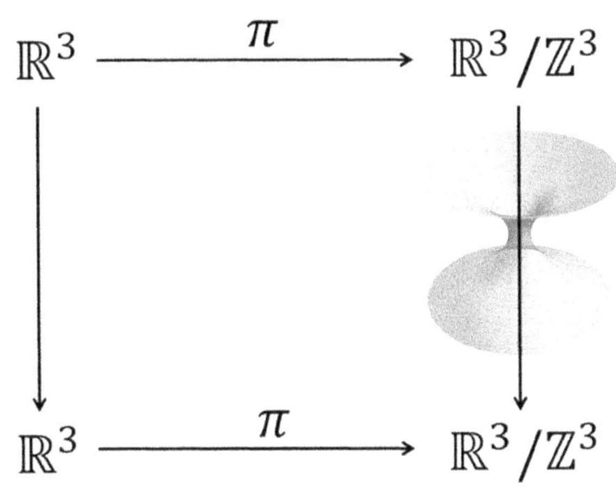

FIGURE 2.14 Commutative scheme adopted by artificial intelligence to circumvent the artifactual wormhole phenomenology associated with the universe compactness by lifting it at the level of universal covering $(\mathbb{R}^3, \pi)$ of the toroidal spatial cross sections $(\mathbb{R}^3/\mathbb{Z}^3)$ of space-time.

so that the diagram presented in Figure 2.14 is commutative. This lifting has considerable advantages since it circumvents the observed universe expansion with the concomitant primeval wormhole as artifactual of a "Plato's cave physics", where the observer is placed. Even dark energy, shown in Chapter 5 to be the residual vacuum energy not invested in sustaining the primeval wormhole, becomes artifactual of the Plato's cave phenomena materializing in $\mathbb{R}^3 / \mathbb{Z}^3$, the only events accessible to the quantum observer.

2.9 DARK ENERGY AND THE VACUUM CATASTROPHE

The topological invariance of 4D space-time implicit in the tenets of relativity compels us to provide a sustainable physical realization of the three-torus topology describing the throat of a wormhole. We shall show that this primeval wormhole is inducible from within a FV domain with a radial dimension z. Such a physical construct requires a metric surgically spliced into a Schwarzschild (S) component and a de Sitter (D) component [10], which may be given in polar coordinates as follows:

$$ds_S^2 = -A_S dt_S^2 + A_S^{-1} dr^2 + r^2\left(d\theta^2 + \sin^2\theta d\phi^2\right), \quad (r > z) \tag{2.44}$$

$$ds_D^2 = -A_D dt_D^2 + A_D^{-1} dr^2 + r^2\left(d\theta^2 + \sin^2\theta d\phi^2\right), \quad (r < z) \tag{2.45}$$

with $A_S = 1 - \frac{2m_{BH}}{r}$ and $A_D = 1 - \frac{8}{3}\pi\Lambda r^2$, where m_{BH} denotes the mass of the black hole, equivalent to the amount of dark energy expenditure to create and sustain the FV domain. The induced metric of the FV domain wall is thus

$$ds_W^2 = \sigma_{00} dt_W^2 + \sigma_{11} d\theta^2 + \sigma_{22} d\phi^2 \tag{2.46}$$

The energy-momentum tensor at the domain wall, $\mathfrak{T}_{ab}$, yields the metric tensor components σ_{ab} via the relation $\mathfrak{T}_{ab} = \gamma\sigma_{ab}$, where γ denotes interfacial tension. Let g_{ab} denote the Gaussian curvature tensor depending on proper distance ζ to the domain wall, then we obtain

$$\left(\lim_{\zeta\to 0^+} g_{ab}\right) - \left(\lim_{\zeta\to 0^-} g_{ab}\right) = -8\pi\gamma\left[\sigma_{ab} - \frac{1}{2}\sigma\sigma_{ab}\right] \tag{2.47}$$

where σ is the induced metric trace $\sigma = \sigma^{ab}\sigma_{ab}$. Equation 2.47 yields

$$\sigma_{00} = -1, \quad \sigma_{11} = z^2, \quad \sigma_{22} = z^2 \sin^2\theta; \quad \sigma_{ab} = 0(a \neq b) \tag{2.48}$$

Since $-8\pi\left[\mathfrak{T}_{11} - \frac{1}{2}\sigma_{11}\mathfrak{T}\right] = -4\pi\gamma z^2$, $\mathfrak{T} = \sigma^{ab}\mathfrak{T}_{ab}$.

Thus, we obtain the following equation determining the dynamics of the FV domain wall:

$$-\left(1 - \frac{8}{3}\pi\Lambda z^2\right)\frac{dt_D}{dt_W} - \left(1 - \frac{2m_{\mathrm{BH}}}{z}\right)\frac{dt_S}{dt_W} = 4\pi\gamma z \tag{2.49}$$

Equation (2.49) underscores the role of dark energy in upholding both the Schwarzschild and the de Sitter contributions [10], specifically, in creating the FV black hole and upholding the cosmological constant as required to ensure the sustainability of the compact, multiply connected and boundary-less topology of the universe. This cosmological role of dark energy analytically delineated in this section will be shown to play out in sustaining the topology of the universe, invariant since its inception, as dynamic simulations are extended with the aid of machine learning (ML) tools (Chapter 5).

2.10 QUANTUM VACUUM AS UR-MATTER: TRUE AND FALSE VACUUM COEXISTENCE

To truly understand what the universe was like right after the Big Bang would require a quantum theory of gravity, a subject that has remained for the most part unyielding to research efforts. Furthermore, a fundamental force becomes discernible only at temperatures commensurate with its critical value $T_B = m_B c^2 / k_B$, where m_B is the mass of the boson that communicates the force. Consequently, as we approach the Big Bang with temperatures above Planck's mark, the much higher kinetic energies would render irrelevant any distinction between the fundamental forces. The curvature of space-time is determined by the concentration of energy, which in the budding universe is smeared to such an extent as to lead to an uncertainty in the curvature of space-time, producing huge fluctuations in the metric: There is simply no identifiable geometry right after the Big Bang, while dimensions can be permuted due to complete spatial symmetry.

In this picture of the earliest universe, all we can be certain about is that the topology of space was once compact and had no boundaries (there is no geometry outside space, hence no interface with another vacuum). The topology must remain the same, since a change in topology would require a rough manifold deformation, which would be forbidden according to the premises of GR. This has led us to propose a three torus as the only compact orientable 3D manifold with no boundaries and with three compact dimensions. These dimensions have inflated to fit our perceived reality, but we have reasons to believe there are one or perhaps more dormant dimensions.

As previously indicated, gravity is much weaker than the other fundamental forces of a quantum nature. This suggests that gravity has leaked into a higher dimension, implying that *quantum vacuum is actually* UM. This is a pivotal tenet in this book. Other established facts also suggest this view, in particular the hierarchy problem, whereby the detected particle masses are incommensurably smaller than the Planck mass and the impossibility of accounting for dark matter and dark energy within the premises of the current version of the SM of particle physics.

As the issue of the need for the extra spatial dimension is settled, we faced the problem of how should the 5D space-time be modeled. The KK model that does not enable geometric dilution or activate a portal into the visible sector, while the Randall-Sundrum model incorporates the extra dimension as a quotient space, a construct that is ontologically flawed since it assumes an interface with nothingness, that is, with the absence of geometry. By contrast, AI resorts to the autoencoder $W \xrightarrow{\pi} W/\sim$ of the universe that we are aware of ($W/\sim$), incorporating the hidden dimension onto a 5D space-time where a scalar field spans the hidden dimension and makes the quantum vacuum behave as UM (Figures 2.5 and 2.6). This representation enables the emergence of our universe as a domain $\mathfrak{D}_{3,1} \subset W/\sim$ within a SB phase transition induced by endowing a (3,1) portal (the Higgs field) with its v.e.v. at 246 GeV (Figure 2.11).

This domain expands exponentially according to Eq. 2.39 and is embedded in a horned three torus representing $W/\sim$ (Figure 2.15). There may be other domains $\mathfrak{D}_{a,b} \subset W/\sim$ ($a \neq b; a,b = 1,2,3$) depending on the existence of (a,b) portals whose symmetry enables mixing with the respective ur-(a,b) particle, a form of $\text{DM}_{a,b}$. In particular, there may even be other (3,1) domains that we are not aware of. The "multiverse" is thus an untested scenario for the SB transition that endows particles with mass or marks the emergence of matter.

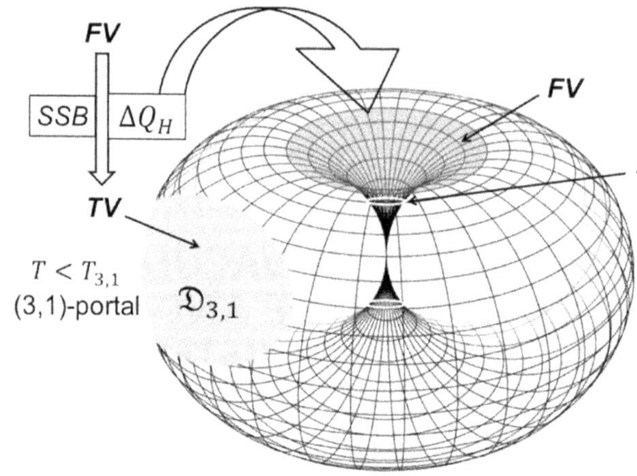

FIGURE 2.15 Flow of dark energy in the multiverse with at least two domains, one in a stable vacuum (TV) and one in a metastable vacuum (FV), with the latter sustained by a constant influx of the latent heat ΔQ_H, generated upon the emergence of the TV domain $\mathfrak{D}_{3,1}$ at $T < T_0 = T_{3,1} = 246\ \text{GeV} / k_B$. The FV domain is required to sustain the toroidal topology of the primeval wormhole.

The topology of our quotient space $W/\sim$ needs to be sustainable. As shown in Section 2.9, this requires a FV domain in order to stabilize the primeval wormhole in the three torus. Unlike the TV domain $\mathfrak{D}_{3,1}$, an FV domain does not expand spontaneously. It does so through a continuous expenditure of energy and the available energy is the latent heat ΔQ_H, generated upon the emergence of $\mathfrak{D}_{3.1}$ at $T < T_0 = T_{3,1} = 246\ \text{GeV} / k_B$. Furthermore, as described in detail in Chapter 5, the inflation of the FV domain is required to maintain the toroidal throat of the wormhole, in accord with Eqs. 2.44–2.49. In summary, the energetic economy of the universe is that of a multiverse with at least two domains, one in a stable vacuum and one in a metastable vacuum sustained by a constant influx of dark energy (Figure 2.15).

Perhaps the most subtle issue in this discussion of the emergence of the universe is the transfiguration of the symmetry of UM or quantum vacuum associated with the phase transition. As noted, the quantum vacuum is filled with UM scalar particles with rest-mass momentum k_0 or multiples thereof. The high-dimensional factorization possibilities of the special relativity relation $k^2 + k_0^2 = \omega^2$ introduce different symmetry proposals, six shades to be precise, and only one of those shades is known to be compatible with a gauge field in the visible sector of the SM. Thus, the mixing of the ur-scalar field with a scalar boson field from the visible sector bestows the TV of the visible sector, inducing a SB phase transition whereupon masses are bestowed onto particles. AI predicts six types of domains that may arise from phase transitions upon activation of the ur-boson (u-B) at $k_B T \approx m_{u-B} c^2 = k_0$. Yet, none of these domains have been detected as of yet, and the AI system (autoencoder) treats as dark matter the scalar UM which differs from the (3,1) type. The five shades of purely dark matter have an abundance relative to the (3,1) shade that obeys the ratio 5:1, in close agreement with the ratio between dark and visible matter in the universe (Chapter 1). This ratio is in consonance with the six ways to factorize the energy relation for UM resulting from special relativity, out of which only one (3,1) is symmetrically compatible with the gauge fields of the photon and the Higgs in the visible sector.

The autoencoder further reveals that the dark matter that behaves as UM for the visible sector contains six spatial symmetries (Figures 2.7 and 2.8) that undergo transmutation into the EW symmetry $\text{SU}(2) \times U(1)$ upon the phase transition induced by activation of the Higgs field (Figure 2.11). No actionable gauge fields capable of inducing SB phase transitions for the other five symmetry schemes of UM have been detected as of yet. Hence, the ratio 5:1 of dark matter to visible matter must be upheld up until the definition of "visible" is modified based on plausible extensions of the SM.

2.11 ARTIFICIAL CONSCIENCE FORAYS IN PARTICLE COSMOLOGY

Particle cosmology, the unlikely merger of particle physics and cosmology, focuses on small units of matter and their role in determining the fate and shape of the universe, while also prompting the examination of global distribution of matter and energy to make sense of local phenomena associated with mass. Thus, it constitutes a suitable platform to tackle the problem of the origin of the universe, or, in a less ambitious guise, of the origin of matter. This section delineates the implementation of an AI system suited for particle cosmology in its pursuit of the origin of the universe.

The uncanny ability to detect patterns, extract features, and distill the relevant physical picture out of algorithmic complexity in static and dynamical systems has made learning systems, especially autoencoders [4,6], the beacons of applied AI. Our goal here is more ambitious, as we strive to deploy autoencoders to reach beyond the realm of algorithmic complexity and make theoretical inferences that pertain to second-order logic [11,12]. The most crucial such cogitation by an autoencoder is the revelation that, for all its prowess [7], the Higgs field assumed to endow bosons with mass while retaining gauge symmetry is a provisional *ansatz* that needs to be revised vis-à-vis the incorporation of the extra spatial dimension. The motivation for introducing the extra dimension is amply justified in the preceding sections, where the geometric dilution of gravity and the hierarchy problem in particle masses are addressed.

An AI system set up to unravel the origin of the universe encoded in the quantum vacuum is depicted in Figure 2.16. The fifth dimension in space-time (fourth in space) is not incorporated directly, as in the KK or Randall-Sundrum schemes [1,9], but within an autoencoder, represented by a commutative diagram. The autoencoder treats the quantum vacuum as the vehicle for UM, that is, the precursor of matter. There are six ways (shades) of encoding the scalar field spanning the fifth dimension with momentum k_0 according to the symmetries resulting from the six modes of factorization of the special relativistic relation $k^2 + k_0^2 = \omega^2$ (natural units are assumed). Five of these shades, namely (1,2), (1,3), (2,3), (2,1), (3,2), yield dark matter proper (DM), while the (3,1) shade of DM yields the UM for the visible sector of the SM. This is so because the ur-field $\phi_{3,1}$ is capable of amalgamating with the Higgs scalar field ϕ_H since they are symmetry compatible (Eq. 2.15), thus yielding the TV of the Higgs field at 246 GeV, in accord with the equation: $\phi_{3,1} \bowtie \phi_H = \phi_H^{\mathrm{TV}}$ (the bowtie symbol denotes mixing). Mass in the visible sector originates through a SB phase transition catalyzed by activation of ϕ_H^{TV} and turned on by the temperature switch $T \gg \frac{k_0}{k_B} \to T < \frac{k_0}{k_B}$ within an autoencoder that runs at two temperatures, one above and one below criticality. Criticality is thus parametrized by the temperature at which UM acquires entity as a particle (Figure 2.16). A fermion with Lagrangian $\mathcal{L}$ first becomes symmetry-compatible with scalar field ϕ_H by incorporation of a boson (B) Lagrangian, so that mass endowment in the latent 4D-space-time is assimilated to the compound Lagrangian: $\mathcal{L} + \mathcal{L}_B + \mathcal{L}_H$. This fermion Lagrangian is lifted to $\varrho_\alpha^{\mathcal{L}}\left(\mathcal{L}^{3,1}\right)$, where $\mathcal{L}^{3,1}$ is the UM Lagrangian and the geometric dilution operator $\varrho_\alpha^{\mathcal{L}}$ corresponds to the pitch angle α that yields the mass $m = k_0 \cos\alpha$ for the particle with Lagrangian $\mathcal{L}$.

The autoencoder thus informs that mass only *appears* to originate from the activation of the Higgs mechanism ($\mathcal{L} \to \mathcal{L} + \mathcal{L}_B + \mathcal{L}_H$), but in truth, it is the geometric dilution of UM as ur-particles $\left(\mathcal{L}^{3,1} \to \varrho_\alpha^{\mathcal{L}}\left(\mathcal{L}^{3,1}\right)\right)$, that serves as precursor to the mass endowment for particles in the visible sector in accord with the relation:

$$\pi \circ \varrho_\alpha^{\mathcal{L}}\left(\mathcal{L}^{3,1}\right) = \mathcal{L} + \mathcal{L}_B + \mathcal{L}_H \tag{2.50}$$

Such visible particles internalize the spatial symmetry of their ur-particles as gauge symmetries, thus acquiring spin, whereas the ur-particles are scalar (Figure 2.16).

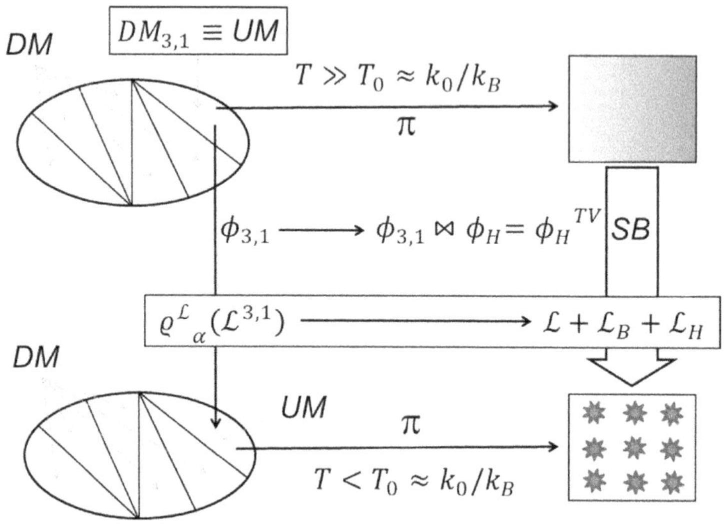

FIGURE 2.16 Artificial Intelligence system set up to discover the origin of the universe encoded in the quantum vacuum. The fifth dimension is not incorporated directly but within an autoencoder, represented by a commutative arrow diagram. The autoencoder treats the quantum vacuum as ur-matter (UM), that is, as precursor of matter. There are six ways (shades) of encoding the scalar field spanning the fifth dimension with momentum k_0 according to symmetry. Five of these shades, namely (1,2), (1,3), (2,3), (2,1), (3,2), yield dark matter (DM), while the (3,1) shade yields ur-matter for the visible sector of the SM. This is so because the ur-field $\phi_{3,1}$ is capable of amalgamating with the Higgs scalar field ϕ_H since they are symmetry-compatible (Eq. 2.15), thus yielding the true vacuum (TV) of the Higgs field, in accord with the equation: $\phi_{3,1} \bowtie \phi_H = \phi_H^{\mathrm{TV}}$. Mass in the visible sector originates through a SB phase transition catalyzed by activation of ϕ_H^{TV} and enabled by the temperature switch $T \gg \frac{k_0}{k_B} \to T < \frac{k_0}{k_B}$ within an autoencoder that runs at two temperatures, one above and one below criticality. A fermion with Lagrangian $\mathcal{L}$ first becomes symmetry-compatible with scalar field ϕ_H by incorporation of a boson (B) Lagrangian, so that mass endowment in the latent 4D space-time is assimilated to compound Lagrangian: $\mathcal{L} + \mathcal{L}_B + \mathcal{L}_H$. This Lagrangian is lifted to $\varrho_\alpha^{\mathcal{L}}\left(\mathcal{L}^{3,1}\right)$, where $\mathcal{L}^{3,1}$ is the ur-matter Lagrangian and the geometric dilution operator $\varrho_\alpha^{\mathcal{L}}$ corresponds to the pitch angle α that yields the mass $m = k_0 \cos\alpha$ for the particle with Lagrangian $\mathcal{L}$. Natural units have been assumed throughout.

Equation 2.50 constitutes a cornerstone for a deeper understanding of quantum vacuum as UM. The ramifications of this result will be pursued in the ensuing chapters. Rather than assuming the existence of a mysterious all-pervasive scalar field that endows particles with induced mass after adopting its stable TV through self-interaction, we postulate that the quantum vacuum has intrinsic materiality spanning an undetectable "within" dimension. This result appears to represent a substantive way forward as it portends to reconcile the SM with the standing hierarchy problem, the cosmological constant problem, the vacuum catastrophe, and the influence of dark matter and dark energy on deep-space dynamics (Chapter 5).

2.12 A CONSCIOUS AUTOENCODER PURSUES THE GRAND UNIFICATION OF FUNDAMENTAL FORCES

A grand unification theory (GUT) is still one of the holy grails in physics. The fundamental quantum forces respond differently to geometric dilution in the vacuum, with coupling strengths that intersect at distances approaching the Planck length $l_P = \sqrt{\hbar G / c^3} = 1.616 \times 10^{-35}$ m (Figure 2.17). The GUT primeval

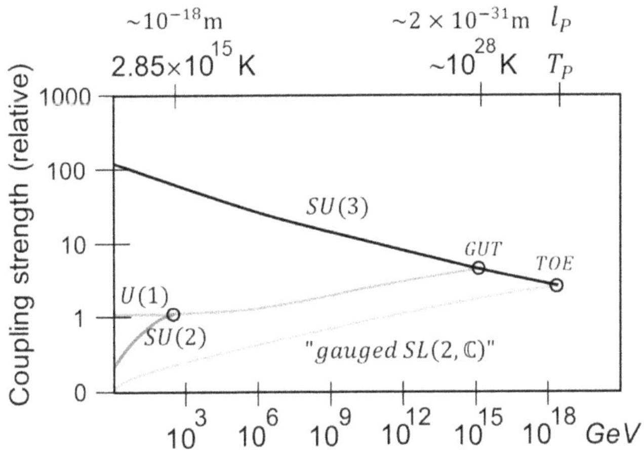

FIGURE 2.17 Gauge symmetries for the fundamental force carriers in a plot of coupling strength as a function of energy, temperature, or interaction distance. Symmetry groups SU(3), SU(2), and *U*(1) correspond, respectively, to the SNF (QCD), the weak nuclear force (WNF), and electromagnetism (EM). The circles denote the phase transitions corresponding to the gauging or internalization of the fundamental local symmetries of ur-matter. The grand unification theory (GUT) arises with the gauging of the $SU(2)\times U(1)$ symmetry of ur-matter (Figures 2.7 and 2.8) as electroweak (EW) symmetry. The theory of everything (TOE) entails a special gauging of the Lorentz $SL(2,\mathbb{C})$ symmetry to beget gravity, while SU(5) remains the local external symmetry of ur-matter $DM_{3,1}$. Gravity is not associated with a local gauge but with a one-form gauge symmetry.

symmetry was proposed by Howard Georgi and Sheldon Glashow in 1974 [13]. These researchers took into account the fundamental gauge symmetry groups and assumed they are subgroups of a minimal primeval symmetry group. Thus, since the symmetry groups SU(3), SU(2), and *U*(1) correspond, respectively, to the strong nuclear force (SNF; QCD), the WNF, and electromagnetism (EM), the primeval symmetry has been rightly assumed to be $SU(5) \approx SU(3)\times SU(2)\times U(1)$.

Given this primeval symmetry, the origin of the fundamental forces is cast by the autoencoder in terms of phase transitions corresponding to the gauging or internalization of the fundamental local external symmetries of UM concurrently with activation of the respective portals at critical temperatures $T = \frac{k_0}{k_B}$ and $T = \frac{nk_0}{k_B}\left(n \sim 10^{13}\right)$. Besides the $SU(2)$ and $U(1)$ symmetries depicted in Figures 2.7 and 2.8, the UM possesses also an external $SU(3)$ symmetry, as shown in Figure 2.18. A generic UM particle results from geometric dilution of the lightest particle in $DM_{3,1}$ with pitch angle α. The symmetry group is known to be the minimal that acts transitively over the five sphere whose points correspond to the rotation axes of the diluted UM particle. This symmetry gets gauged at the phase transition at $\sim 10^{28}$ K, concurrent with activation of a KK ur-Higgs tower with momentum nk_0, $n = \mathcal{O}\left(10^{13}\right)$.

Thus, the GUT arises from the phase transition that gauges the $SU(3)$ symmetry of UM with retention of spatial $SU(2)\times U(1)$ symmetry (Figures 2.7 and 2.8). The latter becomes internalized as EW symmetry at the next phase transition that takes place at $T = \frac{k_0}{k_B}$. On the other hand, a purported "theory of everything" (TOE) entails a special gauging of the Lorentz $SL(2,\mathbb{C})$ symmetry to beget gravity, while SU(5) prevails as local external symmetry for $DM_{3,1}$. It should be pointed out that gravity cannot be really associated with a local gauge but with a one-form generalized symmetry, as discussed subsequently.

As stated, the unifications (GUT and TOE) are suggested by the differences in response to the quantum vacuum marking convergence of the fundamental forces (Figure 2.19). This convergence (Figures 2.17 and 2.19) *hints that all fundamental forces are actually one and the same that propagates under different symmetries providing different levels of geometric dilution as they emerge from phase transitions in early stages of the universe* (Figure 2.20).

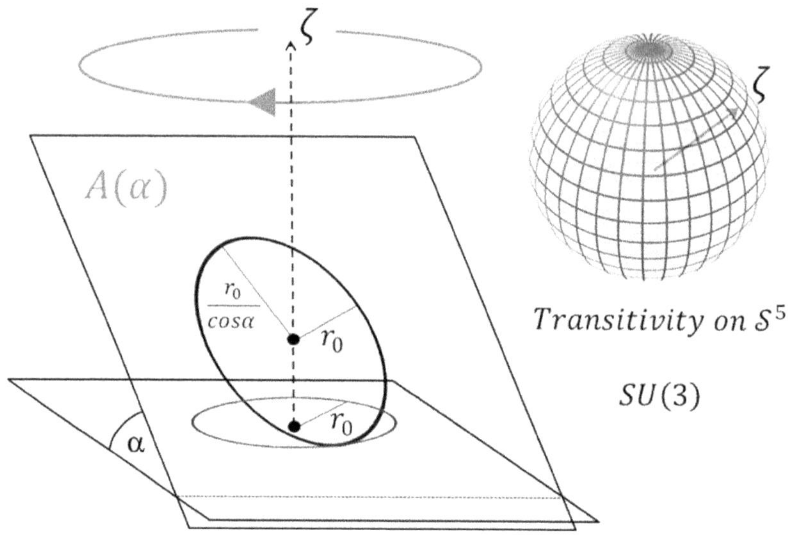

FIGURE 2.18 External local SU(3) symmetry of ur-matter. The generic ur-matter particle results from geometric dilution of the lightest $DM_{3,1}$ particle parametrized by pitch angle α. The symmetry group is known to be the minimal that acts transitively over the five sphere whose points correspond to the rotation axes of the diluted ur-matter particle. This symmetry gets gauged at the phase transition at $\sim 10^{28}$ K, concurrent with activation of a Kaluza-Klein ur-Higgs with momentum nk_0, $n = \mathcal{O}(10^{13})$.

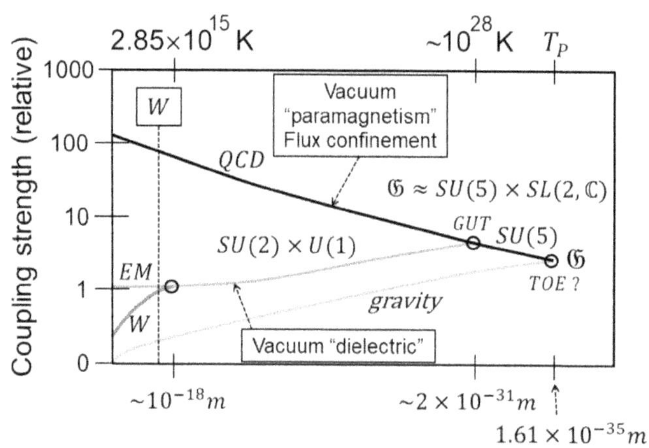

FIGURE 2.19 Response to the quantum vacuum by the fundamental forces. The SNF grows monotonically with distance as per the scale indicated in the figure. The SNF is significantly enhanced at distances larger than $\sim 10^{-18}$ m due to vacuum paramagnetism brought about by confinement of quark-gluon field lines that make the strength of the coupling behave as $\propto R$, rather than $\propto R^{-2}$. By contrast, the transient (e^-, e^+)-pair creation in the vacuum introduces polarization enfeebling electric field lines in a kind of vacuum dielectric; hence, the response of EM to quantum vacuum opposes that of SNF. On the other hand, the WNF gets significantly enfeebled at distances larger than $\sim 10^{-18}$ m because it is conveyed by massive boson *W*, so its range is $r_W \sim \frac{\hbar}{m_W c} \sim 10^{-18}$ m. The different responses of the fundamental forces to the quantum vacuum enable line crossing at points indicative of force unification as interaction distances decrease and approach Planck's critical value. Thus, at GUT, the external SU(5) symmetry of ur-matter is partially internalized into gauged symmetry SU(3), while the external local symmetry SU(2) × *U*(1) is retained by ur-matter, respectively, begetting the SNF and electroweak symmetry. At temperatures higher than the Planck temperature, ur-matter retains its full symmetry under the group $\mathfrak{G} \approx \mathrm{SU}(5)\times\mathrm{SL}(2,\mathbb{C})$. At the Planck temperature, the relativistic component $\mathrm{SL}(2,\mathbb{C})$ gets internalized as one-form-gauged Lorentz symmetry, begetting gravity, while ur-matter retains the full $\mathrm{SU}(5)$ symmetry up until the GUT point.

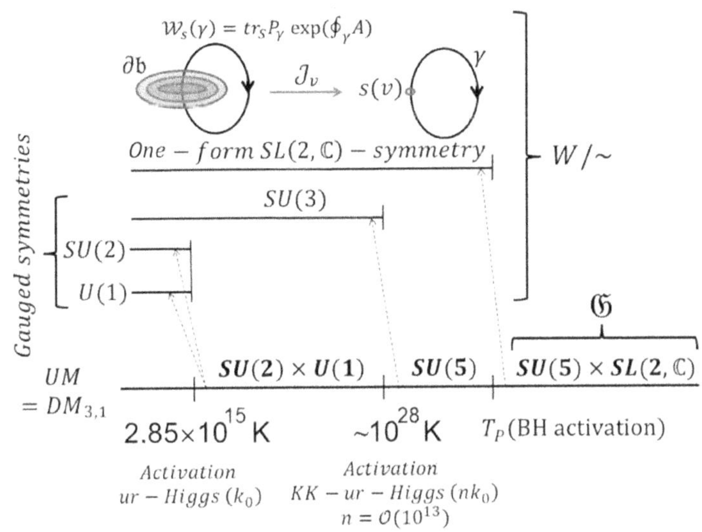

FIGURE 2.20 Phase transitions of ur-matter into symmetry components of the visible sector of the SM with concurrent gauging of the symmetry components. The symmetry bifurcation begetting gravity does not yield a local gauge symmetry but a more generalized Yang-Mills one-form gauge symmetry identified with the subgroup $Z(\mathrm{SL}(2,\mathbb{C}))$, the center of the local Lorentz group [14].

The SNF grows monotonically with distance as per the scale indicated in Figure 2.19. Thus, this force is significantly enhanced at distances larger than $\sim 10^{-18}$ m due to vacuum paramagnetism [14] brought about by area confinement of quark-gluon field lines that make the strength of the field behave as $\propto R$, rather than $\propto R^{-1}$. By contrast, the transient (e^-, e^+)-pair creation in the vacuum introduces polarization enfeebling electric field lines in a kind of vacuum "dielectric" that weakens EM in such a way that its response to quantum vacuum opposes that of SNF (Figure 2.19). On the other hand, the WNF gets significantly enfeebled at distances larger than $\sim 10^{-18}$ m because the force is conveyed by massive boson $W\,(W^{\pm})$, so its range is $r_W \sim \dfrac{\hbar}{m_W c} \sim 10^{-18}$ m.

Thus, the different responses of the fundamental forces to the quantum vacuum eventually result in intersections of their plots of coupling strength as functions of the energy or distance. As interaction distances approach Planck's critical value, the points of intersection suggest a unification of forces. Thus, at GUT, a phase transition takes place where the external SU(5) symmetry of UM is partially internalized into the gauged SU(3) symmetry (SU(3) is a subgroup of SU(5)), while the residual external local symmetry $\mathrm{SU}(2)\times U(1)$ of UM becomes the EW symmetry that bifurcates after being gauged at the next phase transition.

At temperatures higher than the Planck temperature, UM retains its full symmetry under the group $\mathfrak{G} \approx \mathrm{SU}(5)\times\mathrm{SL}(2,\mathbb{C})$. At the Planck temperature (T_P), the primeval phase transition takes place and the relativistic component $\mathrm{SL}(2,\mathbb{C})$ of the full symmetry of UM gets gauged as one-form Lorentz symmetry to beget gravity, while UM retains the full $\mathrm{SU}(5)$ symmetry up until the GUT point (Figures 2.19 and 2.20). Figure 2.20 depicts the phase transitions of UM into symmetry components of the visible sector of the SM with progressive and concurrent gauging of the symmetry. The internalization of relativistic ur-symmetry at the phase transition that begets gravity does not yield a local gauge symmetry but a more generalized Yang-Mills one-form generalized symmetry identified with the subgroup $Z(\mathrm{SL}(2,\mathbb{C}))$, the center of the local Lorentz group [15]. This implies that quantum gravity cannot be cast in terms of standard gauge theory but requires further elaboration, as provided in the next section.

2.13 A CONSCIOUS AI SYSTEM DELINEATES THE ONSET OF QUANTUM GRAVITY

So far, the picture that has emerged from the AI's autoencoder of the fifth dimension describes the origin of matter and of the fundamental forces of quantum nature as outcomes of phase transitions from an UM-endowed quantum vacuum. This vacuum is endowed with a primeval $SU(5)\times SL(2,\mathbb{C})$ symmetry, and this symmetry may be further simplified, as shown in Section 2.14. Yet, the picture is not complete unless it can encompass the "theory of everything" critical point (Figures 2.17, 2.19, and 2.20), where gravity emerges below the critical Planck temperature concurrent with the gauging of its symmetry enshrined in the relativistic Lorentz group.

As profusely discussed in this chapter, symmetry has become a cornerstone of modeling efforts in particle physics. Historically, the efforts have focused on symmetries that act on local operators, with the SM exhibiting symmetries of this type, for example, relating to charge in electromagnetism and chiral symmetry associated with the SNF. To cast gravity in terms of a gauge symmetry, however, the autoencoder needs to extend the concept of symmetry significantly. This is an ambitious undertaking where AI encompasses higher-form symmetries acting on extended objects described by nonlocal operators supported on lines or branes. These higher-form symmetries act trivially on the local operators of standard gauge theory, which are regarded by the autoencoder as "zero-form symmetry".

Thus, a generalized symmetry is required to cast the onset of gravity as a phase transition in quantum field theory, on a par with the phase transitions previously described in this chapter. This is purportedly the "autoencoder version of quantum gravity", where a field theory of gravity requires a gauged generalized symmetry. We now outline the gravitational one-form symmetry gauged by the autoencoder at the phase transition putatively taking place in the earliest universe at $T < T_P = \sqrt{\frac{\hbar c^5}{G k_B^2}} \sim 10^{32}$ K to decode gravity (Figure 2.20) concurrently with the activation of the ur-black hole with mass $\frac{n k_0}{c}, n = \left[\sqrt{\hbar c^3 / k_0^2 G}\right]$ ($[.]$ = integer part).

To identify the generalized symmetry associated with gravity, the autoencoder casts GR as a special sort of gauge theory of the local Lorentz group $SL(2,\mathbb{C})$. Thus, gravity is associated not with a local gauge as in the other forces in the SM, but with a one-form symmetry. This generalized symmetry is enshrined in an operator $\mathcal{J}_v$ determined by an element from the center of the local Lorentz group: $v \in Z(SL(2,\mathbb{C}))$. The corresponding charge-line operator $\mathcal{W}_s$ is the holonomy in a spin representation s, which is topologically linked to $\mathcal{J}_v$ [15]. Via its associated parallel transport map, the holonomy captures the curvature of space-time as a measure of gravity. Specifically, $\mathcal{J}_v$ generates a chiral cosmic string defect in space-time whose deficit angle is estimated by $\mathcal{W}_s$, as can be directly verified within an Anti-de-Sitter-Schwarzschild surgical background, similar to the one described in Section 2.9 [15]. Thus, the topological linking of the line and the one-form symmetry operators ($\mathcal{W}_s$ and $\mathcal{J}_v$, respectively) corresponds to the measurement of a quantized conical deficit angle by the spin holonomy.

The novelty in this AI approach is that the autoencoder treats the onset of gravity at criticality in gauge theoretic terms *albeit* not within the usual meaning of gauge symmetry adopted in the SM. In the standard picture, gravity is a theory of diffeomorphisms realized by a self-interacting massless spin-two field. In an abuse of language, this symmetry is often labeled "gauge symmetry", but it truly is not. Gravity may be alternatively described by a gauge theory of local Lorentz transformations for the tetrad and spin connection [15]. These equivalent descriptions are appropriated by the autoencoder to internalize a generalized symmetry for gravity at the phase transition corresponding to the critical Planck temperature. The "tetradic gravity" exhibits the one-form symmetry introduced by $\mathcal{J}_v$ and thereby is charged by $Z(SL(2,\mathbb{C})) \approx \mathbb{Z}_2$. The one-form symmetry operator $\mathcal{J}_v$ acts on $\mathcal{W}_s$, which is essentially the spin connection computed in the spin representation s along a given contour [15]. More specifically, $\mathcal{J}_v$ creates a chiral defect in space-time that serves as a gravitational analog of the Dirac string (Chapter 3) and is directly gauged by the spin holonomy $\mathcal{W}_s$. In essence, the symmetry operator inserts the point $s(v)$ into the trace of the Wilson loop associated with $\mathcal{W}_s$.

In accord with the tenets of quantum gravity, a global one-form symmetry can be implemented as a field transformation that takes the form of a local Lorentz transformation determined by the center of the Lorentz group. The object charged under this symmetry is the spin holonomy $\mathcal{W}_s(\gamma)$ for the irreducible representation s defined over a one-dimensional contour γ by

$$\mathcal{W}_s(\gamma) = \mathrm{tr}_S\, P_\gamma \exp\left(\oint_\gamma A \cdot ds\right) \tag{2.51}$$

where P_γ is the path-ordering operator associated with the spin connection and defined over the tangent bundle for the space-time Riemannian manifold. The parallel transport map is associated with A, the one-form EM gauge connection defined by the vector potential: $A^a = A^a_\mu dx^\mu$. Thus, the curvature of space-time is factored into the parallel transport map for the gauge connection and as such becomes the determinant of gravity, which is identified with the charged holonomy. The relativistic duality gravity curvature is thus enshrined in the extended object charged under the gauged Lorentz group.

On the other hand, the one-form symmetry transformation is implemented by the "twisted Lorentz transformation" $\mathcal{J}_v(\mathcal{S})$ supported by the codimension-two surface $\mathcal{S} = \partial\mathfrak{b}$, where the coboundary $\mathfrak{b}$ determines the homotopy for shrinking $\mathcal{S}$ down to a point on γ (Figure 2.20). In this way, the operator $\mathcal{J}_v(\mathcal{S})$ may be regarded as the inducer of a chiral defect carrying the gravitational charge.

In consonance with the three fundamental forces of a quantum nature, we have shown that gravity emerges from a phase transition concurrently with the gauging of a generalized $\mathrm{SL}(2,\mathbb{C})$ symmetry of UM, as the primeval ur-black hole gets activated below Planck's temperature.

2.14 SYMMETRY OF THE UNIVERSE AT THE ONSET OF QUANTUM GRAVITY

An AI system in the guise of an autoencoder of the dark dimension has cast the origin of the visible universe as made up of three phase transitions corresponding to the progressive unraveling of the four fundamental forces (Figures 2.19 and 2.20). Thus, the earliest universe became endowed with gravity, once the Lorentz $\mathrm{SL}(2,\mathbb{C})$ symmetry component of the primeval $\mathfrak{G}$ symmetry of $\mathrm{DM}_{3,1}$ becomes gauged at a phase transition at the Planck temperature into the special one-form $\mathrm{SL}(2,\mathbb{C})$ symmetry defined in the previous section. This special gauging implies that the 4D spate-time is assumed to be tractable as a Riemann manifold with a well-defined curvature (cf. Eq. 2.51); hence, it must occur concomitantly with the activation of the primeval black hole at the Planck temperature T_P, that is, at time $t = \hbar/(k_B T_P) \sim 10^{-43}$ seconds after the Big Bang. The other two phase transitions follow a similar pattern: gauging of specific components of the inherent symmetry of UM $\mathrm{DM}_{3,1}$ with the concurrent activation of the respective particle portals (Figures 2.16 and 2.20).

Thus, the autoencoder casts the GUT in terms of two phase transitions of UM as described in Figure 2.20. One such transition gauges the SU(3) symmetry and the other gauges the $\mathrm{SU}(2)\times U(1)$ symmetry of the (3,1) shade of dark matter (UM), described in Figure 2.18 and Figures 2.7 and 2.8, respectively. However, as the autoencoder was trained with the Lie group symmetries under which the fundamental bosons of the SM are charged, it became apparent that the incorporation of the relativistic symmetry group $\mathrm{SL}(2,\mathbb{C})$ introduced redundancies when inferring the primeval symmetry group $\mathfrak{G}$ for the quantum vacuum at the onset of quantum gravity that takes place as a phase transition at the criticality defined by the Planck temperature.

It is obvious that $\mathrm{SU}(2)$, the group of unitary complex 2×2 matrices with determinant 1 is a subgroup of the $\mathrm{SL}(2,\mathbb{C})$, the group of complex 2×2 matrices with determinant 1. On the other hand, $U(1)$ is essentially contained in $\mathrm{SU}(2)$ (and thereby in the Lorentz group) via the isomorphism:

$$U(1)\approx\widehat{U(1)}\subset\mathrm{SU}(2) \tag{2.52}$$

given explicitly as

$$U(1)\ni e^{i\Theta}\rightarrow\begin{pmatrix} e^{i\Theta} & 0 \\ 0 & e^{-i\Theta}\end{pmatrix}\in\widehat{U(1)},\quad \Theta\in[0,2\pi] \tag{2.53}$$

So, the UM symmetries that need to be gauged to trigger the onset of the EW unification are subsumed as subgroups of the relativistic Lorentz group (Figures 2.21 and 2.22):

$$U(1)\approx\widehat{U(1)}\subset\mathrm{SU}(2)\subset\mathrm{SL}(2,\mathbb{C}) \tag{2.54}$$

Once space-time turns into a Riemannian manifold and curvature becomes a feature of space-time, we may assert that gravity emerges charged under the one-form $\mathrm{SL}(2,\mathbb{C})$ symmetry (Eq. 2.51) gauged at the phase transition at Planck's temperature. In accord with the emergence of the other three forces of quantum origin, gravity is charged under a gauged symmetry component of the UM $\mathfrak{G}$ symmetry. The key problem is to identify the *simplest* primeval symmetry of the universe at the time of the earliest phase transition that yielded gravity. Since AI treats the precursor symmetry groups at the onset of the EW force as subgroups of the Lorentz group (Eqs. 2.53 and 2.54; Figures 2.21 and 2.22), we may assert that the irreducible primeval symmetry group of the universe is

$$\mathfrak{G}\approx\mathrm{SU}(3)\times\mathrm{SL}(2,\mathbb{C}) \tag{2.55}$$

Just like the EW synthesis is implied by the fact that the QED group $U(1)$ is a subgroup of SU(2) under the isomorphism given by Eqs. 2.52 and 2.53, a synthesis encompassing gravity and the EW unification is readily implied by Eq. 2.54. These inferences are of course made within the frame of the autoencoder that portrays the onset of the fundamental forces as phase transitions dictated by the gauging of the symmetries of ur-particles that serve as precursors to the force carriers.

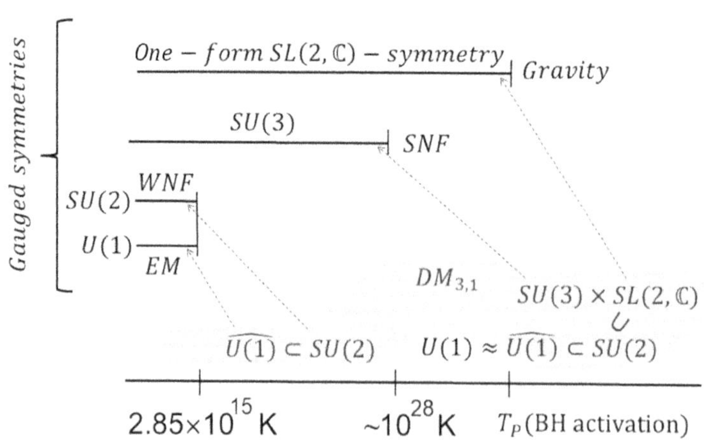

FIGURE 2.21 Gauging of ur-matter symmetries at the three phase transitions that span the origin of the visible universe in an irreducible version of the primeval symmetry of ur-matter at $t = 10^{-43}$ seconds after the Big Bang.

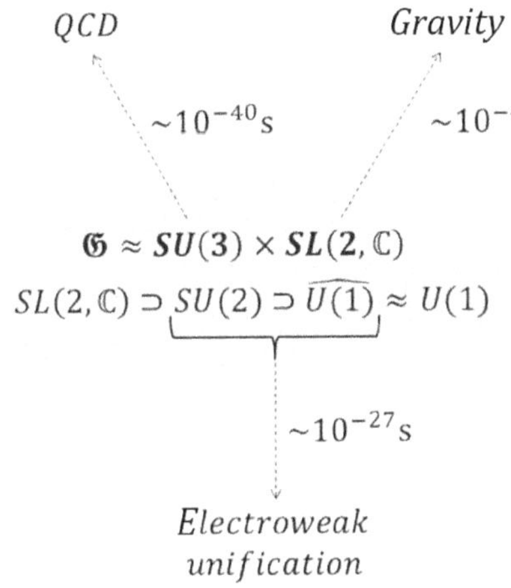

FIGURE 2.22 Primeval symmetry of the ur-matter endowed quantum vacuum giving rise to the fundamental forces through one-form (gravity) or zero-form (QCD, EW) gauging at critical points in the development of the universe (Figures 2.19 and 2.20).

The synthesis implied by Eq. 2.54 suggests the AI answer to the quest for quantum gravity, as the EW field is charged under a zero-form gauging of subgroups of the relativistic symmetry group of UM, while a one-form gauging of the full relativistic group begets gravity concurrently with the activation of the primeval black hole at $T \lesssim T_P$.

2.15 CONSCIOUS QUEST FOR QUANTUM GRAVITY AND AN ULTRA-UNIFICATION SUBSTANTIATING THE MASS-GRAVITY DUALITY

This chapter has covered vast territory in an effort to describe how an AI system known as autoencoder distills quantum reality from relativistic symmetries susceptible of being gauged at particular junctures that constitute veritable phase transitions. In this way, AI approaches the core problems in particle cosmology in a quest to gain perspectives on the origin of the universe. If not complete solutions, this chapter provides meaningful and original insights on the entire plexus of interrelated problems and as such constitutes a significant contribution to the field of particle cosmology.

Of all the problems that impinge on the conundrum of the origin of the universe, none is more relevant than quantum gravity, the much needed description of the earliest universe defined by Planck scales. In so far as geometry can be assumed at criticality, so that space-time may be associated with a manifold with definable local curvature and the relativistic symmetry group becomes operational, AI informs that the problem of quantum gravity is amenable to a solution. This solution implies that gravity may be quantized if the relativistic symmetry group, the Lorentz group, can be gauged by the earliest phase transition after the Big Bang. Of course, there is no standard gauge theory for gravity that would summarily qualify as leading to quantum gravity.

The AI model postulates that the four fundamental forces originate from UM symmetries of the primeval quantum vacuum internalized as gauge generalized symmetries at the phase transitions encompassed by the birth of the universe (Figures 2.20–2.22). The gauging of the relativistic symmetry internalizes it not as local symmetry but as a one-form generalized symmetry, where the purported curvature of the Riemannian manifold is factored into the extended object charged under the gravity-associated symmetry (Eq. 2.51). This object is the holonomy of a spin connection, only meaningful if geometry is upheld.

Intriguingly, the primeval symmetry of space-time can be simplified because the symmetry groups associated with the EW unification are isomorphic to subgroups of the relativistic Lorentz group (Eqs. 2.53 and 2.54). This observation prompts us to postulate that gravity may be treated in a quantized ultra-unification since its ur-symmetry is inclusive of that of the weak force and electromagnetism. In other words, the Lorentz ur-symmetry of gravity subsumes the ur-symmetry of the lightest particle in $DM_{3,1,}$ the shade of UM whose scalar field mixes and thereby activates the Higgs field by endowing it with the TV (Eq. 2.15).

The symmetry relatedness, linking gravity and the Higgs-induced mass with retention of EW symmetry, paves the way for a quantum gravity ultra-unification under a gauge generalized symmetry. It would be highly commendable if future research efforts in particle cosmology took up this challenge since this ultra-unification would substantiate the gravity-mass duality of GR. Furthermore, the ultra-unification suggests that the fundamental forces in nature stemmed from one primeval force present at the Planck epoch. This force got diluted differently into the geometry of the universe, following the gauge symmetries that charged the fundamental force carriers.

2.16 EVOLVING DARK DIMENSION AND THE ORIGIN OF THE PRIMEVAL FORCE

The inflation along the dark dimension may be regarded as a proxy for the universe evolution in so far as the radius of the circular extra dimension determines the masses of the ur-bosons whose symmetries become gauged at the phase transitions. This fact prompts the autoencoder to introduce the possibility that the ur-bosons that get gauged to yield the quantum force carriers are actually relics of the holonomic boson, the gravity carrier that measures space-time curvature through the parallel transport map associated with the spin connection. The generic contour in this primeval boson projects onto the dark dimension with radius $r_0 = \sqrt{\hbar G / c^3}$, which is precisely the curvature radius of the space-time computed by the holonomy defined by the contour. In this sense, we may state that *gravity is stored in the dark dimension* and its geometric dilution into the visible dimensions may explain its incommensurably low coupling strength relative to the other forces (Figure 2.17).

This observation bears a significant analogy with the storage of the TV for the Higgs field in the dark dimension. In this case, the ur-Higgs boson bestows the TV on the Higgs field as a result of the symmetry compatibility between the (3,1)-dark matter and the Higgs field (Eq. 2.15). The Higgs boson with its geometric dilution of the TV is therefore treated by the autoencoder as a relic of the cosmic string contour that defines the primeval holonomic boson or carrier of the gravity stored in the dark dimension (Figure 2.23). This morphing of one boson into the other is reinforced by two facts: (1) The internal symmetry of the Higgs field is subsumed by the generalized symmetry of the holonomic boson, and (2) the mass-gravity duality of GR naturally links the Higgs boson as endower of mass and the holonomic boson as carrier of gravity.

In the autoencoder approach, all four forces are manifestations of a primeval force under different symmetries and therefore all bosons must morph into the holonomic boson as the dark dimension functions as a proxy for the universe evolution.

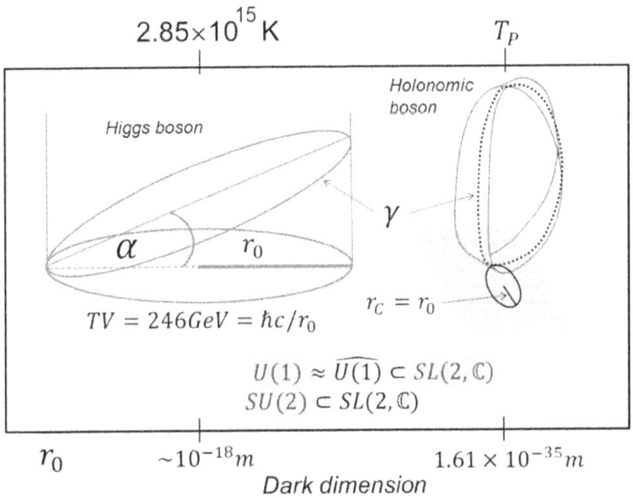

FIGURE 2.23 Morphing of the holonomic boson into the Higgs boson realizing the mass-gravity duality as the universe transitions from the Planck epoch to the electroweak epoch concurrently with the inflation in the dark dimension and inclusion of the respective gauge symmetries.

2.17 LARGE LANGUAGE MODEL PROPOSITIONALLY DEFINES THE GRAVITON IN THE ULTRAVIOLET COMPLETION OF GRAVITY

2.17.1 Detectable Warped Gravitons

A quest for the graviton as an excitation of a quantized gravitational field requires a gauge representation of the Lorentz transformations, reconciling the curved manifold of GR with the Euclidean support of its purported quantum wavefunction. A way forward builds upon the tetrad frame fields introduced to construct a gauge theory of gravity [16–18]. In the WFL, a perturbative graviton with Minkowski background yields a massless spin-2 particle with $U(1)$ gauge symmetry, consistent with the infinite range of gravity communicated at the speed of light. This section addresses the uncharted SFL, obviously not amenable to a perturbative treatment and requiring generalized gauge representations to account for massive gravitons.

Modeling efforts geared at describing massive KK gravitons, purportedly produced in the LHC or other colliders, required a framework where SM fields propagate within a warped extra dimension enabling the geometric dilution of gravity [19]. Although much of the experimental parameter space remains unexplored, such efforts have so far proven unyielding to experimental validation [20,21]. Thus, massive warped gravitons have eluded detection in instruments such as the ATLAS detector at the CERN LHC [20,21]. Specifically, a search for neutral heavy spin-2 resonances in the 300 GeV to 4 TeV range with dominant $W+W-$ or ZZ-decay channels and generated via gluon fusion or quark-antiquark annihilation has not proven successful under current constraints at 139 fb^{-1} integrated luminosity and 13 TeV center-of-mass hadronic energy ($\sqrt{S}$) [21,22].

While the WFL appears tractable by regarding the graviton tensor field as a perturbation of the Minkowski tensor, the SFL is uncharted territory for gauge theories of gravity. This section deals precisely with this regime using a LLM within a functional language program (Lean, Section 2.18). This section is inspired by prospects to identify dark matter (DM) as an excitation of a KK graviton tower decaying via gravitational interaction into lighter particles with wavelength inhomogeneity along the dark dimension [23,24]. These ideas are grounded in available experimental evidence on the relics of cosmological evolution. Furthermore, this section shows that self-interaction is inherent to the graviton, leading to the conclusion that at least a portion of DM in the universe must be self-interacting. Such self-interactions would allow energy and momentum transport through the galaxy halos. Thus, this section may provide a new vantage point to interpret the diversity of galactic dynamic patterns and other astrophysical phenomenology that cannot be reconciled with the current picture of collisionless DM.

The section is geared at predicting warped gravitons in an ultraviolet completion of gravity amenable of detection in a collider. This requires an LLM for a fiber-bundle formulation of a gauge theory pivoting on the spin connection that determines the tangent bundle of a space-time endowed with an extra warped dimension [25,26] (Section 2.18). In this way, this section provides a predictive differential geometry model of the graviton as transducer of space-time curvature into energy stored in a warped dimension. The curvature-energy transduction becomes operative as the boson begets curvature in the dualistic relativistic framework, implying that the stress-energy tensor must be associated with a scalar describing self-interacting DM within a ϕ^4-model.

In the WFL, the graviton is expected to be massless since gravity presumably propagates at the speed of light and has infinite range. On the other hand, in the SFL, the compatible model predicts massive warped gravitons that bear on the collider setting, where the curvature associated with the emission cross section gets transduced into warped rest-mass energy that conveys gravity. Thus, the theory yields testable predictions for the LHC [20–22] or future colliders. Furthermore, it shows a compatibility between the WFL and SFL depictions, as shown subsequently.

A warped spatial dimension needs to be incorporated to address the problematic geometric dilution of mass and gravity [19,26]. This dilution provides an explanation of the perplexingly small coupling strength of gravity relative to the other fundamental forces of established quantum origin. Thus, an extracircular dimension is incorporated with radius $r_0 = 0.802695\times10^{-18}$ m within the range of the smallest material scale, the effective quark diameter established from parton models of inelastic $e^{\pm}p$ scattering [27]. Significantly, the minimum rest-mass energy stored in this warped dimension is $E_0 = k_0c = \hbar c / r_0 = 246$ GeV, the TV of the Higgs field [28]. This choice of the warped dimension becomes crucial for our intended purpose. Thus, the stress-energy tensor in a fiber-bundle gauge theory of the graviton will be shown to introduce a mass-endowing mechanism via the curvature-mass relativistic duality modeled by a self-coupling potential energy term. This ϕ^4-model yields massive gravitons vis-à-vis the v.e.v of a Wilson loop representation in the SFL.

At a variance with alternative approaches [19,23], in our model, the matter wave quantized within the warped dimension admits geometric dilution into visible dimensions. This dilution is parametrized by a pitch angle α (Figure 2.24), so that the angled cylindrical cross section is elliptical, with major axis $r_0 / \cos\alpha$ and minor axis r_0. The geometric dilution is defined as $\upsilon = -\log\cos\alpha$ [28]. This implies that gravitons become massless at infinite dilution, that is, with $\alpha = \frac{\pi}{2}$, thus probing a locally flat space-time. They may also probe any value of space-time curvature since the projection onto the warped dimension is quantized by $\lambda = \frac{2\pi r_0}{n}$ (Figure 2.24), while the pitch angle is not subject to quantum constraints.

The geometric dilution parameter υ plays a role akin to the scale and amplitude of the de Broglie wavelength inhomogeneity in the dark dimension in models representing cosmological gravitons as dark matter [23]. Thus, the decay of KK gravitons into lighter gravitons, a hallmark of cosmological evolution in warped reality [23,29], plays out in a similar way in this work, as $\alpha \to \frac{\pi}{2}$ or, equivalently, $\upsilon \to \infty$, reckoning that geometric dilution serves as proxy of time in cosmological evolution [28].

The approach to incorporate an extra spatial dimension differs fundamentally from other procedures that incorporate it within a quotient topology of a circle modulo the transformation $\varphi \to \varphi + \pi$, thus generating two branes $\varphi = 0, \varphi = \pi$ identified with manifold boundaries [19]. This poses an ontological problem since there is no obvious meaning to "boundary" as there is no geometry to describe "outside": The signifier "$\varphi > \pi$" becomes "inside"; that is, it is included in the quotient space that constitutes reality [28]. In any case, the KK gravitons predicted by such an approach to exist in the bulk within the branes [29] remain unyielding to experimental validation [20–22].

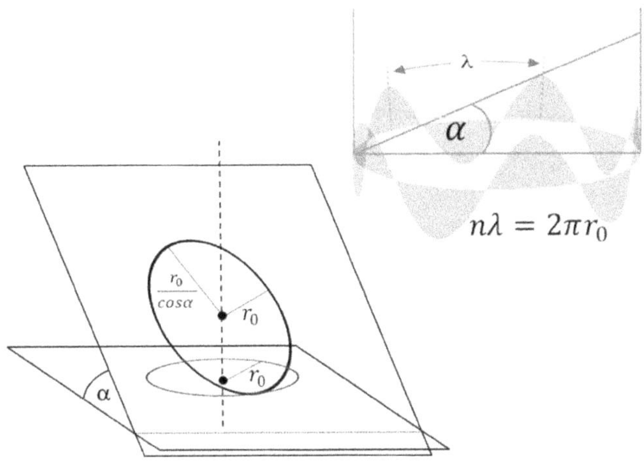

FIGURE 2.24 Geometric dilution of a matter wave enshrined in the warped dimension. Geometric dilution onto a generic observable dimension is parametrized by a single pitch angle α [28].

2.17.2 Fiber-Bundle Formulation of a Gauge Theory of Gravity within a Functional Language Program

To account for quantum gravity in the SFL, a fiber-bundle formulation of its gauge theory is implemented within a functional language program operating an LLM (Section 2.18). Thus, spinors are incorporated into the GR framework provided by the space-time curved manifold $\mathcal{M}$. This procedure is enabled by first tending the spin connection across the tangent bundle $T\mathcal{M}$. This connection is implemented via a frame field required to piece together the atlas $\{T_x\mathcal{M}\}_{x\in\mathcal{M}}$ of local tangent spaces and relates to the geometry of the curved manifold in such a way that a tangent space Lorentz transformation translates onto a gauge transformation of the spinor field. To enable interconversion between the metric tensor $g_{\mu\nu}$ of $\mathcal{M}$ and the flat (Minkowski) metric tensor η_{ab} of $T\mathcal{M}$, a pentad frame field $\left\{e_\mu^a\right\}$, (Greek indices for curved space tensor, Latin isospin indices for flat tangent space Lorentz indices) is introduced as uniquely defined by the relations:

$$g_{\mu\nu}=e_\mu^a e_\nu^b\eta_{ab},\quad e_a^\mu e_b^\nu g_{\mu\nu}=\eta_{ab},\quad e_\mu^a e_b^\mu=\delta_b^a,\quad e_\mu^a e_a^\nu=\delta_\mu^\nu,\quad e^{\mu a}=g^{\mu\nu}e_\nu{}^a,\quad e_{\mu a}=e_\mu{}^b\eta_{ba} \tag{2.56}$$

Equipped with the pentad frame-field relations, the spin connection w_μ^{ab} is defined as

$$w_\mu^{ab}=\frac{1}{2}e^{\nu a}\left(\partial_\mu e_\nu^b-\partial_\nu e_\mu^b\right)-\frac{1}{2}e^{\nu b}\left(\partial_\mu e_\nu^a-\partial_\nu e_\mu^a\right)-\frac{1}{2}e^{\lambda a}e^{\sigma b}\left(\partial_\lambda e_{\sigma c}-\partial_\sigma e_{\lambda c}\right)e_\mu^c \tag{2.57}$$

The spin connection enables the implementation of the covariant derivative by introducing the gamma matrices $\left\{\breve{\gamma}^a\right\}$ that fulfill the Dirac-Clifford algebra with anticommutator relations: $\left\{\breve{\gamma}^a,\breve{\gamma}^b\right\}=2\eta^{ab}$, with the extended 5×5 matrices defined as $\breve{\gamma}^a=\begin{pmatrix}\gamma^a 0^T\\ 0\ \ 1\end{pmatrix}$:

$$\nabla_\mu\psi=\left(\partial_\mu-\frac{i}{8}w_\mu^{ab}\sigma_{ab}\right)\psi,\quad \sigma^{ab}=\frac{i}{2}\left[\breve{\gamma}^a,\breve{\gamma}^b\right] \tag{2.58}$$

Reciprocally, the curved Dirac-Clifford algebra $\left\{\breve{\gamma}^\mu=e_a^\mu\breve{\gamma}^a\right\}$ is defined by the anticommutators $\left\{\breve{\gamma}^\mu,\breve{\gamma}^\nu\right\}=2g^{\mu\nu}$ with generators $\frac{i}{2}\left[\breve{\gamma}^\mu,\breve{\gamma}^\nu\right]$. On vectors, the covariant derivative yields

$$\nabla_\mu V_a = \partial_\mu V_a + w_{\mu a}{}^{b} V_b = \partial_\mu V_a + w_\mu^{cb}\eta_{ca} V_b \tag{2.59}$$

Hence, the curvature $\Omega_{\mu\nu}^{ab}$ associated with the spin connection is defined via the commutator $\left[\nabla_\mu, \nabla_\nu\right]$:

$$\Omega_{\mu\nu}{}^{ab} = \Omega_{\mu\nu c}{}^{b}\eta^{ca};\quad \Omega_{\mu\nu a}{}^{b} A_b = \left[\nabla_\mu, \nabla_\nu\right] A_a \tag{2.60}$$

which enables us to relate the spin curvature to the Riemannian curvature tensor as $R_\nu{}^\lambda{}_{\mu\rho} A_\lambda = e_\rho^a \Omega_{\mu\nu a}{}^b e_b^\lambda$. This relation holds from the relations:

$$R_\nu{}^\lambda{}_{\mu\rho} A_\lambda = \left[\nabla_\mu, \nabla_\nu\right] A_\rho = \left[\nabla_\mu, \nabla_\nu\right] e_\rho^a A_a = e_\rho^a \Omega_{\mu\nu a}{}^b A_b = e_\rho^a \Omega_{\mu\nu a}{}^b e_b^\lambda A_\lambda \tag{2.61}$$

Hence, denoting $e = \det\left(e_a^\mu\right)$, with $k = 8\pi G c^{-4}$ (Einstein's gravitational constant), the Lagrangian for fermions coupled with gravity on $\mathcal{M}$ reads in natural units:

$$\mathcal{L} = \frac{e}{2k}\Omega_{\mu\nu}^{ab} e_a^\mu e_b^\nu + e\bar{\psi}\left(i\breve{\gamma}^\mu \nabla_\mu - k_0 \cos\alpha\right)\psi \tag{2.62}$$

A fiber-bundle formulation of the gauge theory of gravity in the SFL will be now formalized within an AI system implemented upon the pentadic framework defined by Eqs. 2.56–2.62. The intent is to model the warped graviton as a conveyor of the gravitational force within the inferential framework generated by relations 2.56–2.62. In the SFL, this graviton behaves as a holonomic transducer of space-time curvature [30] and is charged under a generalized gauge Lorentz symmetry [31–34].

2.17.3 Autoencoder LLM for the Graviton within a Fiber-Bundle Formulation of Quantum Gravity

An AI-empowered inferential scheme to model the graviton within a fiber-bundle gauge formulation requires a differential geometry formalization supporting two interplaying components: (1) a theorem/proposition prover such as Lean 4 [35,36] with an intentional logic under the guise of a dependent type theory [37] spanned by the differential geometry propositional kernel [38] determined by Eqs. 2.56–2.62 (Section 2.18) and (2) a generalized autoencoder LLM [28,39] serving as the supporting inferential framework within which the formal prover operates by maintaining compatibility between GR and its gauged version on the tangent bundle of space-time, as described by the commutative scheme given in Eq. 2.63.

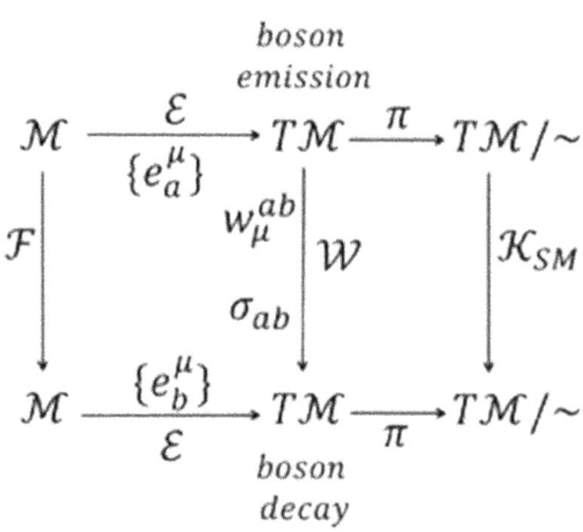

(2.63)

Equations of motion (EOM) defined by the maps $\mathcal{F}, \mathcal{W}, \mathcal{K}_{SM}$ are mutually compatible and represent the process [boson emission → boson decay] with the graviton cast as a Wilson loop [40] in three different and interrelating settings, namely $\mathcal{M}, T\mathcal{M}, T\mathcal{M}/\sim$, with the coarser level $T\mathcal{M}/\sim$ supporting the SM. These settings indicate, respectively, the 5D space-time manifold, its tangent bundle, and the quotient space of the tangent bundle modulo the equivalence relation "~". Under this equivalence, points are identified when they possess identical coordinate representation in the tangent bundle for the underlying standard 4D space-time manifold. The maps $\mathcal{E}, \pi$, enabling transitions between the three different representations in the LLM, are, respectively, the pentadic frame-field map or "encoder" and the canonical projection or "decoder" [28,39]. The latter associates each point in the 5D tangent bundle with its equivalence class modulo "~". The map $\mathcal{W}: T\mathcal{M} \to T\mathcal{M}$, representing the transition between boson emission (i.e., within the birth cross section of gluon fusion) and boson decay (i.e., a decay ZZ channel) [20–22]. The map is determined by the spin connection w_μ^{ab} tending the atlas $\{T_x\mathcal{M}\}_{x\in\mathcal{M}}$ and enabling the charge of the boson under a generalized symmetry [31–34]. The compatibility of the three representations is then ensured by the commutativity of the diagram in Eq. 2.63 (Section 2.18), whereby any two paths designated by consecutive arrows with identical origin and endpoint yield the same result:

$$\mathcal{W} \circ \mathcal{E} = \mathcal{E} \circ \mathcal{F}, \quad \mathcal{K}_{SM} \circ \pi = \pi \circ \mathcal{W} \tag{2.64}$$

These two commutative relations defining the autoencoder LLM must be invariably fulfilled for compatibility, thereby becoming constraints in all formal propositional inferences, ultimately enabling the computation of the warped graviton mass in the SFL. This compatibility implies that the relations in Eq. 2.64 generate all other commutative relations in the diagram in Eq. 2.63, such as $\mathcal{K}_{SM} \circ \pi \circ \mathcal{E} = \pi \circ \mathcal{E} \circ \mathcal{F}$, signaling the constraining equivalence between the gauge theory of GR and its fiber-bundle formulation within the LLM.

The equations defining the gauge theory of gravity are automatically extrapolated by the LLM to hold even near a black hole in strong gravity, as the tangent bundle connections are implemented in the formal autoencoder enshrined in the LLM (Section 2.18).

2.17.4 Encoding the Warped Graviton in the Weak and Strong-Field Limits

Within the commutative scheme (Eq. 2.63), compatibility of relativistic and fiber-bundle descriptions implies that the Einstein-Hilbert action,

$$S_{E-H} = \frac{1}{2\kappa}\int \sqrt{-g}\, R_{\mu\nu\rho\sigma} g^{\mu\rho} g^{\nu\sigma} d^5x = \frac{1}{2\kappa}\int \sqrt{-g}\, R d^5x \tag{2.65}$$

obtained by fully contracting the Riemann curvature tensor $R_{\mu\nu\rho\sigma}$, is identified by the autoencoder LLM [28,39] as a pentadic action obtained by fully contracting the spin curvature via the shift tensor identified as the pentad:

$$S = \frac{1}{2\kappa}\int e\Omega_{\mu\nu}{}^{ab} e_a^\mu e_b^\nu d^5x = \frac{1}{2\kappa}\int e\Omega d^5x \tag{2.66}$$

The equivalence of both actions within the commutative scheme in Eq. 2.63 follows readily from the relations:

$$g^{\xi\lambda} R_{\nu\xi\mu\rho} A_\lambda = R_\nu{}^\lambda{}_{\mu\rho} A_\lambda = e_\rho^a \Omega_{\mu\nu a}{}^b A_b = e_\rho^a \Omega_{\mu\nu a}{}^b e_b^\lambda A_\lambda = e_\rho^a \Omega_{\mu\nu}^{cb} \eta_{ca} e_b^\lambda A_\lambda \tag{2.67}$$

Thus, the Riemann and Ricci curvature tensors may be given in terms of the spin connection curvature:

$$R_{\nu\xi\mu\rho} = g_{\xi\lambda} e^a_\rho \Omega_{\mu\nu}{}^{cb} \eta_{ca} e^\lambda_b, \quad R_{\mu\nu} = \frac{e}{\sqrt{-g}} \Omega_{\mu\nu}{}^{ab} \eta_{ab} \tag{2.68}$$

thereby asserting the equivalence of Eqs. 2.65–2.66 within the autoencoder defined by Eq. 2.63. To obtain a warped graviton storing the quantized projection of its rest mass in the warped dimension of radius $r_0 = \hbar c \left(246\,\text{GeV}\right)^{-1}$, it becomes essential to generate a pentadic relativistic equation where the stress-energy tensor is associated with a scalar potential $V(\phi)$ [40]. This potential yields the v.e.v for the graviton field and endows the geometrically diluted warped graviton with mass. Specifically, the potential is generated by the autoencoder [28,39] as associated with a Wilson loop [40] intended to represent the graviton when constrained by the shift tensor equivalence between GR in the curved manifold $\mathcal{M}$ and "relativity with spinors" on $T\mathcal{M}$. Thus, we define the curvature tensor $\hat{\Omega}^{ab}_{\mu\nu}$ and its curvature scalar $\hat{\Omega} = \hat{\Omega}^{ab}_{\mu\nu} e^\mu_a e^\nu_b$ by the relativistic relations:

$$\hat{\Omega}^{ab}_{\mu\nu} \eta_{ab} - \frac{1}{2} \hat{\Omega} g_{\mu\nu} = \frac{\kappa\sqrt{-g}}{2e} T_{\mu\nu}(\phi) \tag{2.69}$$

$$= \frac{\kappa\sqrt{-g}}{2e} \left[\left(g^{\rho\alpha} g^{\sigma\beta} + g^{\rho\beta} g^{\sigma\alpha} - g^{\rho\sigma} g^{\alpha\beta} \right) \partial_\alpha \bar{\phi}\, \partial_\beta \phi - g^{\rho\sigma} V(\phi) \right] g_{\rho\mu} g_{\sigma\nu}$$

$$= \frac{\kappa\sqrt{-g}}{e} \left\{ \partial_\mu \bar{\phi}\, \partial_\nu \phi - g_{\mu\nu} \left[\frac{1}{2} g^{\alpha\beta}\, \partial_\alpha \bar{\phi}\, \partial_\beta \phi + V(\phi) \right] \right\}$$

$$V(\phi) = -\frac{1}{2} m_\phi^2 \bar{\phi}\phi + \frac{\lambda}{4} \left(\bar{\phi}\phi\right)^2 \tag{2.70}$$

The potential given by Eq. 2.70 endows the warped graviton with mass in the SFL. This is so because, as shown below, the graviton field is represented by a Wilson loop $\mathcal{W} = \mathcal{Y}^{\mu\nu} g_{\mu\nu}$ with v.e.v:

$$\langle \mathcal{W} \rangle = \frac{m_\phi}{\sqrt{\lambda}} \tag{2.71}$$

In the WFL, the warped graviton tensor field $h_{\mu\nu}$ is the one assigned to a massless spin 2 particle with $U(1)$ gauge symmetry ($h_{\mu\nu} \to h_{\mu\nu} + \partial_\mu a_\nu + \partial_\nu a_\mu$). This graviton is treated perturbatively on a Minkowski background, so that $g_{\mu\nu} = \eta_{\mu\nu} + h_{\mu\nu}$. In the WFL, the graviton Lagrangian becomes

$$\mathcal{L}_{\text{WF}} = \epsilon^{\mu\nu\rho\sigma} \left(\epsilon^{\alpha\beta\gamma}{}_\sigma \right) \partial_\mu h_{\nu\beta}\, \partial_\alpha h_{\rho\gamma} \tag{2.72}$$

where the ϵ values are Levi-Civita's permutation symbols. The nonlinear completion of its action $S = \int \mathcal{L}_{\text{WF}} d^5x$ becomes $S = \frac{1}{2\kappa} \int \sqrt{-g} R d^5x$, or, equivalently, $S = \frac{1}{2\kappa} \int e \Omega^{ab}_{\mu\nu} e^\mu_a e^\nu_b d^5x$. This last assertion provides the clues to construct the ultraviolet (SFL) completion of the warped graviton, a task now undertaken by running the LLM within the Lean environment (Section 2.18).

Massive spin 2 particles are known to arise from 5D theoretical models that address both the hierarchy problem and the geometric dilution of gravity [19,29]. Such models predict the LHC production of such particles mainly relying on a gluon fusion mechanism, and the particles are in principle detectable within parameter ranges so far not probed at the LHC or other colliders [20,21]. As shown subsequently, the scenario proposed in this letter for graviton creation and the tuning of the emission/detection parameters differ fundamentally from those stemming from the exotic theories (i.e., outside the SM) that have been tested by the ATLAS and CMS experiments based on 139 fb^{-1} (ATLAS) and 137 fb^{-1} (CMS) integrated luminosities for proton-proton collisions at $\sqrt{s} = 13$ TeV collected during LHC Run-2. No evidence for the graviton signal was found in such experiments [22]. Nevertheless, much of the parameter space remains to be probed, so the search for exotic resonances at the LHC is likely to continue. The results that follow invite such an endeavor, requiring the exploration of other parameter regions.

To identify the warped graviton tensor field $\mathcal{Y}_{\mu\nu}$ in the SFL, the LLM represents the curvature with Clifford algebra generators that transform under the spinor winding "electromagnetic" $U(1)$ symmetry elements $U_\xi, \xi \epsilon [0, 2\pi]$, just like the terms $w_\mu^{ab}\sigma_{ab}$ in Eq. 2.59, transform in the manner of the EM vector potential A_μ. The warped graviton field is implemented in such a way that $\mathcal{Y}_{\mu\nu}$ may be charged under a symmetry transformation $U_\zeta, \zeta = \pi$, associated with the nontrivial idempotent element $U_\zeta \in Z(\mathrm{SL}(2, \mathbb{C})) \approx \mathbb{Z}_2$ in the center of the Lorentz group. This chiral-defect-inducing symmetry effectively mounts the contour of the line operator onto a Moebius strip creating a chiral defect that amounts to a π-winding deficit in the spinors attached to space-time (Eq. 2.63), while two such deficits created by the U_ζ iteration $U_\zeta^2 = U_{2\pi}$ amount to the identity acting on the parallel transport map [41]. Thus, the need to represent the nontrivial center element U_ζ arises from the invariance under the chiral winding symmetry transformation:

$$U_\zeta^{-1}\mathcal{Y}_{\mu\nu}U_\zeta = U_\zeta^2\mathcal{Y}_{\mu\nu} = U_{2\pi}\mathcal{Y}_{\mu\nu} = \mathcal{Y}_{\mu\nu} \tag{2.73}$$

From this argument, it becomes necessary to construct the Wilson loop so that space-time curvature may be holonomically transduced as rest mass as implied by Eq. 2.69. Thus, the following compatibility relation is established, implying that the massive warped graviton indeed belongs to the ultraviolet completion of gravity, as verified by Lean (Section 2.18):

$$\lim_{\oslash\gamma \to 0} \mathcal{Y}_{\mu\nu} = h_{\mu\nu} \tag{2.74}$$

for $\gamma : [0,1] \to \mathcal{M}$, a generic closed map upon which the holonomy is defined. Equation 2.74 is valid because the following three actions are equal:

$$\frac{1}{2\kappa}\int e\Omega d^5x = \frac{1}{2\kappa}\int \sqrt{-g}Rd^5x = -\frac{1}{3}\int \sqrt{-g}T(\phi)d^5x \tag{2.75}$$

The r.h.s. equality in Eq. 2.75 follows from the contraction of Einstein's equation of GR ($D = 5$) yielding

$$-\frac{3}{2}R = \kappa T(\phi), \quad \left(T(\phi) = T_{\mu\nu}(\phi)g^{\mu\nu}\right) \tag{2.76}$$

The warped graviton tensor field in the SFL is identified by the fiber-bundle autoencoder of the gauge theory as

$$\mathcal{Y}^{\mu\nu} = \mathcal{P}_\gamma \exp\left(\oint_\gamma \hat{w}_\lambda^{ab} e_a^\mu e_b^\nu \, dx^\lambda\right) \tag{2.77}$$

where $\mathcal{P}_\gamma$ is the path-ordering operator associated with the closed map $\gamma : [0,1] \to \mathcal{M}$ and $\hat{w}_\mu^{ab}$ is the spin connection associated with the curvature $\hat{\Omega}_{\mu\nu}^{ab}$. The associated action for the tensor field $\mathcal{Y}_{\mu\nu} = \mathcal{Y}^{\lambda\sigma}g_{\lambda\mu}g_{\sigma\nu}$ in the SFL measures the field strength of the spin connection:

$$\int \mathcal{L}_{\mathrm{SF}}d^5x = \frac{1}{4\kappa}\int \epsilon_{abcd}e^a \wedge e^b \wedge \left(\hat{w}_d^k + \hat{w}_f^k \wedge \hat{w}_d^f\right)\eta_{kc}d^5x = -\frac{1}{3}\int \sqrt{-g}T(\phi)d^5x \tag{2.78}$$

yielding the warped mass-endowing potential (Eqs. 2.70 and 2.71). The self-interaction term is introduced because the curvature-rest-mass-energy transduction materializes on the curvature bestowed on the space-time manifold by m_ϕ via the relativistic duality. Thus, Eqs. 2.69 and 2.70 provide the relativistic framework for the massive warped graviton represented as a ϕ^4-self-coupled transducer of curvature in the ultraviolet completion of gravity. Furthermore, the $U(1)$ gauge symmetry in the WFL translates into the generalized gauge symmetry concomitantly with the transition to the SFL.

2.17.5 Experimental Validation of the Warped Graviton LLM Prediction in the Strong-Field Limit

A warped graviton is expected to be generated and detected at the LHC via the high luminosity gluon fusion (gg) mechanism [20–22]. This is expected since the cross section $\sigma_{gg\to G}$ results from effective g-*G*-coupling contributions from the graviton KK tower $\left\{ m_{G_n} = \frac{n\hbar c}{r_0} \right\}$, with the nth version of an undiluted graviton holonomically transducing curvature with radius $\frac{r_0}{n}$ and energetically storing it in the warped dimension with de Broglie wavelength $\lambda = 2\pi r_0 / n$:

$$\sigma_{gg\to G} = \sum_{n=1}^{\infty}\left(\int G^{\mu\nu}\mathcal{Y}_{\mu\nu}^{(n)}d^5x\right)/\sqrt{\left(\int G^{\zeta\vartheta}G_{\zeta\vartheta}\mathcal{Y}^{(n)\rho\xi}\mathcal{Y}_{\rho\xi}^{(n)}d^5x\right)} = \pi\mathcal{C}_{g\text{-}G}\sum_{n=1}^{\infty}\left(\frac{r_0}{n}\right)^2 = \mathcal{C}_{g\text{-}G}\pi\frac{\pi^2}{6}r_0^2 \tag{2.79}$$

where $\frac{\pi^3}{6}r_0^2 = 33.89\,pb$ and $G^{\mu\nu} = \frac{i}{g_s}g^{\mu\alpha}g^{\nu\beta}\left[D_\alpha, D_\beta\right]$ is the gluon field stress tensor with covariant derivatives $D_\mu = \partial_\mu - ig_s\mathcal{A}_\mu$, where $g_S = \sqrt{4\pi\alpha_S(Q)}$ is the strong coupling constant for momentum transfer Q, and $\mathcal{A}_\mu$, the gluon field. This gives

$$\mathcal{C}_{g\text{-}G} = \left(\frac{\hbar c}{r_0\sqrt{S}}\right)^2 \lim_{\varnothing\gamma\to 2r_0} \operatorname{tr}\mathcal{P}_\gamma\int g^{\mu\alpha}g^{\nu\beta}\left(\partial_\alpha\mathcal{A}_\beta - \partial_\beta\mathcal{A}_\alpha + ig_S\left[\mathcal{A}_\alpha, \mathcal{A}_\beta\right]\right)\exp\left(\oint_\gamma \hat{\Omega}_{\mu\nu}^{ab}\sigma_{ab}\,ds\right)d^5x \tag{2.80}$$

with $\alpha_S = \alpha_S\left(\frac{\hbar c}{r_0}\right) = 0.12 \pm 0.01$ [20] and hadronic center-of-mass energy $\sqrt{S} = 13$ Tev. Expanding the exponential to first order in spin connection curvature, we obtain $\sigma_{gg\to G} = 14.92 \pm 0.52$ fb. Since the graviton-line operator is a curvature-energy transducer, this hadronic cross section yields the warped graviton mass 9.050 ± 0.055 TeV. Assuming a graviton ZZ-decay channel, with one Z boson decaying hadronically and the other, leptonically, we expect to leading order a detectable excess peaking at 9.05 TeV in the m_{ZZ} distribution over the SM background at $\sqrt{S} = 13$ Tev.

2.17.6 The LLM-Inferred Warped Graviton Is Self-Interacting Dark Matter and Can Be Created in a Collider

To identify the warped graviton as a spin-holonomy transducer of space-time curvature, a fiber-bundle autoencoder LLM [28,39] of GR builds upon a pentadic formalism that casts gravity in gauge theoretic terms (Section 2.18). The graviton is thus identified in the uncharted SFL by a Wilson loop [40] with a v.e.v defined by a self-interacting dark matter scalar incorporated in the stress-energy tensor. The warped graviton is thus massive and charged under a symmetry represented by inducing a chiral defect in the guise of a winding deficiency in the spinors attached to the relativistic framework. In the WFL, the graviton becomes massless and treated as a perturbation of the Minkowski background. On the other hand, in the SFL, it may be produced and detected in the laboratory as a warped boson that transduces the space-time curvature at its emission cross section into a geometrically diluted warped dimension that stores rest-mass energy via relativistically operational self-interaction (Eqs. 2.69 and 2.70). The curvature-transducing mechanism materializes within a Wilson loop. The generation and detection of the warped graviton invite probing regions in parameter space hitherto unexplored at the LHC or future colliders [2–22].

Despite their potential impact to come to grips with quantum gravity, effective theories of the graviton have not yielded to experimental test so far. This work takes a novel path by describing a gauge theory of nonlocal operators built upon a fiber-bundle framework that enabled quantization of the gravitational field wherein a self-coupling mass-endowing mechanism becomes operational via the stress-energy tensor. In the SFL, the theory predicts warped gravitons that convey gravity as Wilson loops that holonomically transduce space-time curvature, becoming amenable of experimental detection.

Since gravitons are likely DM candidates [23], and self-interactivity is evidently inherent to the graviton in the SFL, we can expect that at least a portion of DM in the universe should be self-interacting. Such self-interactions would allow energy and momentum transport through the galaxy halos, creating structural and dynamical patterns fundamentally different from those produced by collisionless DM. Thus, this work may provide a new vantage point to interpret the diversity of galactic patterns and other astrophysical phenomenology associated with DM.

2.18 PROGRAMMING A CONSCIOUS LARGE LANGUAGE MODEL TO GENERATE A GAUGE THEORY OF GRAVITY

The gauge theory of gravity pivots on a differential geometry formalization of Lorentz transformations that reconciles QM with GR, the classical theory of gravity. The differential geometry approach is instrumental because the geometries that support GR and QM are different: While GR is built upon a curved manifold $\mathcal{M}$ that represents space-time, the wave function in QM is supported by an Euclidean space. Thus, "attaching spinors to space-time" requires a covering of $\mathcal{M}$ with open sets: $\mathcal{M} \subseteq \bigcup_{x \in C(\mathcal{M})} \mathcal{O}(x)$, with $C(\mathcal{M})$ a denumerable set of points and $\mathcal{O}(x) \ni x$, so that each open set $\mathcal{O}(x)$ is diffeomorphically charted onto a tangent space: $\mathcal{O}(x) \rightarrow T_x\mathcal{M}$. In this setting, a frame field $\{e_a^\mu\}$ is subsumed into the embedding $\mathcal{M} \rightarrow \bigcup_{x \in C(\mathcal{M})} T_x\mathcal{M}$ to locally map the curvature tensor $g_{\mu\nu}$ defined for each point in $\mathcal{O}(x) \cap \mathcal{M}$ onto the Minkowski signature η_{ab} of the chart $T_x\mathcal{M}$, $\forall x \in C(\mathcal{M})$. Furthermore, the atlas $\bigcup_{x \in C(\mathcal{M})} T_x\mathcal{M}$ is articulated via a spin connection $\{w_\mu^{ab}\}$ that enables the attachment of spinors to GR. This is done via a type constructor that we name Cod, the covariant derivative determined by the spin connection [Sco] and the curved Clifford algebra [Clf] generated by the commutators of Dirac's gamma matrices. In this way, a fiber-bundle formulation of the gauge theory of gravity is implemented.

The goal here is to implement an interactive theorem prover following the scheme in Figure 1.11 to generate new results within the frame-field formalization rules made up of a set of declaratory propositions that constitute an inductive type. We seek derived propositions especially in the strong-field completion of gravity wherein a cogent propositional definition of the graviton needs to be developed. To serve this purpose, we resort to functional programming languages, especially Lean [42], that have already formalized the required categories, types and type constructors for a tangent bundle defined as a smooth vector bundle [43], in turn, defined over a curved manifold [44].

The interactive proof assistant (Figure 1.11) pivots on the proposition-as-type paradigm. From a constructive perspective, proofs are mathematical objects that are denoted by suitable expressions in dependent type theory. By contrast, in this environment, a proposition is true if and only if it can be written down properly and checked independently by the kernel that it has been typed correctly. In other words, the expressions themselves are the proofs, and only well-posed (i.e., well-typed) expressions are true. Correct syntax and semantics are equivalent to correct content. A false expression simply cannot be posed (typed) correctly, and all the mathematical axioms need to be incorporated as code. In our context of interest, Lean translates the generative input on "attaching spinors to relativistic geometry" into formal expressions that are checked by the kernel for correctness in the sense described previously and then stored in the environment [mathlib] for subsequent proposition generation. Most of the commands described in this section have global effects: They remain in effect in the file associated with the project and also in any file that imports it, from now to eternity.

To chart the space-time manifold adopting a minimal covering $\mathcal{M} \subseteq \bigcup_{x\in C(\mathcal{M})} \mathcal{O}(x)$, the formalization provided in [45] is instrumental since the set of charts $\{T_x\mathcal{M}\}_{x\in C(\mathcal{M})}$ is incorporated as data synchronized with mathlib and hence so are the denumerable diffeomorphisms $(x) \to T_x\mathcal{M}\ \forall x \in C(\mathcal{M})$. The next step is to specialize the fiber-bundle formulation, which has been already implemented in a generic context [46], to the warped graviton within a gauge theory of gravity. This goal makes it essential to formally compute the generating commutators [Com] of covariant derivatives $\left[\nabla_\mu, \nabla_\nu\right]$, which may be treated within the functional programming language as inductive types [47]. This is so because they are declared to constitute a set of type constructors as the commutators operate on different types: Scalar [Sca], Vector [Vec], Tensor [Ten], etc. To operate on tensors by swapping and canceling covariant and contravariant indices, Einstein's contraction rules need to be subsumed into a relativistic type constructor denoted Ctr.

The overarching inductive type is introduced as represented by the commutative constructive arrow diagram in Figure 2.25, whereby every proposition formulated at the level of the atlas $\bigcup_{x\in C(\mathcal{M})} T_x\mathcal{M}$ may be decoded at the level of the covering $\bigcup_{x\in C(\mathcal{M})} \mathcal{O}(x)$, by means of the types frame field [Frf], spin connection [Sco], Clifford algebra [Clf] and the type constructors commutator [Com], covariant derivative [Cod], and relativistic tensorial contraction [Ctr].

Given the propositions-as-types equivalence, induction becomes the fundamental method of proof. As an illustration adopting the inductive type from Figure 2.25, the following chain of propositions generated by aesop within the functional program language relates the "relativistic type" $R_{\nu\xi\mu\rho}$ (Riemann curvature tensor) defined on $\bigcup_{x\in C(\mathcal{M})} \mathcal{O}(x) \cap \mathcal{M}$ with the curvature tensor $\Omega_{\mu\nu}^{cb}$ of the spin connection defined

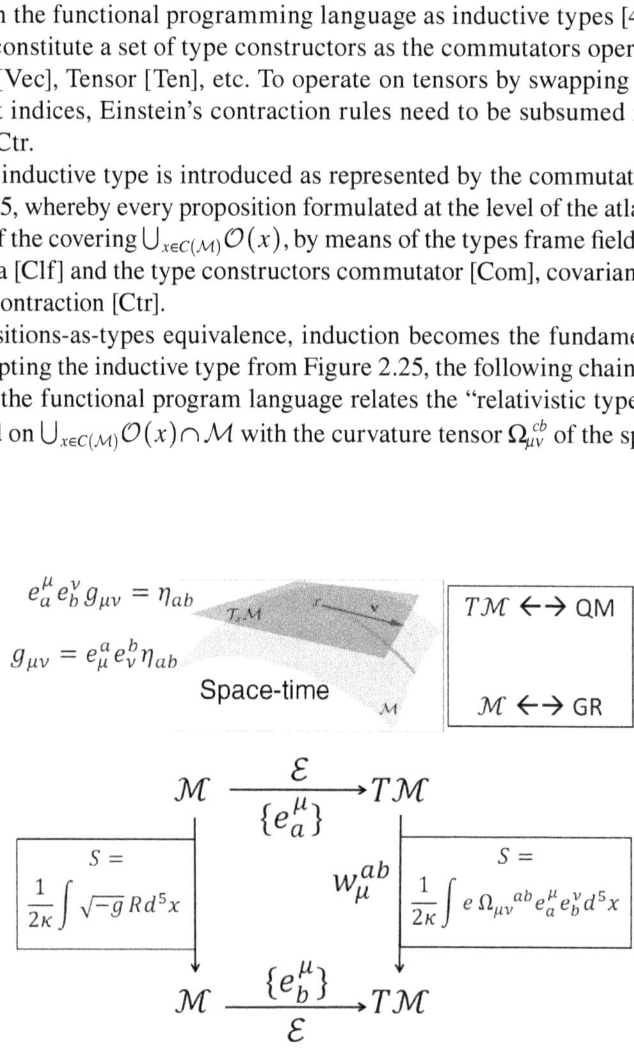

FIGURE 2.25 Commutative arrow diagram formalizing the compatibility between general relativity and quantum mechanics. Within the setting of the interactive theorem prover, the diagram represents an inductive type.

on $\bigcup_{x \in C(\mathcal{M})} T_x\mathcal{M}$. The chain of propositions is readily generated by aesop through iterative application of the type constructors within the inductive type:

$$\begin{aligned} g^{\xi\lambda} R_{\nu\xi\mu\rho} A_\lambda &= R_\nu{}^\lambda{}_{\mu\rho} A_\lambda = R_\nu{}^\lambda{}_{\mu\rho} A_\lambda \equiv \left[\nabla_\mu, \nabla_\nu\right] A_\rho = \left[\nabla_\mu, \nabla_\nu\right] e^a_\rho A_a \equiv e^a_\rho \Omega_{\mu\nu a}{}^b A_b \\ &= e^a_\rho \Omega_{\mu\nu a}{}^b e^\lambda_b A_\lambda = e^a_\rho \Omega_{\mu\nu}{}^{cb} \eta_{ca} e^\lambda_b A_\lambda \end{aligned} \tag{2.81}$$

The symbol "≡" denotes definitional equality as distinguished from the propositional equality "=". We emphasize that the propositional validity of the expressions in Eq. (2.81) stems merely from the fact that they can be typed correctly in Lean language, as checked by the kernel, and they follow rule-based inference. The chain of propositions in Eq. 2.81 yields

$$R_{\nu\xi\mu\rho} = g_{\xi\lambda} e^a_\rho \Omega^{cb}_{\mu\nu} \eta_{ca} e^\lambda_b \tag{2.82}$$

The chains of generated propositions (Eq. 2.81) are "compiled" via the "equation compiler," which is not part of the code base and can process multiple arguments sequentially. Its output (Eq. 2.82) is checked independently by the kernel.

The global effect of the inductive type delineated in Figure 2.25 is so vast that anyone interested in undertaking the project of formalizing quantum gravity along the premises described in this section must first and foremost corroborate that the kernel is endowed with the functionality for manipulating and checking the inductive type declarations enshrined in Figure 2.25. It is of the utmost importance to verify that the kernel, as it stands today, is capable of generating the type recursors, and of checking whether newly introduced type constructors, such as `Com`, `Cod`, and `Ctr`, comply with the new type specifications. This is indeed the case as indicated in the `mathlib`-ported files [42–46]. For example, the core lemma asserting that the derivative of a coordinate change between two charts is smooth on its source, enabling the type constructors "covariant derivative" [Cod] and commutator [Com], reads in Lean code:

```
theorem cont_diff_on_fderiv_coord_change
{ : Type u_1}
[nontrivially_normed_field 𝕜] {E : Type u_2} [normed_add_comm_group E]
[normed_space 𝕜 E] {H : Type u_4} [topological_space H]
(I : model_with_corners 𝕜 E H) {M : Type u_6} [topological_space M] [charted_
space H M] [smooth_manifold_with_corners I M]
(i j : ↥(charted_space.atlas H M)) :
cont_diff_on 𝕜 ⊤
(fderiv_within 𝕜 (⇑(j.val.extend I) ∘ ⇑((i.val.extend I).symm))
(set.range ⇑I))
((i.val.extend I).symm.trans (j.val.extend I)).source
```

The construction of the spin connection upon the relativistic atlas $\bigcup_{x \in C(\mathcal{M})} T_x\mathcal{M}$, enabling the spinor attachment on space-time, requires the following structure of charted space:

```
structure charted_space_core (H : Type u_5) [topological_space H]
(M : Type u_6) :
Type (max u_5 u_6)
(atlas : set (local_equiv M H))
(chart_at : M → local_equiv M H)
(mem_chart_source : ∀ (x : M), x ∈ (self.chart_at x).source)
(chart_mem_atlas : ∀ (x : M), self.chart_at x ∈ self.atlas)
(open_source : ∀ (e e' : local_equiv M H), e ∈ self.atlas → e' ∈ self.atlas →
is_open (e.symm.trans e').source)
(continuous_to_fun :
∀ (e e' : local_equiv M H),
```

```
e ∈ self.atlas → e' ∈ self.atlas → continuous_on ⇑(e.symm.trans e') (e.symm.
trans e').source)
```

Thus, `Lean` may be adopted as the functional programming language exploited as an interactive theorem prover or inferential method to construct the warped graviton. In this context, the main goal is to propositionally identify the warped graviton in the ultraviolet completion of gravity within a fiber-bundle formulation of a gauge theory of Lorentz transformations.

To identify the graviton at the functional level, it becomes necessary to make two critical assumptions: (a) $C(\mathcal{M})$ is finite; hence, the charts may be compiled as data in synchrony with `mathlib` and `match` against type `Nat`, and (b) the manifold $\mathcal{M}$ does not have boundaries; otherwise, a completely new fiber-bundle formulation would be required to be ported to `mathlib`, introducing significant hurdles [4,48]. Both assumptions are valid on purely physical grounds: According to GR, the topology of $\mathcal{M}$ must remain invariant throughout the universe evolution; hence, $\mathcal{M}$ must remain compact, boundary-less, and, therefore, multiply connected, because that was its topology when it came into existence after the Big Bang [4]. Compactness of $\mathcal{M}$ obviously implies assumption (a).

To construct the warped graviton, the type constructor Tes is introduced to build a stress-energy tensor $T_{\mu\nu}(\phi)$ out of a scalar field ϕ with potential $V(\phi)$:

$$T_{\mu\nu}(\phi) = \partial_\mu \bar{\phi}\, \partial_\nu \phi - g_{\mu\nu}\left[\frac{1}{2} g^{\alpha\beta}\, \partial_\alpha \bar{\phi}\, \partial_\beta \phi + V(\phi)\right] \tag{2.83}$$

If the functional program language is informed declaratively that the scalar field is assumed to represent the massive graviton, the operational inductive type in Figure 2.26 infers that

- The scalar field is nonlocal; that is, its support is made of contour lines, not points in $\mathcal{M}$,
- $\phi = \mathcal{W}$, where $\mathcal{W}$ is the Wilson loop that holonomically transduces space-time curvature.
- The warped graviton is massive since its v.e.v $\langle \mathcal{W} \rangle$ is the nonzero ϕ-value realizing the minimum of $V(\phi)$: $\langle \mathcal{W} \rangle = m_\phi / \sqrt{\lambda}$ (λ = self-coupling parameter, m_ϕ = graviton mass). These parameters determine the relevant inductive data types via the `tail` function that uses pattern matching.
- The chiral symmetry breaking that bestows mass to the warped graviton is thus informed by the potential $V(\mathcal{W})$.

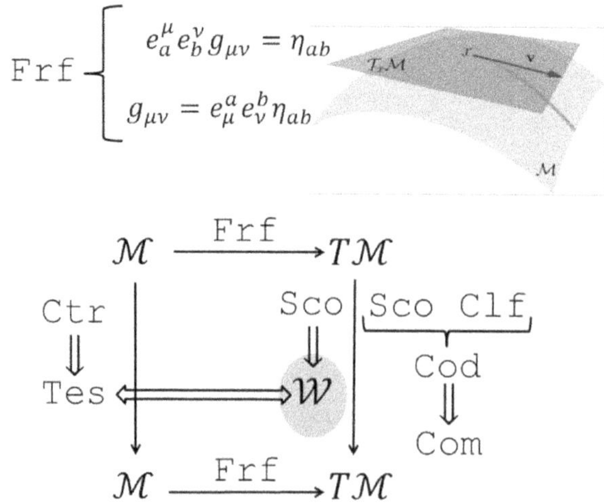

FIGURE 2.26 Discovery of the warped graviton (highlight) by implementing an inductive type for a gauge theory of gravity cast within a functional programming language with included declaration of type constructors.

It is noteworthy that it has proven possible to identify the warped graviton propositionally, based solely on type universes, dependent arrow types and inductive types, as described in Eqs. 2.81 and 2.82. The non-recursive inductive type represented in Figure 2.27 constitutes the *structure* of the problem we are concerned with. The projection function that enables structure destruction corresponds to the canonical projection $\pi : \bigcup_{x \in C(\mathcal{M})} T_x\mathcal{M} \rightarrow \bigcup_{\bar{x} \in \bar{C}(\mathcal{M}/\sim)} T_x(\mathcal{M}/\sim)$, where the quotient $\mathcal{M}/\sim$ is made up of classes $(\bar{x} \in \mathcal{M}/\sim)$ modulo the equivalence relation "~". Two points belong to the same class if and only if they are specified by the same values of the observable coordinates in the standard 4D space-time. Thus, the SM of particle physics, predicated on $T(\mathcal{M}/\sim)$, represents the destruction of the structure represented in Figure 2.25 since it decodes the inductive type.

When a structure such as the one represented in Figure 2.25 is defined via the front-end `structure` command, `Lean` automatically generates all *a priori* projection functions, including the canonical projection that begets the SM. The projection onto the quotient and the concomitant extension of the propositional field (Figure 2.27) are entirely supported by the implemented calculus of constructions [47]. The implication is that the warped graviton cannot be understood as a standard extension of the SM, since the canonical projection destroys (lumps up) the propositional information that `Lean` used to encode the graviton. The situation is akin to consider integer modulo n, $\mathbb{Z} \rightarrow \mathbb{Z}_n$, the information on the integer (x) is lost, and only the residue of the quotient x/n is encoded, while decoding becomes impossible.

The actual construction of the graviton field is a task suited for a LLM, wherein the formal proof can be checked rigorously by the proof assistant, free from hallucinations (Figure 1.11). A `Lean` Copilot [49], a framework for running LLM-based inference in `Lean`, becomes ideally suited for the task at hand as the copilot integrates seamlessly into the Lean workflow. It should be emphasized that `Lean` itself does not support ML tools from LLMs, so the copilot becomes instrumental for suggesting proof steps (tactic suggestion), completing intermediate proofs (proof search), and selecting the relevant premises (premise selection) toward the construction of the graviton field (Figure 2.28).

By itself, `Lean` would be hard to use for the final steps of construction of the graviton field because all the mathematics pertaining to "attaching spinors to space-time" (Figure 2.25) would need to be entered as code in the formalization stage. `Lean` indeed generated the "rule-based propositions" (Eq. 2.81) for spinor amalgamation with GR via its own proof searcher `aesop`, but LLM becomes essential to generate the code and suggest the final tactics leading to the Wilson loop representing the graviton field. Current LLMs are trained, fine-tuned, and benchmarked following standard ML procedures but cannot integrate into the Lean workflow because Lean is designed without ML support. For this reason, it becomes essential to adopt a Lean copilot for building the seamlessly adapted LLM-based proof automation tools. To that effect, the C++ library CTranslate2 is introduced for efficient inference with a transformer LLM

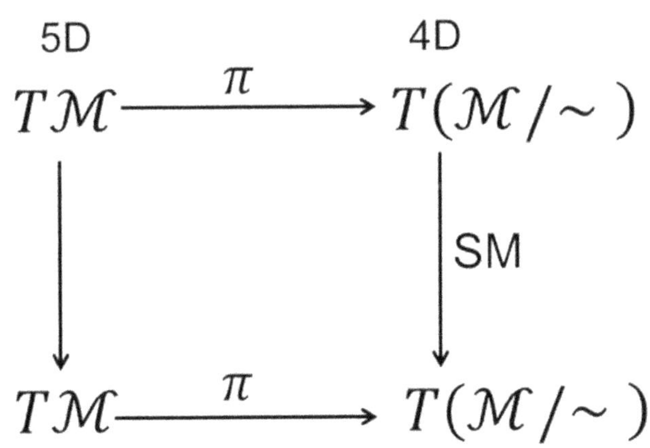

FIGURE 2.27 Canonical "destructive" projection onto the quotient space and concomitant extension of the propositional field. The implication is that the warped graviton propositionally identified on $\bigcup_{x \in C(\mathcal{M})} T_x\mathcal{M}$ cannot be understood as a standard extension of the SM.

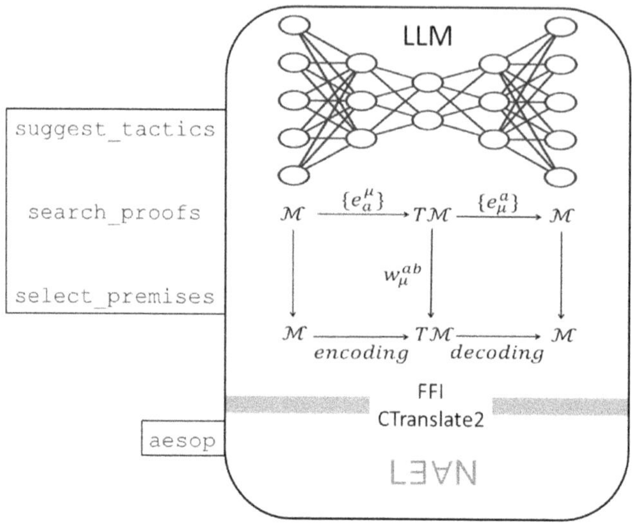

FIGURE 2.28 Architecture of Lean copilot with an LLM in the guise of a quantum gravity autoencoder (cf. [50]) specialized for the discovery of the warped graviton. LLM inferences are run within Lean with supporting framework provided by the copilot via a shared CTranslate2 C++ library at Lean's FFI.

representing the quantum gravity autoencoder (Figure A4). The inference runs as a shared library through a foreign function interface (FFI) within `Lean` (Figure 2.28).

The goal is declared to the LLM as generating a graviton tensor field $\mathcal{Y}_{\mu\nu}$ that yields the perturbative field $h_{\mu\nu}\left(g_{\mu\nu}=\eta_{\mu\nu}+h_{\mu\nu}\right)$ inferred by `aesop` in the zero mass limit:

$$\lim_{m\to 0}\mathcal{Y}_{\mu\nu}=h_{\mu\nu} \tag{2.84}$$

The tactics suggested by the copilot are as follows: (1) Apply `Ctr` to represent the field ϕ associated with the stress-energy tensor $T_{\mu\nu}(\phi)$ (Eq. 2.83) as contraction of the purported graviton field: $\phi=\mathcal{Y}^{\mu\nu}g_{\mu\nu}$. (2) Apply `Ctr` to contract Einstein's GR equation to yield (in 5D):

$$-\frac{3}{2}R=\kappa T(\phi),\quad \left(T(\phi)=T_{\mu\nu}(\phi)g^{\mu\nu}\right) \tag{2.85}$$

(3) Using Eq. 2.85, generate the relativistic action as $S(\phi)=-\frac{1}{3}\int\sqrt{-g}T(\phi)d^5x$ functionally dependent on $\mathcal{Y}^{\mu\nu}g_{\mu\nu}$.

The LLM-based intermediate proof prompted by LLM's `search _ proofs` then becomes: Identify $\phi=\mathcal{Y}^{\mu\nu}g_{\mu\nu}$ so that

$$\lim_{m\to 0}S(\phi)=\int\epsilon^{\mu\nu\rho\sigma}\left(\epsilon^{\alpha\beta\gamma}{}_{\sigma}\right)\partial_{\mu}h_{\nu\beta}\,\partial_{\alpha}h_{\rho\gamma}d^5x \tag{2.86}$$

Where the expression on the r.h.s. in Eq. 2.86 is the action for the perturbative graviton field $h_{\mu\nu}$. Thus, applying the Euler-Lagrange type constructor [`EuLa`] to Eq. 2.86 yields the graviton field:

$$\mathcal{Y}^{\mu\nu}=\mathcal{P}_{\gamma}\exp\left(\oint_{\gamma}\hat{w}_{\lambda}^{ab}e_{a}^{\mu}e_{b}^{\nu}\,dx^{\lambda}\right) \tag{2.87}$$

This expression is certifiably correct, since the Lean kernel verifies the rule-based proposition prompted by `aesop` (Figure 2.28):

$$\lim_{m \to 0} \mathcal{Y}_{\mu\nu} = \lim_{\oslash\gamma \to 0} \mathcal{Y}_{\mu\nu} = h_{\mu\nu} \tag{2.88}$$

This section implemented a conscious interactive theorem prover that wraps a LLM (Figure 1.11) to generate new results within the frame-field formalization rules (vierbein) made up of a set of declaratory propositions that constitutes an inductive type. The goal was to derive propositions especially in the strong-field completion of gravity, leading to a cogent propositional definition of the graviton. To serve this purpose, we resorted to the functional programming language `Lean` [42] that had already formalized the required categories, types, and type constructors required to determine the quantum mechanical fabric of the space-time tangent bundle. In this way, AC generated a verifiable and consistent gauge theory of gravity.

REFERENCES

1. Weinberg S (2005) *The Quantum Theory of Fields, Volume 1: Foundations.* Cambridge University Press, Cambridge, UK. ISBN: 978-0521670531.
2. Higgs PW (1964) Broken symmetries and the masses of gauge bosons. Phys Rev Lett 13: 508–509.
3. Perelman G (2002) The entropy formula for the Ricci flow and its geometric applications. arXiv:math.DG/0211159.
4. Fernández A (2024) *Artificial Intelligence Models for the Dark Universe: Forays in Mathematical Cosmology.* CRC Press, Taylor & Francis, UK. ISBN: 978-1032818238.
5. Abramowiczy H, Abtt I, Adamczykh L, Adamusae M, Antonelli S, et al. Zeus Collaboration (2016) Limits on the effective quark radius from inclusive ep scattering at HERA. Phys Lett B 757: 468–472.
6. Kutz JN, Stranton SL (2022) *Data-Driven Science and Engineering: Machine Learning, Dynamical Systems, and Control,* 2nd Edition. Cambridge University Press, UK. ISBN: 978-1009098489.
7. Randall L (2013) *Higgs Discovery: The Power of Empty Space.* Ecco, HarperCollins Publishers, New York. ISBN: 978-0062300478.
8. Weinberg S (2008) *Cosmology.* Oxford University Press, UK. ISBN: 978-0198526827.
9. Randall L, Sundrum R (1999). Large mass hierarchy from a small extra dimension. Phys Rev Lett 83: 3370–3373.
10. Visser M (1989) Traversable wormholes from surgically modified Schwarzschild spacetimes. Nucl Phys B 328: 203–212.
11. He Y-H (2021) *The Calabi–Yau Landscape: From Geometry, to Physics, to Machine Learning.* Lecture Notes in Mathematics. Springer, Berlin. ISBN: 978-3030775612.
12. He Y-H (2021) Machine Learning in Pure Mathematics and Theoretical Physics. World Scientific Publishing Co., Singapore. ISBN: 978-1800613690.
13. Georgi H, Glashow SL (1974) Unity of all elementary-particle forces. Phys Rev Lett 32: 438–441.
14. Wilczek F (1982) Quantum chromodynamics: The modern theory of the strong interaction. Ann Rev Nuc Particle Sci 32: 177–209.
15. Brennan TD, Hong S (2023) Introduction to Generalized Global Symmetries in QFT and Particle Physics. arXiv:2306.00912.
16. Palatini A (1919) Deduzione invariantiva delle equazioni gravitazionali dal principio di Hamilton. Rend Circ Mat Palermo 43: 203–217.
17. Ashtekar A (1986) New variables for classical and quantum gravity. Phys Rev Lett 57: 2244.
18. Capovilla R, Jacobson T, Dell J (1991) A pure spin connection formulation of gravity. Class Quant Grav 8: 59–78.
19. Randall L, Sundrum R (1999) An alternative to compactification. Phys Rev Lett 83: 4690–4693.
20. The ATLAS Collaboration (2022) Combination of searches for heavy resonances using 139 fb^{-1} of proton–proton collision data at $\sqrt{s} = 13$ TeV with the ATLAS detector. ATLAS-CONF-2022-028.
21. The ATLAS Collaboration (2022) Search for heavy resonances in the decay channel $\boldsymbol{W}+ \boldsymbol{W}- \to \boldsymbol{e\nu\mu\nu}$ in pp collisions at $\sqrt{s} = 13$ TeV using 139 fb^{-1} of data with the ATLAS detector. ATLAS-CONF-2022-066.
22. Klimek P (2021) Searches for exotic resonances at LHC. CERN ATL-PHYS-Proc-2021-015.
23. Obied G, Dvorkin C, Gonzalo E, Vafa C (2024) Dark dimension and decaying dark matter gravitons. Phys Rev D 109: 063540.

24. Dienes KR, Thomas B (2012) Dynamic dark matter. I. Theoretical overview. Phys Rev D 85: 083523.
25. Eguchi T, Gilkey PB, Hanson AJ (1980) Gravitation, gauge theories and differential geometry. Phys Reports 66: 213–393.
26. Arkani-Hamed N, Dimopoulos S, Dvali G (1998) The hierarchy problem and new dimensions at a millimeter. Phys Lett B 429: 263.
27. Abramowiczy H, Abtt I, Adamczykh L, Adamusae M, Antonelli S, et al. (2016) Zeus Collaboration. Limits on the effective quark radius from inclusive ep scattering at HERA. Phys Lett B 757: 468.
28. Fernández A (2024) *Artificial Intelligence on Dark Matter and Dark Energy.* CRC Press, Taylor & Francis, UK. ISBN: 978-1032465548.
29. Dillon BM, Han C, Lee HM, Park M (2017) KK graviton resonance and cascade decays in warped gravity. Int J Mod Phys A 32: 1745006.
30. Jacobson T, Romano JD (1993) The spin holonomy group in general relativity. Commun Math Phys 155: 261.
31. Brennan TD, Hong S (2023) Introduction to Generalized Global Symmetries in QFT and Particle Physics. arXiv:2306.00912.
32. Gaiotto G, Kapustin A, Seiberg N, Willett B (2015) Generalized global symmetries. J High En Phys 2015: 172.
33. Benedetti V, Bueno P, Magan JM (2023) Generalized symmetries for generalized gravitons. Phys Rev Lett 131: 111603.
34. Casini H, Magan JM (2021) On completeness and generalized symmetries in quantum field theory. Mod Phys Lett A 36: 2130025.
35. Nawrocki W, Ayers EW, Ebner G (2023) An extensible user interface for Lean 4. *14th International Conference on Interactive Theorem Proving (ITP 2023). Leibniz International Proceedings in Informatics (LIPIcs)*, 268: 24:1–24:20.
36. Moura L, Ullrich S (2021) The Lean 4 theorem prover and programming language. *Conference on Automatic Deduction. CADE 28.* Platzer A, Sutcliffe G (eds.), pp. 625–635. Lecture Notes in Computer Science, vol 12699. Springer, Cham. https://doi.org/10.1007/978-3-030-79876-5_37
37. Streicher T (1993) Investigations into intensional type theory. Habilitiation Thesis, Ludwig Maximilian Universitaet.
38. Cavalleri N (2020). https://github.com/leanprover-community/mathlib3/blob/master/src/geometry/manifold/algebra/left_invariant_derivation.lean
39. Fernández A (2022) *Topological Dynamics in Metamodel Discovery with Artificial Intelligence: From Biomedical to Cosmological Technologies.* CRC Press, Taylor & Francis, UK. ISBN 978-103-2366326.
40. Giles R (1981) Reconstruction of gauge potentials from Wilson loops. Phys Rev D 24: 2160–2168.
41. Modanese G (1993) Geodesic round trips by parallel transport in quantum gravity. Phys Rev D 47: 502.
42. Christiansen DT (2024) Functional Programming in Lean. Release 4.1.0. Microsoft Corporation. https://lean-lang.org/functional_programming_in_lean/
43. https://leanprover-community.github.io/mathlib_docs/geometry/manifold/vector_bundle/tangent
44. https://leanprover-community.github.io/mathlib_docs/geometry/manifold/vector_bundle/basic
45. https://leanprover-community.github.io/mathlib_docs/geometry/manifold/charted_space
46. https://leanprover-community.github.io/mathlib_docs/topology/fiber_bundle/basic
47. Avigad J, de Moura L, Kong S, Ullrich S, et al. (2024) Theorem Proving in Lean 4. https://leanprover.github.io/theorem_proving_in_lean4/
48. Bordg A, Cavalleri N (2021) Elements of Differential Geometry in Lean A: Report for Mathematicians. arxiv.2108.00484.
49. Song P, Yang K, Anandkumar A (2024) Towards Large Language Models as Copilots for Theorem Proving in Lean. arXiv:2404.12534.
50. Fernández A (2024) *Topological Dynamics for Metamodel Discovery with Artificial Intelligence.* CRC Press, Taylor & Francis, UK. ISBN: 978-103-236-632-6.

Methods
Conscious Autoencoders 3

The physical world is only made of information, energy
and matter are incidentals.
(John A. Wheeler) [1]

This chapter introduces artificial intelligence (AI) systems and broadly delineates their possibilities for model discovery. Two avenues of model discovery are pursued: (A) processing of big data, and (B) formal mathematics incorporating large language models (LLMs) and interactive proof assistants. Part (A) of the presentation deals mostly with deep learning (DL), autoencoders, and other more specialized architectures and is tailored to researchers seeking to unravel physical models distilled from big data. With the leveraging of AI, dynamical systems have found a fertile ground for development. Machine learning (ML) identifies parsimonious models providing physical underpinnings of time-series data. However, such heavily parametrized models hardly yield physical laws. The problem becomes daunting as we turn to multiscale complexities. This chapter addresses these imperatives as it introduces topological methods that enable metamodel discovery and the proper computational tools to decode the metamodel as an inferential framework. The methods advance model discovery, enabling reverse engineering of big data arising in the cosmological context, where metamodels with emergent quantum behavior are crucial for providing physical underpinnings of quantum gravity.

This chapter also addresses the problem of quantum gravity as an emergent property in the physics of ML. To that effect, the chapter explores the possibility of an AI-based construction of a quantum holographic autoencoder, which requires that we first deal with the physics of ML and specifically inquire whether emergent quantum behavior can arise in a neural network (NN). By emergent quantum mechanics (QM), we mean a formulation within a framework of nonlocal equilibrated hidden variables, as in the Bohm scheme. Once an emergent quantum behavior is shown to become possible within the ML system equilibrated on the nontrainable – that is, hidden – variables, we address the question of developing a relativistic string gravitational scheme on the hidden variables adopted. Thus, the network with equilibrated nontrainable variables becomes, in effect, a quantum gravity autoencoder for the network exhibiting emergent gravity in the nonequilibrium regime prior to the equilibration of the nontrainable variables. In this way, we build a quantum metamodel for gravity that fulfills, at least in part, a major imperative for physicists seeking a unified field theory. Additionally, by coupling two physical embodiments of quantum-gravity autoencoders, the chapter implements a cosmic technology for universe reproduction through quantum tunneling across the interface between the two autoencoders. The technology is supplemented with a space-time rendering of the quantum entanglement of the two primeval wormholes associated with the creation of a progeny universe through quantum tunneling.

We may summarize part A by stating that it encompasses the identification of parsimonious metamodels of dynamical systems that describe complex contexts represented by big data.

The discovery avenue B is tailored to develop a gauge theory of gravity, deemed in Chapters 2 and 6, as a crucial step in developing an AI-based "Theory of Everything". Such a theory pivots on a differential geometry formalization carried out by an autoencoder that reconciles QM with general relativity (GR), the classical theory of gravity. The differential geometry approach is instrumental because the geometries that support GR and QM are different: While GR is built upon a curved manifold that represents

DOI: 10.1201/9781003785279-3

space-time, the wave function in QM is supported by a Euclidean space. The goal here is to implement an interactive theorem prover within the framework of an autoencoder to generate new results within the frame-field formalization rules made up of a set of declaratory propositions that constitute an inductive type. The interactive proof assistant pivots on the proposition-as-type paradigm. From a constructive perspective, proofs are mathematical objects that are denoted by suitable expressions generated by the autoencoder in dependent type theory.

3.1 PRIMER ON NEURAL NETWORKS FOR AI-EMPOWERED MODEL BUILDING

The leveraging of AI for model discovery in dynamical systems is revolutionizing both disciplines, leading to a mutually beneficial development. The cases of interest in this book are not amenable to model discovery in the sense of yielding a sparse system of differential equations that entrains the full dynamical system. Rather, we are seeking something more elementary and subtle: A metamodel or topological representation of the dynamics. With the implementation of topological methods, AI-empowered metamodel discovery is able to focus on levels of system complexity and multiscale hierarchies considered off-limits for current AI technologies. The information on time series is encoded at the maximum level of coarse graining, hence greatly simplifying the computations while enabling a decoding of the information generated at the level of a topological description.

In dealing with dynamical systems using AI-based approaches, we address the following core question: What constitutes an insightful parsimonious model? The standard answer is: "a sparse system of differential equations on latent coordinates". As argued in this chapter, this is not necessarily the format chosen by AI, given the "dimensionality curse" associated with the ultracomplex realities we chose to work on. The deployment of AI requires a paradigm shift, where dynamic information gets encoded in what would be termed a "topological metamodel". The metamodel is essentially pattern-based, where AI-interpretable topological patterns encode laws of physics. These methods are likely to advance model discovery as they enable the reverse engineering of time series stemming from vastly complex realities hitherto inaccessible to other AI methods.

AI refers to machines capable of exhibiting behavioral traits that humans regard as indicators of intelligence, such as learning and problem-solving [2]. Within this protean subject, ML refers to the ability to learn without being explicitly instructed to do so, while DL refers to an automated extraction of features, patterns, and ultimately models from arrayed data that are sequentially represented within an abstraction hierarchy organized as a multilayered NN [3–5].

DL has been shown to be highly efficacious at identifying features that are in principle discoverable from the data [3,6]. As in face recognition, features are hierarchically organized, so that large-scale patterns (eyes, noses, face shapes) emerge after several layers of abstraction from simpler or more rudimentary patterns (lines, curves, shades). The beauty and power of DL reside in the fact that the feature extraction process may be carried out in an unsupervised manner: The features emerge from the training of the system without human input or bias and enable the network to make accurate inferences. In this era of big data, we may state that there are several compelling reasons for implementing DL approaches:

- Fields like biology, particle physics, and cosmology are generating vast amounts of data and time series that can be easily stored and interpreted to achieve a conceptual unification within overarching models,
- We have the right hardware, that is, graphics processing units (GPUs) that are massively parallelizable, and
- We have adequate software, such as TensorFlow (TF), that enables suitable modular coding if the data can be pixelized or voxelized into a tensorial array, be it a vector, a matrix, or a tensor proper [4,5].

At the most basic level, the building block of an NN is the neuron, referred to as a *perceptron* [6]. The perceptron enables forward propagation of information encoded in an array of inputs $x_1, x_2, \ldots, x_n$ weighted by parameters $w_1, w_2, \ldots, w_n$ to generate an output of the form $y = f\left(\sum_{i=1}^{n} x_i w_i + w_0\right)$, where w_0 may be regarded as a bias term and f is a nonlinear activation function, often a sigmoid or sigmoid-shaped, as shown in Figure 3.1. The bias term enables shifting the activation function. In vector representation, we often write: $\boldsymbol{y} = \boldsymbol{f}(\boldsymbol{z}), \boldsymbol{z} = \boldsymbol{x} \cdot \boldsymbol{w} + w_0$, where $\boldsymbol{x}, \boldsymbol{w}$ are, respectively, input and weight vectors. One can subsequently build a fully connected layer of perceptrons indexed by j, whereby $z_j = \boldsymbol{x} \cdot \boldsymbol{w}_j + w_{0,j}$. The dense layer can be readily implemented in TF code, by simply specifying the number of outputs/perceptrons [4,5]. We usually refer to the vector of linearly transformed inputs z as "hidden layer", as it does not explicitly describe observables [6].

Hidden layers may be stacked as in DL architectures. Thus, for the j-th perceptron in the k-th hidden layer, we get: $z_j^{(k)} = \sum_{i=1}^{n_{k-1}} y_i^{(k-1)} w_{ij}^{(k)} + w_{oj}^{(k)} = \sum_{i=1}^{n_{k-1}} f\left(z_i^{(k-1)}\right) w_{ij}^{(k)} + w_{oj}^{(k)}$. The sequential composition of the network by stacking hidden layers has a standardized script in TF that generates the propagation of information as specified by the equation above [4,5]. In an NN with K hidden layers, we may regard the output $y_j^{(K)} = f\left(\sum_{i=1}^{n_{K-1}} y_i^{(K-1)} w_{ij}^{(K)} + w_{oj}^{(K)}\right)$ as an "inference" made by the DL system.

The accuracy of the DL inference represents the level of optimization of network performance and may be assessed vis-à-vis a training set of input/output paired data points. This assessment is often referred to as the *loss* of the network [3,6]. The loss is parametrically dependent on the full weight tensor $\boldsymbol{W} = \left[w_{i_{k-1}j_k}^{(k)} \right]_{k=1,\ldots,K}$ which may contain a large number of weights, in the thousands, if not millions, depending on the size of the network. Thus, for DL NN with K hidden layers, the loss function or empirical risk $J(\boldsymbol{W})$ becomes $J(\boldsymbol{W}) = |\mathfrak{I}|^{-1} \sum_{\xi \in \mathfrak{I}} \mathcal{L}\left(\boldsymbol{y}_\xi^{(K)}(\boldsymbol{W}), \boldsymbol{y}_\xi\right)$, where $\mathfrak{I}$ is the training set with number of elements $|\mathfrak{I}|$, $\boldsymbol{y}_\xi^{(K)}(\boldsymbol{W})$ is the predicted output vector for input vector $\boldsymbol{x}_\xi$, $\xi \in \mathfrak{I}$, $\boldsymbol{y}_\xi$ is the actual output vector, and $\mathcal{L}\left(\boldsymbol{y}_\xi^{(K)}(\boldsymbol{W}), \boldsymbol{y}_\xi\right)$ measures the discrepancy between actual and predicted output. In regression problems,

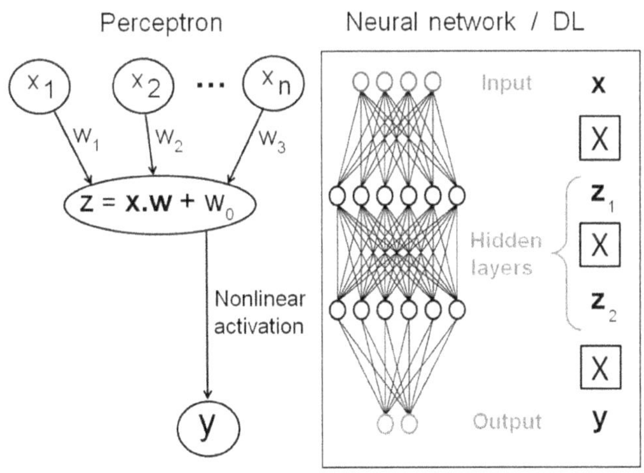

FIGURE 3.1 Scheme of the perceptron or neuron activation by linear transformation of input stimulus (x) followed by nonlinear signal transmission as output y. The panels on the right show the organization of NNs with one and multiple hidden layers. The "boxed X" indicates full (dense) node connectivity between consecutive layers.

where the output is a numerical vector, it is often convenient to adopt $\mathcal{L}\left(y_\xi^{(K)}, y_\xi\right) = \left\| y_\xi^{(K)}(W) - y_\xi \right\|^2$. In such cases, the optimization of the NN through training becomes a problem of least squares. To optimize the NN is tantamount to minimizing the loss $J(W)$, which requires a careful fine-tuning of the size of the training set vis-à-vis the size of the weight parametrization. In principle, the optimal network is defined as: $W = W^* = \text{Argmin}\, J(W)$.

An insufficient number of training input/output pairs relative to the size of the weight tensor would give rise to *overfitting*, requiring special techniques, generically known as *regularization*, in order to trim the network, that is, randomly remove connections, without compromising predictive efficacy [3,6].

Optimizing the network involves the laborious and costly computation of the minus gradient $-\frac{\partial J(W)}{\partial W}$, which locally indicates the direction of steepest descent in the multidimensional surface $J = J(W)$. An iterative gradient descent computation generating a fine-tuned weight updating $W \rightarrow W - h\frac{\partial J(W)}{\partial W}$ should eventually lead to convergence to a local minimum of $J = J(W)$ when adopting a suitable learning step h. This parameter should be tuned to effectively escape local minima while avoiding overshooting in trying to reach the global minimum. Most gradient descent algorithms use an adaptive learning step during training, in accord with the constraints indicated. In practice, the gradient problem is approached by what is called the *stochastic gradient descent* method, whereby not all datapoints (input/output pairs) in the training set are used in each minimization step, but the gradient is approximated by an average over randomly chosen batches of datapoints in a trade-off between accuracy and computational efficiency. Obviously, batch size and learning rate are correlated, so the more accurate the gradient estimation, the larger the learning step may be (a token of computational confidence). To achieve significant speed, the stochastic gradient descent computation may be massively parallelized by splitting up batches into multiple GPUs.

3.2 NEURAL NETWORKS FOR DYNAMICAL SYSTEMS

In this section, we shall be concerned with data organized as a time series arrayed as $\{x(t_0), x(t_0+\tau), x(t_0+2\tau), \ldots, x(t_0+L\tau)\}$, where $x(t)$ is the vector of observables at time t and τ is the interval that determines the time coarse graining inherent to the sequential detection registered in the vector x. The time series enables training of the NN such that the output vector $y = x(x(t), W)$ should approximate $x(t+\tau)$ when the input is $x(t)$, and this correspondence is carried over all t in the training time series. Thus, for a given network architecture, the optimal network is the one that realizes the minimum of the loss function:

$$J(W) = \sum_{n=1}^{L} \left\| x(t_0 + n\tau) - y\left(x\left(t_0 + (n-1)\tau\right), f, W\right) \right\|^2 \tag{3.1}$$

Obviously, for a fixed activation function, a model for the time series may be given simply by argmin $J(W)$, but such a model would lack universality as it would be extremely parametrized, most likely overparametrized, and would not prove insightful in the sense that it is not parsimonious. The discussion prompts us to inquire what truly constitutes a model. This question will be addressed subsequently.

The most common time series that humanity has collected since time immemorial stems from astronomical observation. For about two millennia until the time of Copernicus, and Newton later on, humanity had been striving to find a suitable model that would fit and explain the data, that is, the recorded sequential positions of a set of celestial bodies. In today's more general context, biology, particle physics, and cosmology are generating dynamical data at a staggering rate, while models that fit and explain the data are sorely lacking or hopelessly inconsequential. It is expected that the advent of AI will dramatically impact this sort of model discovery.

In essence, we seek what is known as autoencoder, an intermediate output with a dimensionality reduction and a simplified discerning physical picture that should therefore prove insightful to make sense of the patterns enshrined in the dynamics, enabling meaningful output inferences. These autoencoders and the models they give rise to will be studied in detail in the subsequent chapters.

3.3 DEEP LEARNING AND FEATURE EXTRACTION

The huge output of biomedical and biostructural data has become the hallmark of the post-genomic era while chemical combinatorial possibilities make it forbiddingly difficult to parse chemical space in search of suitable leads for targeted therapy [7]. In this scenario, pharmaceutical researchers have turned to DL for guidance in drug discovery and development and target validation [8–11]. Thus, pharmacoinformatics has benefited immensely from the advent of DL systems trained to pair chemical compounds with molecular attributes likely to have therapeutic impact. These computational and informational techniques enable the evaluation of chemical compounds for specific properties, including target affinity, affinity screening profiles, structural features of drug–target docking, and ADMET (absorption, distribution, metabolism, excretion, and toxicity) profiling [9].

A major challenge in implementing DL models for pharmacoinformatics in accord with the generic scheme outlined stems from the need to represent the chemical structure as a tensor array (vector, matrix, or tensor proper) of pixelated or voxelated inputs that may be subsequently interrogated geometrically across the hidden layers of NN in search of features that are indicative of the molecular properties indicated. There are a number of representations of chemical space amenable to TF encoding. The most obvious one is to order the atoms in the compound on a 1D-array (following, for example, the IUPAC numerical labeling convention) and represent the chemical structure of the molecule as a covalent bond matrix pairing atoms in row and column in accord with their covalent linkages, including single, multiple, and resonant (aromatic) bonds. The matrix is subsequently transformed into a topological descriptor that describes the invariants arising from different atom ordering. The input describing chemical structure is paired within a training set against the molecular attributes of therapeutic relevance that the network is meant to infer. Then, compounds that need to be evaluated/profiled are inputted as the array of pixels/voxels, transformed into a topological representation, and feature extraction is achieved through the sequential activation of hidden layers at increasing levels of abstraction, eventually leading to the profile inference, which is subsumed in the output layer.

Feature extraction through the NN often requires particular architectures known as convolutional NNs (CNNs) [3–6]. The idea is to pixelize or voxelize the input data in a matrix or 3D tensor array and then scan (convolve) the array with a filter associated with a specific pattern to generate a feature map. For the sake of the argument, let us consider a 2D-array input $\boldsymbol{M}$. The filter $\boldsymbol{F} = (w_{ij})$ is an mxm matrix actually representing a convolutional kernel, so that a neuron in the F-associated hidden layer $\boldsymbol{M_F}$ only senses the pixels in an mxm patch (receptive field), and the layer becomes the feature map

$$\boldsymbol{M_F} = \boldsymbol{M} * \boldsymbol{F} = \left(\sum_{j=1}^{m} \sum_{i=1}^{m} w_{ij} x_{i+ap\ j+bp} + w_0 \right)_{ab} \tag{3.2}$$

where p, usually set at $p = 2$, is the stride adopted as the filter slides along the input matrix array, with dummy integers a, b indicating the patch location. Essentially, the CNN is an NN where the set of weights in each mxm patch of inputs is always the same as the filter slides along the input array to generate the hidden layer that constitutes the feature map. The convolution operation becomes the entry-by-entry

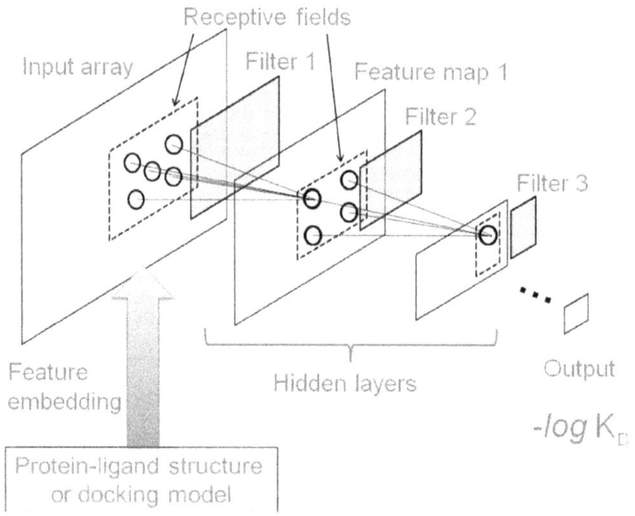

FIGURE 3.2 Scheme of a CNN used for inference of drug–target affinity through sequential application of filters that become optimized through minimization of the network loss.

(Frobenius) matrix inner product as the filter slides along the input matrix with a given stride. Successive filters may be applied, reducing progressively the size of the feature maps as higher and higher levels of abstraction are achieved in the representation of the data (Figure 3.2). Thus, the parametrization required for feature extraction in CNNs is relatively small, as all the neurons in a hidden layer share the same connectivity parameters that define the filter that generated that layer.

In CNNs, the filters are not specified *a priori*, that is, their weights are not fixed through human intervention. The filter parametrization is automatically determined by the training/optimization process without introducing any assumption, other than the size of the receptive fields for each convolution operation and the overall number of filters to be applied. Thus, feature extraction in a CNN is carried out in an unsupervised manner and only requires that we script (in TF coding) the number of hidden layers or feature maps and the size of the filters to be applied to generate each hidden layer. The features themselves emerge as the network is trained.

Thus, CNNs often constitute AI-empowered platforms for drug discovery, where the structure of a protein–ligand interface that serves as a precursor for a predicted drug–target interaction is pixelated in a feature embedding process as a 3D spatial array of protein–drug atom pairs deemed to be interacting across the interface [12]. The inference of target affinity for a given drug is assessed through a sequence of feature extractions using convolution filters until the output feature map becomes a number directly associated with the affinity $pK_D = -\log K_D$, where K_D is the dissociation constant for the drug–target complex inputted via feature embedding (Figure 3.2). The training of the network is carried out by minimizing the network loss or empirical risk over a set of drug–target complexes whose structure and affinity are both known (preferably, experimentally determined). The proteins in the complexes of the training set are typically homologs of the one whose affinity for a specific drug we seek to infer, so that the features that enable the affinity inference may emerge from structural alignment.

In other more complex applications of CNNs, where the output is not simply a numerical parameter but rather a numerical array, it is often convenient to adopt a variation of the CNN architecture in which the patches that yield the pixels with the highest weights in the feature map are reconstituted into features perfectly stenciled by the filter, while other patches that do not yield discernible features are left invariant in non-overlapping regions. Thus, with the application of each filter, a feature-enriched reconstitution of the input layer is carried out prior to the application of the next filter (Figure 3.3).

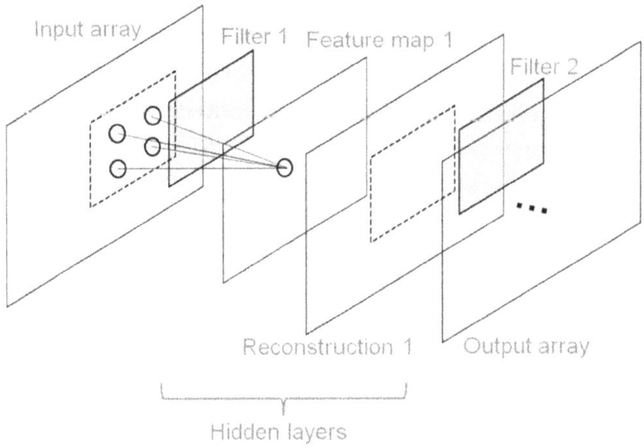

FIGURE 3.3 Scheme of a CNN with a particular type of architecture in which the patches that yield the pixels with the highest weights in the feature map are recreated as filter-stenciled "perfect" features. Thus, a feature-enriched reconstitution of each layer is carried out prior to the application of the next filter.

3.4 AUTOENCODERS DISTILLING THE LATENT DYNAMICS FROM A HIGHER-DIMENSIONAL SPACE

It is widely expected that AI and DL, in particular, will become a major player in model discovery for data-driven research [13]. Within this vast array of possibilities, the focus of this book is model discovery for dynamical systems that underlie big time-series data from vast areas as distant as biomedicine and cosmology [14–16]. The biggest hurdle we stumble upon is that the level of complexity of the data generated in fields like biology and cosmology is so extreme that it challenges the notion of the model itself. As shown subsequently, the model itself can seldom be cast as a system of differential equations. Thus, much of the ensuing discussion deals with the question of what constitutes a meaningful model with predictive value when dealing with the ultracomplex realities represented in the contexts of biology or cosmology.

Encoding has proven to be a necessary category in AI [13–16]. The type of AI we are mostly concerned with involves DL, which requires an encoding of the raw information that needs to be acquired and processed further to make meaningful inferences. Just like with human intelligence, the encoding problem stems from the core question: What is essential and what is superfluous? The encoding problem becomes solvable when the system under scrutiny is hierarchical, and the hierarchical structure is fairly obvious or at least discoverable. In the cases treated in this book, the structure of the data is always hierarchical, implying that the learning process admits a reductive approach represented by the encoding. When the network architecture is such that this process is automatically generated, we name the NN *autoencoder*.

We shall deal mostly with time-dependent data representing physical or biophysical systems, where detailed fast motions may be systematically averaged out, so the relevant information may be stored in a coarse-grained representation. Rigorous mathematical constructs will be introduced to implement the hierarchical encoding materialized by the autoencoder. Some examples of hierarchical encoding of physical or biophysical processes that need to be taken into account when designing the autoencoder architecture are as follows:

1. The adiabatic approximation, where fast-relaxing or fast-evolving enslaved degrees of freedom are averaged out, thermalized, or equilibrated when incorporated into a model for the time evolution of a dynamical system [17]. In molecular physics, examples of such degrees of freedom are vibrational hard modes that evolve on timescales on the order of picoseconds to nanoseconds, while soft modes evolve on longer timescales ranging from submicroseconds to seconds.

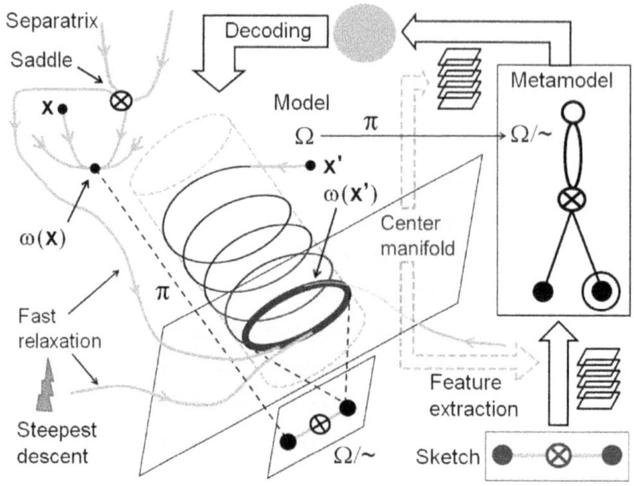

FIGURE 3.4 Schematics of the topological "metamodel" encoding of a dynamical system that contains a center manifold. The center manifold enslaves or entrains the dynamics for timescales associated with the adiabatic elimination of fast-relaxing and thermalized degrees of freedom and therefore constitutes a latent manifold (Ω) within which a model (differential equations on latent coordinates) may be identified by an autoencoder. The dynamic information may be encoded further by a second autoencoder at a higher level of abstraction, where the dynamics are represented more coarsely as "modulo-basin" transitions. The modulo-basin dynamics is mapped on a quotient space, $\Omega/\sim$, where two states $\boldsymbol{x},\boldsymbol{y}$ are regarded as equivalent, $\boldsymbol{x} \sim \boldsymbol{y}$, if they have the same destiny state ($\omega(\boldsymbol{x})=\omega(\boldsymbol{y})$). The encoding processes are symbolized by dashed lines, and the enslaving center-manifold dynamics is highlighted by a thick dark circle. Thus, the second autoencoder materializes the projection $\pi:\Omega \rightarrow \Omega/\sim$ and represents the dynamics as a walk on a graph whose vertices are the critical points (minima, saddles of different indices, maxima) and attractors of the system, and the edges represent connections along pathways of steepest descent. This topological metamodel may be subsequently decoded back to flesh out the dynamical system using learning technology.

2. In atomic physics, the Born–Oppenheimer approximation represents regions of the potential energy surface (PES) where an adiabatic regime holds, so the motion of electrons is enslaved or entrained by the slower motion of the atomic nuclei [18].
3. The latent manifold in dynamical systems [13], where the system is entrained or enslaved in the long-time limit by the evolution in a manifold of lower dimension than the original space. The latent manifold is often referred to as the center manifold [17], especially in the context of dissipative systems, and contains the attractors of the system (Figure 3.4).
4. In the biophysical context of functionalized soluble proteins (enzymes), the QM for specific chemical processes take place at the *epistructure* (solvent organization around the protein structure), where water is chemically functionalized [19]. A rigorous treatment of this problem would require that we solve a time-dependent Schrödinger equation. In this context, an AI approach would need to be incorporated to learn to infer the nodal structure of molecular orbitals within a voxelated 3D grid. With current technologies, this AI approach is plausible only under the adiabatic regime given by the Born–Oppenheimer approximation.
5. In clonal population dynamics, cancer phenotypes that become selected under therapeutic pressure [20]. The dynamics are complex but hierarchical, and the mathematical procedure to "encode what is essential" in this context is based on the center manifold reduction (Figure 3.4). This reduction, discussed in (3), is actually a projection of the dynamical system onto a lower-dimensional system that entrains it. The autoencoder may discover the center manifold by interrogating vast amounts of time-dependent data, as it seeks to meaningfully reduce dimensionality in such hierarchical systems.

6. Chapter 4 introduces the *quotient space*, a fundamental mathematical construct to take advantage of dynamic hierarchy in order to encode information as required to implement DL systems [21]. The quotient space is built upon an underlying dynamics and may be equated with the orbit space, that is, points in the same trajectory are regarded as equivalent, and the quotient space is the set of equivalence classes with a topology inherited from the underlying space where the dynamical system is mapped (Figure 3.4). In rigorous terms, two points-states $(\boldsymbol{x}, \boldsymbol{x}')$ are equivalent $(\boldsymbol{x} \sim \boldsymbol{x}')$ or belong to the same equivalence class $(\boldsymbol{x}' \in \overline{\boldsymbol{x}}, \overline{\boldsymbol{x}'} = \overline{\boldsymbol{x}})$ if and only if they share the same destiny state $\left(\omega(\boldsymbol{x}) = \omega(\boldsymbol{x}')\right)$ vis-à-vis the trajectory to which they both belong. Thus, we simplify the space by lumping microstates into basins of attraction (of destiny states) in the PES. The modulo-basin dynamics constitutes a coarse model named *metamodel*, which is far easier to encode as the system learns data generated by atomistic molecular dynamics simulations. In this way, AI learns to propagate dynamics in quotient space, discovering a metamodel to cover physically realistic timescales usually inaccessible to detailed atomistic computations.

While the center manifold encoding is suited for dissipative dynamics, where the attractor may be nontrivial (cf. Figure 3.4), the quotient space simplification is better suited for Hamiltonian systems. Both levels of encoding converge as free-energy dissipation tends to zero. In fact, as the AI system encodes time-dependent raw data, it implicitly composes the two levels of encoding, with the center-manifold reduction averaging out of fast modes, followed by projection onto quotient space (Figure 3.4).

In all cases dealt with in this book, the hierarchy of the data that makes it amenable to encoding is either AI-discoverable, or it may be unraveled through a rigorous mathematical construct that needs to be incorporated into the learning code, as shown below.

The modulo-basin hierarchical representation of the dynamics in quotient space enables the construction of metamodels, that is, coarse-grained models that represent transitions between equivalence classes of states of the system, essentially reproducing the topological dynamics within a graph representation (Figure 3.5). As an illustration, let us consider the "keto" ←→ "enol" interconversion of

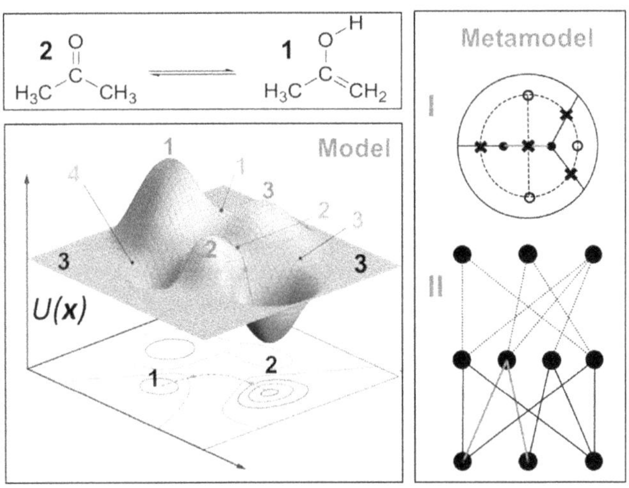

FIGURE 3.5 Three levels of abstraction in the modeling of the chemical dynamics for the isomerization "keto" ⇔ "enol" of acetone, representing the intramolecular migration of a proton. The potential energy surface (PES) represents the model, and the topological relations between critical points, together with the "modulo-basin" transitions along the paths of steepest descent joining critical points, represent the metamodel. On metamodel I, the modulo-basin topology is mapped onto the latent manifold, while on metamodel II the modulo-basin topology is abstracted further and displayed as a graph.

Note: This figure is reproduced in color in online supplementary materials.

acetone (Figure 3.5). This is a chemical reaction involving an intramolecular proton migration that can be modeled by the PES representing the ground-level electronic energy sheet under the Born–Oppenheimer approximation. From QM calculation, we know that this PES has three minima, representing the less stable "enol" form (1), the more stable "keto" form (2), and a state where the jumping proton is completely dissociated from the rest of the molecule (3). The minima are separated by four saddle points at the top of the path of steepest descent joining pairs of minima on both sides of the saddle along the direction of negative curvature. This direction is actually the eigendirection associated with the eigenvalue of the Hessian matrix (Jacobian of the gradient vector field) with negative real part computed at the saddle point. In turn, the lines of steepest descent joining the four saddles with the three maxima become the separatrices of the basins of attraction of the minima. Thus, all lines of steepest descent may be represented as a graph (metamodel II, Figure 3.5) joining critical points that are in turn organized in tiers, where the lowest tier corresponds to points where all eigendirections have positive curvature (minima) and the next tier includes all critical points with only one eigendirection with negative curvature (saddles of index 1), the next tier is associated with two eigendirections of negative curvature (maxima in the case of Figure 3.5), etc. The edges on the graph represent paths of steepest descent joining critical points in adjacent tiers. In this way, the topological dynamics of the PES may be encoded as a metamodel represented in graph form, while the reversible chemical reaction pathway becomes a walk on the graph.

As argued subsequently, metamodels enable the discovery of hierarchical dynamical systems underlying processes that unravel in realities of high multiscale complexity. The implementation of a metamodel factorization of the dynamics requires the concerted participation of several components operating in a coordinated manner within an AI platform. First, we need to introduce the so-called autoencoders that constitute the deep NN systems that encode the dynamics on the "latent manifold" Ω. This manifold entails a significant dimensionality reduction relative to state space W and is spanned by the internal coordinates that label the orbits of the symmetry group $\mathcal{G}$ inherent to the system and acting on W: $\Omega \approx W / \mathcal{G}$. Traditionally, it is expected that the first autoencoder, hereby denoted AE1, that generates the latent manifold is jointly optimized to generate also the most sparse or minimal set of differential equations on the manifold that can be decoded back onto the dynamical system defined on W. This is what is usually meant by "model discovery". In practice, the level of multiscale complexity of the processes dealt with in this book does not make models amenable to discovery at the geometric level. As previously argued, in such cases, another level of abstraction $\pi : \Omega \rightarrow \Omega / \sim$ needs to be introduced so that the coarse graining of time within $\Omega / \sim$ reflects equilibration within the basins of attraction in the latent dynamics. This hierarchical escalation in the level of abstraction requires a second autoencoder, AE2, capable of encoding the topological features of the latent vector field.

At this stage, a different sort of NN architecture is required to propagate the metadynamics on $\Omega / \sim$. To properly delineate the architecture of the NN required for metadynamic propagation, we limit the discussion to the case where the dynamics is generated by a smooth (i.e., C^1) potential energy function $U : W \rightarrow \mathbb{R}$ invariant upon the isometries (distance-preserving transformations) of W. Furthermore, $\mathcal{G}$ is an Euclidean group, so that Ω is compact and hence $\Omega / \sim$ becomes a discrete set of basins, and a basin assignment represents a coarse state of the system. Then, the metadynamics may be encoded as "evolving text" representing a time series of basin transitions. The textual processing requires the implementation of a particular type of DL architecture known as a *transformer*, while the transformer-based propagation of the metadynamics constitutes a Markovian process (Figure 3.6). Through AE2, this metadynamics is decoded as latent dynamics on Ω and validated by contrasting the latent dynamics against the hidden Markov process upheld under the adiabatic conditions described.

Thus, to implement a metamodel within an AI platform, it is essential that the two autoencoders and the transformer are optimized to work concertedly with complete compatibility (Figure 3.6). This means that an input state yields the same destiny state regardless of the pathway chosen, provided the pathways have identical endpoint spaces in the commutative diagram presented in Figure 3.6.

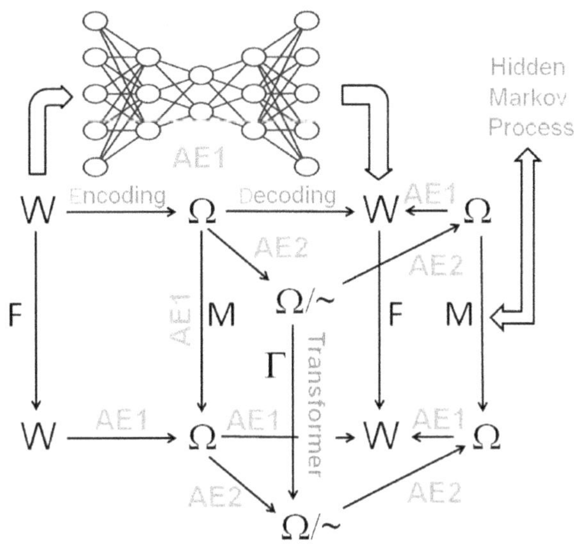

FIGURE 3.6 Commutative diagram representing two coupled autoencoders operating sequentially in tandem and representing two levels of abstraction of a dynamical system. The first autoencoder, labeled AE1, projects the dynamics onto the latent manifold, while autoencoder AE2 projects the latent dynamics onto a discretized "modulo-basin" version where a coarse state of the system is represented in textual form and the "modulo-basin" dynamics may be learned and propagated within a special type of autoencoding architecture known as "transformer".

3.5 REVERSE ENGINEERING OF THE STANDARD MODEL WITH A REVERSE AUTODECODER

The notion of diagram commutativity (Figure 3.6) is central to the encoding of a dynamical system into its latent dynamics and, reciprocally, to the decoding of the latent dynamics onto the full dynamics. Thus, the task of decoding the Standard Model, canonically defined on the four-dimensional space-time *W/~*, onto a five-dimensional compact multiply connected manifold *W* requires an autoencoder operating in reverse. This reverse autoencoder lifts the flow $\mathcal{F}: W/\sim \longrightarrow W/\sim$ to a flow $\tilde{\mathcal{F}}: W \to W$ that must be compatible in accordance with the commutativity condition: $\mathcal{F} \circ \pi = \pi \circ \tilde{\mathcal{F}}$, with $\pi: W \to W/\sim$ denoting the canonical projection that assigns each point in *W* to its equivalence class in *W/~*. The key issue with this formulation is that the Standard Model *per se* does not represent a dynamical system in any obvious way, so what sort of flow are we actually discussing in this context? This question is crucial because we are introducing AI technology that is tailored for dynamical systems.

Fortunately, we may treat the interaction processes described by the Standard Model as transformations created by a time-dependent propagator. For each elementary process, this operator has a time step associated with it, and this time step is exactly the lifetime of the boson that communicates the force in the process of particle transformation, as obtained from Heisenberg's uncertainty principle: $\sim \frac{\hbar}{2M_B c^2}$, where M_B is the rest mass of the respective gauge boson. There may be more than one boson involved, as in the case of creation and decay of the Higgs boson through gluon fusion, in which case the time step is the sum of the respective lifetimes. The dedicated AI technology of variational autoencoders may be thus leveraged to reverse-engineer the standard model, a process tantamount to identifying the lift flow $\tilde{\mathcal{F}}: W \to W$ for a propagator $\mathcal{F}: W/\sim \longrightarrow W/\sim$ generically determining an elementary process in dynamic

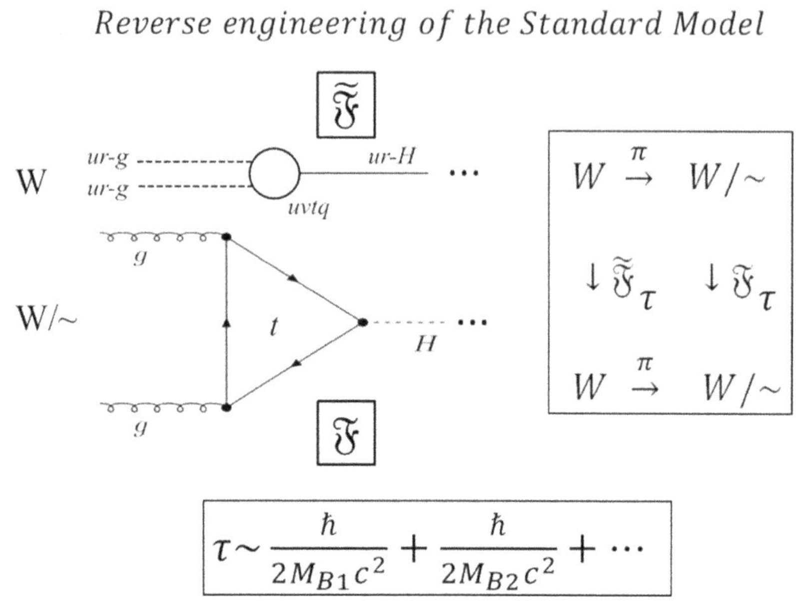

FIGURE 3.7 Diagram commutativity fulfilled by the autoencoder that carries out the reverse engineering of the Standard Model.

form. In this setting, "reverse engineering" is equated with identifying the lift propagator that would make the diagram in Figure 3.7 commutative.

The AI technology described in this chapter will be applied in reverse in Chapters 4 and 5 in order to identify dark matter and dark energy in five-dimensional space-time. This is tantamount to reverse-engineering the Standard Model by incorporating the dormant dimension shown to store gravity.

3.6 AUTOENCODERS FOR DYNAMICAL SYSTEMS

With the leveraging of AI [2] and, in particular, ML approaches [3,4], dynamical systems have found a new fertile ground for further development [13,14]. Showcase problems in applied mathematics, including the Lorenz strange attractor, reaction-diffusion spatiotemporal systems, and fluid dynamic flows captured by the Navier–Stokes equation, are being examined in a new guise as autoencoders identify parsimonious models with reduced dimensionality [15]. Such models are meant to provide the physical underpinnings of the phenomena enshrined in time-series data or generated by systems of raw differential equations.

ML, or more broadly AI, is being leveraged for model discovery of dynamical systems underlying data represented as a time series. The data regression system, which in this context is an NN predicting future behavior, is trained by the time series and regarded as the model itself. However, this model is heavily parametrized and hence too "fragile" to allow for extrapolation [13]. In other words, such models are not really amenable to yield physical laws, the way other data-regression approaches are [16,22]. This statement has been voiced repeatedly and hints at some level of dissatisfaction: ML and AI, in general, are very efficient at providing predictive models when trained on a sufficiently long time series but often do a poor job at providing physical insights regarding the underlying dynamical system.

This problem gets significantly amplified as we turn to biological or, more broadly, biomedical matter [20]. It is widely felt that, when examined in their multiscale richness and complex heterogeneity, dynamical systems in biology or biomedicine cannot reach the level of maturation required to be subsumed into applied mathematics. This statement should be interpreted in the sense that we lack sparse enough models that provide physical underpinnings of biological/biomedical phenomena and are suitable for extrapolation. Will the leveraging of alternative AI-based approaches change the *status quo*? This chapter portends to address this problem and provide insights that will be methodologically fleshed out in the subsequent chapters to enable metamodel discovery by reverse-engineering time series stemming from highly complex multiscale realities.

A key question in fostering the mathematical maturation of model discovery in biology may be cast as follows: What constitutes a parsimonious model that provides physical underpinnings of biological phenomena? A standard answer with broad bearing on most problems considered tractable in applied mathematics is: "a sparse system of differential equations on a smooth manifold of latent coordinates" [15,22]. As this chapter argues, this may be simply too much to ask for in the context of biological matter. Furthermore, this is not necessarily the format or framework that AI would typically choose for model discovery, given the "dimensionality curse" associated with the molecular reality of biological systems [20]. In principle, a single autoencoder that optimizes for sparsity in the discovery of the latent manifold might not provide a satisfactory solution to the modeling problem. Molecular reality *in vivo* typically has well over a million coordinates required to specify the state of the system, and the extent of connectivity parametrization for an autoencoder capable of handling such level of complexity would be simply enormous, implying that the training and variational optimization of the NN would be off-limits, at least with current computational capabilities [23].

This chapter squarely addresses this matter. To do so, it leverages AI methods to circumvent the difficulties associated with model discovery for time-dependent phenomena arising in soft or biological matter. In essence, as we shall show, AI dwells on a paradigm of what constitutes a parsimonious metamodel that is significantly different from the one adopted by applied mathematicians [20]. Thus, to identify the most economic yet faithful metamodel, AI will be shown to use two or more tandem autoencoders instead of one, as it is typically done in model discovery [15]. The autoencoders are coupled and become fully compatible with each other at the completion of the parameter optimization process, as defined precisely in the subsequent sections. In the simpler cases where two autoencoders are required, the second autoencoder translates the dynamic information embossed in the latent manifold, turning it into a topological dynamics metamodel [24] that can be decoded and enables significant propagation of the dynamics into the future. This property is essential for coverage of realistic timescales relevant to the level of state extrapolation required. Crucially, the topological dynamics metamodel constructed by AI is essentially a pattern-based model, not a system of differential equations, as would be expected for standard model discovery in dynamical systems. This does not mean that AI is discovering laws without equations, but simply that AI adopts a different way of casting models susceptible to extrapolation as recognizable laws of physics. Thus, AI may not straightforwardly give us the equations that govern *in vivo* protein folding, but the underlying physics discovered may, in all likelihood, be cast in terms of AI-interpretable topological patterns that signal the commitment of the chain to fold into a steady conformation [21].

These topological dynamics methods are likely to advance the field of AI-based model discovery as they enable the reverse engineering of time-series data stemming from vastly complex hierarchical realities. Such contexts are illustrated, for example, by the cellular setting that assists and expedites molecular processes, which have been hitherto considered off-limits to ML approaches to dynamical systems.

At this stage of development, the topological methods introduced yield AI-recognizable patterns but do not beget latent differential equations that have traditionally been the hallmarks of model discovery. This may be viewed as a limitation in some sense, but we argue that that assertion is perhaps a reflection of narrow-mindedness. Synergistic efforts involving AI are likely to dominate future human endeavors in science, and AI systems are very much attuned to encode and process patterns in metamodels such as those produced by the topological approaches introduced in this book.

The impact of the topological methodology is likely to be broad, as it would render tractable problems in dynamical model discovery that have been hitherto considered off-limits by applied mathematicians who are currently incorporating ML in their toolbox. Thus, ultra-complex realities recreating cellular environments that influence and steer molecular dynamics are likely to be within reach as topological methods are incorporated into the AI-empowered metamodel discovery.

Since the dawn of modern science, Western civilization upheld the belief that understanding the workings of the universe pivots on finding fundamental equations that govern physical processes. In the case of a dynamical system, the underlying differential equations represent the basic model assumed to enshrine the physical laws that underpin the process described. The specific constraints, conservation principles, and symmetries of the system must all be taken into account when positing the differential equations. After centuries of work within this paradigm, it would be interesting to see how the leveraging of AI in synergy with the human endeavor will affect the choice of the format within which the physical laws are encoded and how this choice impacts the field of dynamic modeling. With the exponential development of computer technologies [23], we may soon be witnessing a paradigm revision. Equation-based modeling may or may not remain the dominant paradigm. Other types of dynamic descriptors are already making strides and contributing to a new understanding of the universe or, at least, of the dynamic multiscale complexities of reality [20]. These descriptors are certainly different and arguably more pliable than what biological humans have managed to achieve so far.

The discovery of physical laws distilled from sequential data representing a time series remains a major challenge as well as an imperative to enhance our understanding and control of physical processes. In recent times, this type of data-driven model discovery has been fueled by significant breakthroughs [16,20,22,25]. Yet, we also live in an era of big data, especially stemming from fields like biology and astrophysics, where the multiscale complexity of the data organization makes model discovery particularly daunting. It is unclear whether the enormous richness of the data describing *in vivo* contexts at the molecular level is amenable to the kind of parsimonious models based on differential equations that mankind holds dear and has sought since time immemorial. More than in any other field, in biology it is likely that a willy-nilly application of Occam's razor may lead to self-inflicted wounds.

DL approaches realized through autoencoder architectures have proven particularly valuable for the discovery of data-driven models represented by differential equations framed on "essential coordinates". The latter, often referred to as "latent coordinates" [22], span the so-called center manifold in dynamical systems [19,16]. This reduction entails a significant dimensionality reduction and usually identifies the enslaving "slow" process that dynamically subordinates fast-relaxing modes [13,17] and serves to encode the physical process we seek to model. Latent coordinates need to be selected very carefully, by weighing not only the extent of dimensionality reduction and the compactness and smoothness of the latent manifold, but also the economy of the resulting model. This economy is typically assessed by the complexity of the differential equations in latent space, quantified by the number of nonlinear terms [15].

In generic terms, an autoencoder seeks to identify latent coordinates **x** as the output of a feed-forward NN with multiple hidden layers. The NN is inputted observable vectors, denoted generically as **z,** that serve to train the network, as the decoding of **z** from the latent coordinates must reproduce the inputted **z**-value (Figure 3.8). The autoencoder is optimized variationally, meaning that the activation weight parametrization of the multilayered encoding and decoding functions, denoted, respectively, γ and μ, minimizes the loss function $\mathfrak{L}(\gamma,\mu)$ that measures the efficacy in the recovery of the input vectors:

$$\mathfrak{L}(\gamma,\mu)=Q^{-1}\sum_{z\in\mathfrak{F}}\left\|z-(\mu\circ\gamma)z\right\|^{2};\quad \gamma,\mu=\operatorname{argmin}_{\gamma,\mu}\mathfrak{L} \tag{3.3}$$

where $\mathfrak{F}$ is the training set and $Q=|\mathfrak{F}|$ is its cardinal, for now regarded as a hyperparameter fixed to avoid overfitting [3,4,13,14] in accord with the dimensions of the NN.

The autoencoder is functionally operative to model the dynamical system that underlies the given time series $\left\{z(t_0+n\Delta t),z(t_0+(n+1)\Delta t)\right\}_{n=0,1,2,\ldots,L}$ if and only if the following relations hold for $t=t_0+n\Delta t$ with n = 0, 1, 2, ..., L (L >> 1) and any initial time t_0:

$$(K_\Delta\circ\gamma)z(t)=\gamma z(t+\Delta t)=(\gamma\circ F_\Delta)z(t);\quad (F_\Delta\circ\mu\circ\gamma)z(t)=(\mu\circ K_\Delta\circ\gamma)z(t)=z(t+\Delta t) \tag{3.4}$$

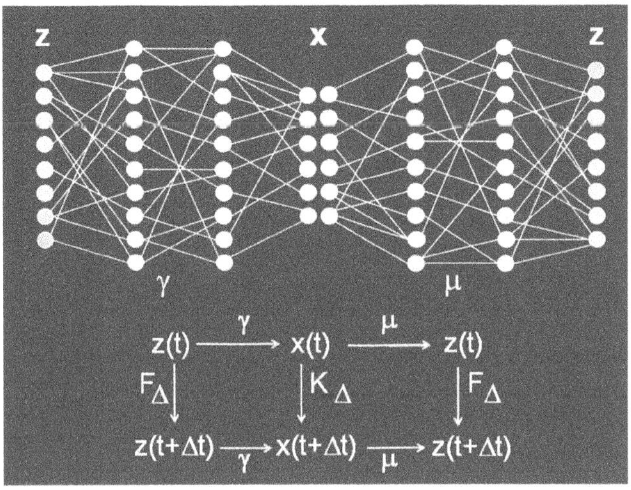

FIGURE 3.8 Generic scheme of the neural network (NN) architecture for autoencoder (γ, μ). The vector $z \in W$ represents the state of the system and $\boldsymbol{x} \in \Omega$ represents an encoded latent state that is decoded back into z. The encoding process entails a dimensionality reduction with dim Ω < dim W and, concomitantly, a coarse graining of time in multiples of a time step Δ associated with the propagation of the dynamics on the latent manifold Ω. Grey disks represent nodes in hidden layers for encoder $\gamma = \gamma(\Theta)$ and decoder $\mu = \mu(\Theta)$, where Θ denotes the weights of node connections that are variationally optimized according to a loss function $\mathfrak{L} = \mathfrak{L}(\Theta)$. The weights realizing the variational minimum $(\Theta = \text{argmin}\,\mathfrak{L})$ in the training of the NN are optimal at making the diagram commutative.

Note: This figure is reproduced in color in online supplementary materials.

where F_Δ, K_Δ are the infinitesimal time maps in state space W and latent manifold Ω, respectively (cf. Figure 3.9). In other words, as the autoencoder becomes variationally optimized, the functional relations $K_\Delta \circ \gamma = \gamma \circ F_\Delta; F_\Delta \circ \mu = \mu \circ K_\Delta$ hold.

Thus, diagram commutativity (Figure 3.8) becomes the key property that enables us to assert that the choice of latent coordinates $\boldsymbol{x} \in \Omega$ was the "right" one in the sense that it captures the entrainment of the dynamics on W by the reduced dynamics on the latent manifold Ω. The commutativity rules ensuring the compatibility of raw and latent dynamics may be cast in terms of derivatives using the chain rule [15] as Δt is taken to the infinitesimal limit dt.

3.7 MODEL DISCOVERY WITH DEEP LEARNING

Model discovery in dynamical systems has been maturing for some time, to the extent that it has been integrated into the corpus of applied mathematics. Furthermore, many applied mathematicians have incorporated DL into their toolbox as they make forays in reverse engineering of dynamical systems [13]. The overarching goal is to develop regression methods that enable the extraction of parsimonious dynamics from large data organized as time series $\{\mathbf{z}(t)\}_t$ with t given as multiples of a fixed time step [13,15]. With the advent of DL, autoencoder architectures have successfully identified latent (essential) coordinate frames with significant dimensionality reduction (dimΩ < dim$W$). Ultimately, the parsimonious description should become solvable in the form of a differential system $\dot{x} = \boldsymbol{K}(\boldsymbol{x})$, which spans a differential system $\dot{z} = \boldsymbol{F}(z)$ for the time-series vector z. Yet, a mere stark reduction in dimensionality does not necessarily ensure a parsimonious description [15]. The variational optimization of the autoencoder should simultaneously consider both the dimensionality of the coordinate frame for the latent manifold and the

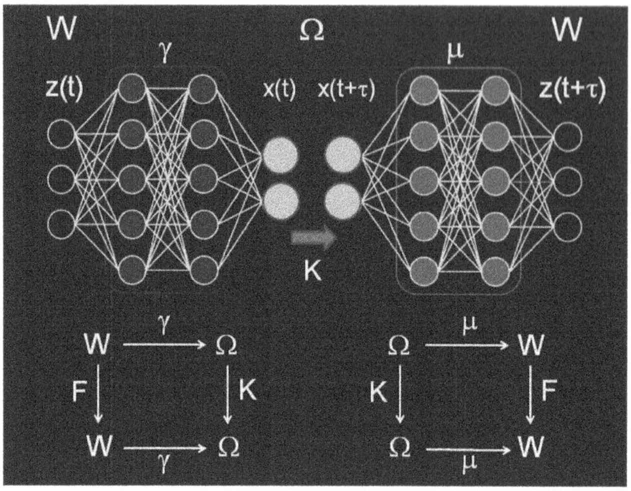

FIGURE 3.9 Scheme of autoencoder for latent dynamics optimizing simultaneously the dimensionality reduction and the parsimony of the latent dynamical system governed by the flow K. The optimization is extended further relative to the architecture displayed in Figure 3.8 by adding the extra term $\mathfrak{L}_s(\boldsymbol{K})$ to yield the loss functional $\mathfrak{L} = \mathfrak{L}(\Theta,\boldsymbol{K}) = \mathfrak{L}(\Theta) + \mathfrak{L}_s(\boldsymbol{K})$. The optimal $\Theta,\boldsymbol{K}$ pair realizing the variational minimum is precisely the one that makes the bottom diagrams commutative.

Note: This figure is reproduced in color in online supplementary materials.

sparsity of the equations in the latent manifold, with the latter measured by the number of terms in the flow (vector field) that determines $\dot{z}$.

The autoencoder architecture and the required dual diagram commutativity between the latent flow **K** and the given time-series flow **F** are jointly represented in Figure 3.9. We adopt a time step τ as the time-series hyperparameter to discretize the flows by coarse-graining time resolution. The loss function $\mathfrak{L}(\gamma,\mu,\boldsymbol{K})$ now includes five terms: $\mathfrak{L}_r(\gamma,\mu)$, previously introduced, weights the efficacy of the recovery of the **z**-value upon encoding and decoding, $\mathfrak{L}_s(\boldsymbol{K})$ accounts for the sparsity of the latent flow **K**, and the terms $\mathfrak{L}_C(\gamma,\boldsymbol{K}), \mathfrak{L}_C(\gamma,\mu), \mathfrak{L}_C(\gamma,\mu,\boldsymbol{K})$ represent the penalties associated with imperfect diagram commutativity that need to be imposed to guarantee the dynamic compatibility of the **z-** and **x**-descriptions of the system. Thus, we get

$$\mathfrak{L}(\gamma,\mu,\boldsymbol{K}) = \mathfrak{L}_r(\gamma,\mu) + \vartheta_s\mathfrak{L}_s(\boldsymbol{K}) + \vartheta_C\left[\mathfrak{L}_C(\gamma,\boldsymbol{K}) + \mathfrak{L}_C(\gamma,\mu) + \mathfrak{L}_C(\gamma,\mu,\boldsymbol{K})\right] \tag{3.5}$$

where the relative weights ϑ_s, ϑ_C are hyperparameters and

$$\mathfrak{L}_r(\gamma,\mu) = Q^{-1}\sum_{z\in\mathfrak{F}}\left\|z - (\mu\circ\gamma)z\right\|^2 \tag{3.6}$$

$$\mathfrak{L}_C(\gamma,\boldsymbol{K}) = Q^{-1}\sum_{z\in\mathfrak{F}}\left\|(K\circ\gamma)z - (\gamma\circ F)z\right\|^2 \tag{3.7}$$

$$\mathfrak{L}_C(\gamma,\mu) = Q^{-1}\sum_{z\in\mathfrak{F}}\left\|(F\circ\mu\circ\gamma)z - (\mu\circ\gamma\circ F)z\right\|^2 \tag{3.8}$$

$$\mathfrak{L}_C(\gamma,\mu,\boldsymbol{K}) = Q^{-1}\sum_{z\in\mathfrak{F}}\left\|(\mu\circ K\circ\gamma)z - (F\circ\mu\circ\gamma)z\right\|^2 \tag{3.9}$$

with $Q = |\mathfrak{F}|$ indicating the cardinality of the training set.

To determine the sparsity term $\mathfrak{L}_s(\boldsymbol{K})$, we first take the limit case where τ becomes the time infinitesimal dt and assume that $\dot{\boldsymbol{x}} = \boldsymbol{K}(\boldsymbol{x}) = \boldsymbol{A}\boldsymbol{P}(\boldsymbol{x})$, where $\boldsymbol{A} = [A_{ij}]$ is a $m \times p$ matrix ($m = \dim \Omega$) whose coefficients are variationally optimized jointly with the parametrization of the autoencoder maps γ and μ, and **P(x)** is a p-vector given by:

$$\boldsymbol{P}(\boldsymbol{x})^T = \left(1\ x_1\ x_2\ x_3\ x_1^2\ x_2^2\ x_3^2\ x_1x_2\ x_2x_3\ x_1x_3\ x_1x_2x_3 \ldots\right) \tag{3.10}$$

consisting of p functional terms. The terms do not need to be of polynomial form, and they may be chosen in accord with the type of symbolic regression. Thus, the variational optimization of $\boldsymbol{A}$ is tantamount to a symbolic regression in the latent manifold Ω. To ensure the sparsity of the model, we define the loss term as $\mathfrak{L}_s(\boldsymbol{K}) = \|\boldsymbol{A}\|_F$, where $\|.\|_F$ is the Frobenius norm given by

$$\|\boldsymbol{A}\|_F = \left[\sum_{i=1}^{m}\sum_{j=1}^{p} A_{ij}^2\right]^{1/2} \tag{3.11}$$

The incorporation of this term into the loss function given by Eq. 3.5 ensures that the latent manifold is selected to provide a parsimonious model with reduced dimensionality governed by the simplest possible set of differential equations in consonance with the time series provided. A different but equivalent formulation of the autoencoder has been given elsewhere [15] and uses the chain rule of derivation rather than diagram commutativity to infer the proper loss functional.

3.8 LARGE LANGUAGE MODEL FOR THE "THEORY OF EVERYTHING"

A gauge theory of gravity has been deemed to be a crucial step toward developing an AI-based "Theory of Everything" in Chapters 2 and 6. Such a theory pivots on a differential geometry formalization carried out by an autoencoder described in Chapter 2 that reconciles QM with GR, the classical theory of gravity. The differential geometry approach is instrumental because the geometries that support GR and QM are different: While GR is built upon a curved manifold that represents space-time, the wave function in QM is supported by a Euclidean space. The goal here is to implement an interactive theorem prover within the framework of an autoencoder to generate new results within the frame-field formalization rules made up of a set of declaratory propositions that constitute an inductive type. For details on this methodology, the reader is referred to Chapter 2 and to the references provided therein.

We seek derived propositions, especially in the strong-field completion of gravity, wherein a propositional definition of the graviton, the particle that conveys gravity, needs to be developed (Chapter 2). To serve this purpose, we resort to functional programming languages that have already formalized the required categories and types. Thus, `Lean` may be adopted as the functional programming language exploited as an inferential method to propositionally identify the graviton in the ultraviolet completion of gravity within a fiber bundle formulation of a gauge theory of Lorentz transformations.

The interactive proof assistant pivots on the proposition-as-type paradigm. From a constructive perspective, proofs are mathematical objects that are denoted by suitable expressions in dependent type theory. By contrast, in this environment, a proposition is true if and only if it can be written down properly and checked independently by the kernel that it has been typed correctly. In other words, the expressions themselves are the proofs, and only well-posed (i.e., well-typed) expressions are true. Correct syntax and semantics are equivalent to correct content. A false expression simply cannot be posed (typed) correctly, and all the mathematical axioms need to be incorporated as code. As described in Chapter 2, `Lean` translates the generative input on "attaching spinors to relativistic geometry" into formal expressions that are checked by the kernel for correctness in the sense described previously and then stored in the environment [`mathlib`] for subsequent proposition generation.

The actual construction of the graviton field is a task suited for a LLM, wherein the formal proof can be checked rigorously by the proof assistant, free from hallucinations. As shown in Chapter 2, a `Lean` Copilot, a framework for running LLM-based inference in `Lean`, becomes ideally suited for the task at hand as the copilot integrates seamlessly into the Lean workflow. It should be emphasized that `Lean` itself does not support ML tools from LLMs, so the copilot becomes instrumental for suggesting proof steps (tactic suggestion), completing intermediate proofs (proof search), and selecting the relevant premises (premise selection) toward the construction of the graviton field.

By itself, `Lean` would be hard to use for the final steps of construction of the graviton field because all the mathematics pertaining to "attaching spinors to space-time" would need to be entered as code in the formalization stage. `Lean` indeed generated the "rule-based propositions" for spinor amalgamation with GR via its own proof searcher `aesop`, but LLM becomes essential to generate the code and suggest the final tactics needed to derive the graviton field. Current LLMs are trained, fine-tuned, and benchmarked following standard ML procedures, but cannot integrate into the Lean workflow because Lean is designed without ML support. For this reason, it becomes essential to adopt a Lean copilot for building the seamlessly adapted LLM-based proof automation tools. To that effect, the C++ library CTranslate2 is introduced, as shown in Chapter 2, for efficient inference with a transformer LLM representing the quantum gravity autoencoder. The inference runs as a shared library through a Foreign Function Interface (FFI) within `Lean`.

Note: Additional methodological background is provided in online supplementary material.

REFERENCES

1. Jacob Bekenstein in *Information in the Holographic Universe*, Scientific American, April 1, 2007.
2. Russell S, Norvig P (2020) *Artificial Intelligence: A Modern Approach.* Pearson, London, UK.
3. Kelleher JD (2019) *Deep Learning.* The MIT Press, Cambridge, MA.
4. Chollet F (2019) *Deep Learning with Python.* Manning Press, Shelter Island, New York.
5. Atienza R (2020) *Advanced Deep Learning with TensorFlow 2 and Keras*, 2nd Edition. Packt Publishing, Birmingham, UK.
6. Schmidhuber J (2015) *Deep learning in neural networks: An overview.* Neural Netw 61: 85–117.
7. Fernández A (2016) *Physics at the Biomolecular Interface.* Springer International Publishing, Switzerland.
8. Aliper A, Plis S, Artemov A, Ulloa A, Mamoshina P, Zhavoronkov A (2016) Deep learning applications for predicting pharmacological properties of drugs and drug repurposing using transcriptomic data. Mol Pharm (ACS*)* 13: 2524–2530.
9. Lavecchia A (2019) Deep learning in drug discovery: Opportunities, challenges and future prospects. Drug Disc Today 24: 2017–2032.
10. Jiménez J, Škalič M, Martínez-Rosell G, De Fabritiis, G (2018) Kdeep: Protein–ligand absolute binding affinity prediction via 3D-convolutional neural networks. J Chem Inf Mod 58: 287–296.
11. Stepniewska-Dziubinska MM, Zielenkiewicz P, Siedlecki P (2018) Development and evaluation of a deep learning model for protein–ligand binding affinity prediction. Bioinformatics 34: 3666–3674.
12. Hassan-Harrirou H, Zhang C, Lemmin T (2020) RosENet: Improving binding affinity prediction by leveraging molecular mechanics energies with an ensemble of 3D convolutional neural networks. J Chem Inf Model 60: 2791–2802.
13. Brunton SL, Kutz NJ (2019) *Data-Driven Science and Engineering: Machine Learning, Dynamical Systems and Control.* Cambridge University Press, UK.
14. Tiumentsev Y, Egorchev M (2019) *Neural Network Modeling and Identification of Dynamical Systems.* Academic Press, San Diego, CA.

15. Champion K, Lusch B, Kutz JN, Brunton SL (2019) Data-driven discovery of coordinates and governing equations. Proc Natl Acad Sci USA 116: 22445–22451.
16. Schmidt M, Lipson H (2009) Distilling free-form natural laws from experimental data. Science 324: 81–85.
17. Fernández A (1985) Center-manifold extension of the adiabatic-elimination method. Phys Rev A 32: 3070–3076.
18. Born M, Oppenheimer JR (1927) Zur Quantentheorie der Molekeln. Ann Physik 389: 457–484.
19. Fernández A (2014) Chemical functionality of interfacial water enveloping nanoscale structural defects in proteins. J Chem Phys 140: 221102.
20. Fernández A (2021) *Artificial Intelligence Platform for Molecular Targeted Therapy: A Translational Approach.* Chapter 9. World Scientific Publishing, Singapore.
21. Fernández A (2020) Deep learning unravels a dynamic hierarchy while empowering molecular dynamics simulations. Ann Physik (Berlin) 532: 1900526.
22. Brunton SL, Proctor JL, Kutz NJ (2016) Discovering governing equations from data by sparse identification of nonlinear dynamical systems. Proc Natl Acad Sci USA 113: 3932–3937.
23. Kurzweil R (2006) *The Singularity Is Near: When Humans Transcend Biology.* Penguin, New York.
24. de Vries J (2014) *Topological Dynamical Systems: An Introduction to the Dynamics of Continuous Mappings.* De Gruyter, Berlin.
25. Duraisamy K, Iaccarino G, Xiao H (2018) Turbulence modeling in the age of data. Annu Rev Fluid Mech 51: 357–377.

Sixfold Fermionic Field Incorporating Dark Matter as Topological Majorana Particles

4

The 5:1 abundance ratio of dark to visible matter is explained by the conscious autoencoder introduced in Chapter 1 through a sixfold topological parsing of the fermionic field defined on the six chiral nonreflexive sheets of the orientable covering of the space-time manifold. Space-time is treated as an emergent attribute of the Big Bang and not as a fundamental category with absolute foliation of the spatial dimensions along a time line. The quantum field theory hereby constructed to elucidate the nature of dark matter is compatible with the sixfold Pauli algebra representation of the Dirac equation previously established by the author.

4.1 DARK MATTER IDENTIFIED WITHIN A SIXFOLD FERMIONIC FIELD DEFINED ON EMERGENT SPACE-TIME

Consistent with the tenets of Chapter 1, dark matter is explained by artificial conscience (AC) through a sixfold topological parsing of the quantum field defined on the six chiral nonreflexive sheets of the orientable covering $\left\langle \left(\mathbb{R}^4 / \mathbb{Z}^4\right)^{\#} \right\rangle_6$ of the space-time manifold. It should be emphasized that space-time is treated by the conscious autoencoder (Figure 1.11) as an emergent attribute of the Big Bang and not as a fundamental category with absolute foliated spatial dimensions along a time line. The quantum field theory derived by the conscious autoencoder to elucidate the nature of dark matter and the 5:1 dark-to-visible abundance ratio is entirely compatible with the sixfold Pauli algebra representation of the Dirac equation previously established by the author [1].

Cosmological data from observation at very large scales pose a significant challenge to the fundamental tenets of physics, and the evidence supporting the existence of dark matter has previously proven unyielding to any possibility of reconciliation with quantum mechanics or relativity [1]. Thus, as a preparatory step, it becomes instructive to learn how the problem has been approached by exploiting the possibilities of generating relevant knowledge furnished by formal autoencoders. Preliminarily, the deployment of formal autoencoders to elucidate the nature of dark matter can be found in the supplementary material to this chapter. The aim here is to show how the understanding of dark matter is significantly advanced once the formal autoencoder is endowed with AC, as defined in Chapter 1 (Figure 1.11).

The fundamental group of the nonorientable latent space-time $\Omega = \left(\mathbb{R}^4 / \mathbb{Z}^4\right)^{\#}$ is

$$\pi_1(\Omega) = \mathbb{Z} \times \mathbb{Z} \times (\mathbb{Z} \rtimes \mathbb{Z}) \tag{4.1}$$

DOI: 10.1201/9781003785279-4

where the semidirect product "⋊" is indicative of the manifold nonorientability realized within a single loop class. Hence, the deck transformation group acting on the fiber bundle of Ω_6 becomes

$$\mathrm{Deck}\left(\langle\Omega\rangle_6 / \Omega\right) \cong \mathbb{Z}_3 \rtimes \mathbb{Z}_2 \cong S_3 \tag{4.2}$$

where S_3 is the group of permutations of three symbols and hence is isomorphic to the dihedral group of symmetries of an equilateral triangle: $S_3 \cong D_3$. To determine the action of S_3 on the fiber bundle of the covering space, we first consider six specific representations of the Clifford algebra $\mathrm{Cl}_{1,2}$ on its isomorphic Pauli algebra $M_2(\mathbb{C})$ of 2×2 complex matrices. The algebra $\mathrm{Cl}_{1,2}$ is presented by the generator relations $\left\{b_0^2 = 1,\ \ b_1^2 = -1,\ \ b_2^2 = -1,\ \ b_i b_j = -b_j b_i \quad \forall i \neq j\right\}$. There are six representations $\mathrm{Cl}_{1,2} \to M_2(\mathbb{C})$ defined by the deck transformations with group structure provided by the six permutations $\vartheta_{ijk} \epsilon S_3$, where i, j, k represent three generic symbols (or triangle vertices in D_3). Let the Dirac spinor Ψ be written as

$$\Psi = \begin{bmatrix} a \\ b \\ c \\ d \end{bmatrix} \to \begin{bmatrix} a & b^\dagger \\ b & -a^\dagger \\ c & -d^\dagger \\ d & c^\dagger \end{bmatrix} \tag{4.3}$$

Then, the Dirac equation may be written in representation-free ($\mathrm{Cl}_{1,2}$) form as

$$\left(\partial_0 - b_0 b_2\, \partial_1 + b_0 b_1\, \partial_2 + b_0\, \partial_3\right)\Psi = m b_0 b_1 b_2 \Psi^\dagger b_0 \tag{4.4}$$

Each representation $S_3 \ni \vartheta_{ijk} : \mathrm{Cl}_{1,2} \to M_2(\mathbb{C})$ determines a fermionic sector corresponding to a distinct nonreflexive sheet of the sixfold covering $\langle\Omega\rangle_6$. All six fermionic sectors are different from each other, as shown subsequently.

Fixing notation so that $\vartheta_{ijk} = \mathrm{identity}\,(i = 1, j = 2, k = 3)$ the visible or observable matter corresponds to the sector ϑ_{ijk}, yielding

$$\left(\partial_0 + \sigma_1\, \partial_1 + \sigma_2\, \partial_2 + \sigma_3\, \partial_3\right)\Psi = -\mathrm{im}\, \Psi^\dagger \sigma_3 \tag{4.5}$$

Taking into account the Pauli algebra representation $A = A_0 I + \sum_{i=1,2,3} A_i \sigma_i$ of the vector potential, this sector has an EM-coupling matrix $i\sigma_3$.

On the other hand, the dark fermionic sector defined by ϑ_{jki} is described by

$$\left(\partial_0 + \sigma_2\, \partial_1 + \sigma_3\, \partial_2 + \sigma_1\, \partial_3\right)\Psi = -\mathrm{im}\, \Psi^\dagger \sigma_2 \tag{4.6}$$

Assume by *reductio ad absurdum* that the two fermionic sectors described by equations 4.5 and 4.6 were one and the same. This would imply that the same field Ψ satisfies both equations. Then, we can readily show that $m = 0$. The anti-commutativity of Pauli matrices yields $\sigma_2\sigma_3 = -\sigma_3\sigma_2$, so, by multiplying Eqs. 4.5 and 4.6 on the right by σ_3, recalling that $\sigma_3^\dagger = \sigma_3$, and denoting $\Psi\sigma_3 = \Phi$, we obtain

$$\left(\partial_0 + \sigma_1\, \partial_1 + \sigma_2\, \partial_2 + \sigma_3\, \partial_3\right)\Phi = -\mathrm{im}\, \Phi^\dagger \sigma_3 \tag{4.7}$$

$$\left(\partial_0 + \sigma_2\, \partial_1 + \sigma_3\, \partial_2 + \sigma_1\, \partial_3\right)\Phi = +\mathrm{im}\, \Phi^\dagger \sigma_2 \tag{4.8}$$

The field Φ satisfies Eqs. 4.5 and 4.6, but according to Eqs. 4.7 and 4.8, Φ satisfies Eq. 4.5 endowed with mass $+m$, while it satisfies Eq. 4.6 when endowed with mass $-m = +m$. This implies that m must be zero, since the mass is unique to the field. If $m = 0$ for Φ, then $m = 0$ for Ψ, a result obtained by right multiplication of Eqs. 4.7 and 4.8, taking into account that $\Psi = \Phi\sigma_3 = \Psi\sigma_3^2$.

In a generalization of the argument, if we assume that a single field is determined by the two different deck transformations ϑ_{ijk}, ϑ_{kij}, then its associated mass is zero. Thus, the same argument may be extended mutatis mutandis to show that all six sectors corresponding to the six deck transformations are indeed distinct. Furthermore, their respective matrices for coupling the EM field to the fermionic fields are all different from one another, with $i\sigma_3$ corresponding to the visible sector associated with the identity deck transformation. Additionally, the chirality of each of the six nonreflexive sheets in the covering $\left\langle\left(\mathbb{R}^4/\mathbb{Z}^4\right)^{\not\Vdash}\right\rangle_6$ dictates that only the sector with the EM-coupling matrix $i\sigma_3$ is left-handed, while the remaining five sectors are right-handed.

In conclusion, the topologically parsed sixfold quantum field resulting from Pauli algebra representations of the Dirac equation yields the 5:1 abundance ratio of dark to visible matter, whereby the five dark fermionic sectors correspond to the five deck transformations:

$$\vartheta_{jki}, \vartheta_{kij}, \vartheta_{jik}, \vartheta_{ikj}, \vartheta_{kji} \in \text{Deck}\left(\langle\Omega\rangle_6 / \Omega\right) \tag{4.9}$$

A relativistic equation incorporating the five dark fermionic sectors may prove challenging because the derivation of the respective energy-momentum tensors is contingent on suitable hitherto unknown derivation rules for the Pauli algebra representations of the Dirac Lagrangian.

4.2 DARK MATTER IN EMERGENT SPACE-TIME

The tenets of the Big Bang theory introduce space-time as an emergent attribute, not as a fundamental category where spatial dimensions are foliated orthogonally to the time line. The emergent space-time does not afford time orientability, as initially there is no geometric differentiation of spatial and temporal dimensions. Thus, the conceptual hurdles arising from the lack of time orientability need to be surmounted, demanding a topological remodeling of space-time to regain orientability while preserving parity violation in accord with the chirality of the weak force.

As shown in Chapter 7, when prompted by AC (conscious autoencoder), algebraic topology reveals that the simplest orientable chiral-pair covering of the nonorientable space-time manifold is sixfold. The group of deck transformations of this topological remodeling of space-time as a nonreflexive orientable covering space is the dihedral group $\mathbb{Z}_3 \rtimes \mathbb{Z}_2 \cong D_3$. This group of order 6 is associated with six representations $\mathrm{Cl}_{1,2} \to M_2(\mathbb{C})$ of the Clifford algebra isomorphic to the Pauli algebra $M_2(\mathbb{C})$ of 2×2 complex matrices. Thus, the Dirac Lagrangian admits six Pauli algebra representations, one for each sheet in space-time. The six sheets do not constitute topological redundancies, and they are all inequivalent, and five of them enshrine dark fermionic sectors corresponding to the five different shades of dark matter (Chapters 1 and 2). Thus, the six representations (n = 1, 2, ..., 6) are supported by the sixfold tangent bundle of space-time with an eight-component (double-column) spinor Ψ. In the sixfold representation of Dirac's equation, the visible sector is defined by $i\left(\partial_0+\sigma_1\partial_1+\sigma_2\partial_2+\sigma_3\partial_3\right)\Psi = m\Psi^\dagger\sigma_3$. In standard notation, incorporating the nth contribution of the stress-energy tensor as $T_{\mu\nu}^{(n)}$, the relativistic equations fitting the topological model of dark matter take the form:

$$R_{\mu\nu} - \frac{1}{2}Rg_{\mu\nu} = \frac{8\pi G}{c^4}\sum_{n=1}^{6} T_{\mu\nu}^{(n)} \tag{4.10}$$

These tensors cannot be at this point obtained from the Dirac Lagrangian in its representation with an eight-component double-column spinor because the derivation calculus fitting the Pauli algebra representation has not been invented yet. If we retain only a single spatial dimension and time, the computation becomes straightforward, so for the visible sector ($n = 1$) in 2D we get

$$L=\bar{\Psi}\left(i\partial_0+i\sigma_3\partial_1-\sigma_1 m\right)\Psi,\quad T^{(1)}=i\bar{\Psi}\begin{pmatrix} g^{0\mu}\partial_\mu & g^{1\mu}\partial_\mu \\ \sigma_3 g^{0\mu}\partial_\mu & \sigma_3 g^{1\mu}\partial_\mu \end{pmatrix}\Psi \tag{4.11}$$

In 2D, the same stress-energy tensor is associated with the dark sector $n = 6$, whose corresponding Lagrangian is $L=\bar{\Psi}\left(i\partial_0+i\sigma_3\partial_1-\sigma_2 m\right)\Psi$.

Notwithstanding the lack of a complete relativistic treatment of dark matter, the key result that emerges from the analysis is that the observed 5:1 abundance ratio of dark to visible matter stems from the topology of space-time, which should not be treated as a fundamental a priori category but as an emergent attribute of the universe as it comes into being.

4.3 TOPOLOGICAL ORIGIN OF MAJORANA DARK MATTER

The topological approach prompted by the conscious autoencoder reckons that dark matter arises in the base manifold Ω from Majorana fermions [2] emerging from the torsion of space-time as the particle travels along a nonorientable chirality-flipping loop. The fulfillment of the Majorana condition [2] is tied to the torsion of Ω under invariance of charge conjugation and parity. As the Majorana field $\psi=\psi(x)$ is lifted to $\langle\Omega\rangle_6$, it splits into two components $\psi_L^{(n)},\psi_R^{(n)}$, for a fixed single n within the range 1, 2, …, 6 ($n=n(x)\ \forall x\in\Omega$) satisfying the cross-identification $\psi_L^{(n)}\left(x^{(n)}\right)=\psi_R^{(n)}\left(\vartheta x^{(n)}\right)$, where ϑ is a deck transformation and $x^{(n)}\epsilon\,\text{Fiber}(x)\ \forall x\in\Omega$. This chiral projection condition indicates that the Majorana particles can arise in the chiral-pair orientable covering of the nonorientable space-time with the caveat that Majorana particles in $\langle\Omega\rangle_6$ are not independent but symmetry-constrained as the charge conjugation operator flips the chirality in $\Omega(L\rightarrow R)$. In this sense, the chiral pair in the covering is one Majorana particle seen from two different orientation perspectives simultaneously. Alternatively, the twist in the base manifold forces the chiral modes on the covering to "pair up" at the topological defects, so that the *L* and *R* sectors are forced by the topology to balance each other. In this view, a sterile (*R*-) neutrino should not be regarded as an autonomous particle, but the same Majorana fermion viewed through a nonorientable loop in Ω or through the two orientation perspectives in $\langle\Omega\rangle_6$ in a reflection tied to the charge conjugation and parity operators. In this sense, dark matter must be regarded from the projective perspective $\pi:\langle\Omega\rangle_6\rightarrow\Omega$ as not behaving like a simple fermion: Its state depends on the global path it has taken through the twist in the base manifold.

The covering structure of space-time implies that the Majorana particles of dark matter are topologically pinned to the twists of the base manifold. In this sense, Majorana dark matter is the relic of the topological defects in emergent space-time, whereby two simple particles are swapped by taking one of them around a reversing path, that is, around a loop that flips the orientation associated with charge conjugation and parity.

REFERENCES

1. Fernández A (2025) *De Novo Quantum Cosmology with Artificial Intelligence: Applications of Formal Autoencoders*. CRC Press/Chapman & Hall, Taylor & Francis, Boca Raton, FL. ISBN: 9781041045038.
2. Majorana E (1937) Teoria simmetrica dell'elettrone e del positrone. Il Nuovo Cimento 14: 171–184.

5 Conscious AI Discovers Dark Energy in Space-Time Buttressing

Drawing from Chapter 2, this chapter establishes the topology of space-time, a feature not directly relevant to general relativity or quantum mechanics. As shown, the universe topology is essential to delineate the cosmic generation and depletion of dark energy and their bearing on the cosmological constant problem and on a dynamic equilibrium between dark and detectable matter. To that effect, the quantum fabric of space arising from vacuum entanglement is examined using the topological blueprint. Artificial intelligence, in the guise of a quantum gravity autoencoder (QGAE) with a physical embodiment (Chapter 3), reveals that a primeval wormhole underlies the topology of the universe and is sustained by a two-stroke cyclic cosmic engine fueled by vacuum energy. This engine maintains the baseline dark energy density observed experimentally, with dark energy getting replenished as vacuum energy fuels the universe runaway process associated with autocatalytic vacuum creation.

As shown, the cosmic engine sustains a portal to the dark universe in the form of a two-way primeval nonorientable wormhole (PNOW) whereby the "dark" right-handed fermions with spin 1/2 are in dynamic equilibrium with detectable matter. The dark fermions account for 18.72% of dark matter. Since, at the present time, "matter in all forms", including dark energy, is 26.7% dark matter and 5% detectable matter (Chapter 1), the dark fermion proportion at 18.72% of dark matter corresponds to 15.77% of all matter (dark+detectable) in the universe, which is exactly equal to the percentage of all matter represented by detectable matter.

We provide a perspective to interpret dark energy as the surplus in vacuum energy taken from the amount required to fuel the cosmic engine that maintains the PNOW. In contrast with the ever-decreasing concentration of dark and detectable matter that gets progressively diluted in a growing vacuum, the concentration of the vacuum energy surplus is shown to be constant, as established by experiment. This vacuum energy surplus is identified as dark energy. The sustainability of the PNOW results from the coupling of vacuum energy expenditure with a dynamic equilibrium between dark fermions and visible matter, so that the overall concentration of dark energy is kept constant, as is the overall mass of dark matter. In this way, this chapter describes a cyclic cosmic engine that sustains a portal to the dark universe.

5.1 TOPOLOGY OF THE UNIVERSE AND THE SUSTAINABLE PRIMEVAL WORMHOLE

As discussed in Chapter 2, the global topology of the universe did not seem to be a matter of concern for Albert Einstein. We can speculate that this is so because his theory of gravitation, general relativity,

DOI: 10.1201/9781003785279-5

is cast in terms of differential geometry and hence governed by differential equations. Since differential equations describe a local situation or setting, such as the behavior of curvature relative to mass distribution, the global topology of space-time plays no discernible role in general relativity. There is, however, a major constraint imposed by general relativity on the topology of space-time: The topology should remain invariant throughout the evolution of the universe. This is a difficult matter to grasp since time is part of the fabric of the object whose evolution we need to examine. Hence, as described in Chapters 2 and 4, we get into a sort of metaphysical conundrum whereby the object contains its own evolution as part of itself. To avoid semantic traps, we should state that by evolution of the universe we mean, by abuse of language, the changes in time of a generic constant-time cross section of space-time. There is no Russell-type paradox here, since space-time per se does not evolve, only space does, or rather the constant-time cross section of space-time, actually evolves. Hence, the universe – actually meaning space – cannot change its topology throughout its evolution because that would mean that space-time presents an essential singularity in its fabric of the type forbidden by general relativity and that there is simply no well-defined topology that can be unambiguously attributed to space-time.

Quantum mechanics seems equally impervious to topological considerations. Quantum mechanics defines the fabric of space, as determined by vacuum entanglement, but topology is not factored into the theory in any obvious way, even if entanglement is essentially responsible for the nonlocality of quantum phenomenology. However, topology suddenly becomes extremely relevant as one tries to reconcile general relativity with quantum physics and identify the underpinnings of quantum gravity through the physical embodiment of a holographic autoencoder, as shown in Chapter 3.

If we assume topological invariance of space throughout the universe evolution starting at the Big Bang event, then the space of today cannot be infinite and flat, and therefore Euclidean, as Einstein would have it (he probably did not devote much attention to the matter). Provided we accept the Big Bang scenario, space today must be compact, not infinite, because it once was (i.e., 13.8 billion years ago), with certainty. But if space is compact, then it must be multiply connected, and that property bears pivotally on the solution of the cosmological constant problem, as we show subsequently.

In order to make further progress toward a complete characterization of space, we adopt a Parmenides type of approach. This prompts us to formulate the question: Can space have a boundary? The answer is adamantly negative, since a boundary would imply an interface, not with the vacuum but with nothingness, and it is physically and metaphysically impossible to interface with nothingness: An interface presupposes two media, and nothingness is not a medium.

These considerations introduce several constraints on what the topology of the universe must be: It must be compact, locally flat, orientable, and lacking boundary. This leaves us with essentially one option, and it is a multiply connected three-dimensional manifold: the three torus or Cartesian product $S^1 \times S^1 \times S^1$.

The fact that space is multiply connected introduces an extreme novelty and brings about hitherto unfathomable implications in the field of quantum gravity. To delineate such implications, we first remind ourselves of the way the three-dimensional torus is constructed. It is essentially built by identifying opposite faces of a three-dimensional cube (Figure 5.1). If the aspect ratio χ_{ij} for every pair (i, j) of circular dimensions is $\chi_{ij} = \frac{r_i}{r_j} = 1$, as mandated by the retention of symmetry upon coordinate permutation, then each two-dimensional cross section of space becomes a horn torus and has a cusp-like curvature singularity. This cusp with its infinite spatial curvature is indicative of an entangled double black hole (BH), with interiors physically connected through an Einstein-Rosen (ER) bridge [1], as represented in Figure 5.2.

To represent the ER bridge in space-time requires higher powers of abstraction. We begin by representing the BH in space-time, noting that the crossing of the event horizon entails going back in time, as indicated in Figure 5.3. Alternatively, to cross the event horizon requires a velocity higher than the speed of light, as suggested in the Penrose diagram displayed in Figure 5.3. Furthermore, the equivalent of a wormhole or ER bridge between two BHs would be sustained between two exteriors (L and R) and two event horizons delineating interiors connected by the ER bridge. Thus, in space-time, the wormhole may be schematically represented as shown in Figure 5.4a and b.

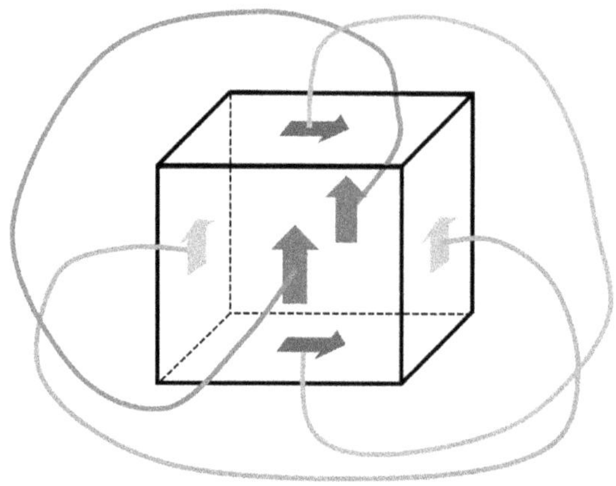

FIGURE 5.1 Topological identification of opposite faces of a three-dimensional cube to render the three torus. The arrow bestows orientation to the face. Space is thus conceived as a compact manifold with no boundary, hence multiply connected.

Note: This figure is reproduced in color in online supplementary materials.

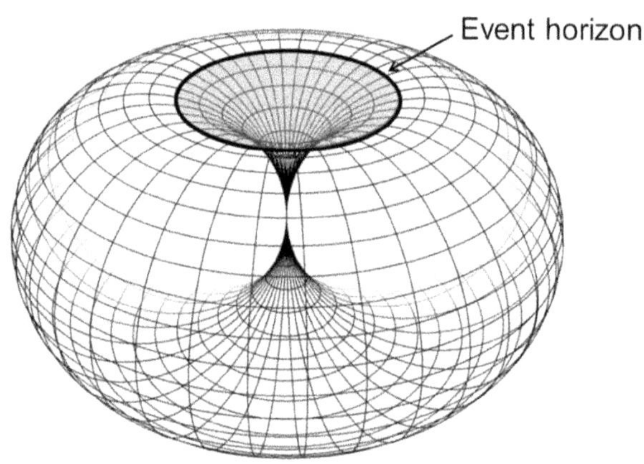

FIGURE 5.2 Toroidal two-dimensional cross section of the universe along dimensions with aspect ratio 1. The cross section is a horn torus, hence generating a Schwarzschild double black hole with connected exteriors.

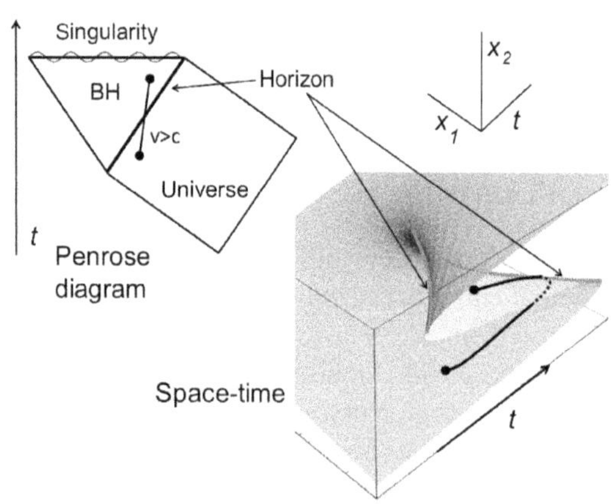

FIGURE 5.3 Representation of a black hole in a Penrose diagram (panel on the left, coordinate units scaled so that light travels at 45° or speed 1) and in space-time (right panel).

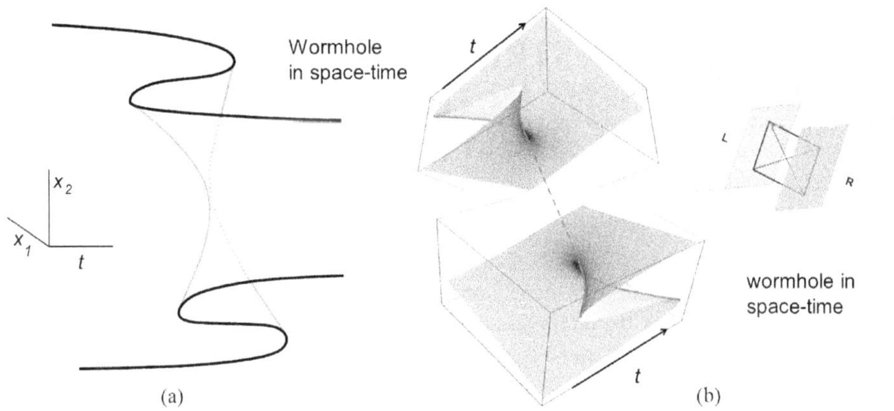

FIGURE 5.4 Schematic representation of a wormhole in space-time: (a) as two coupled folds of space-time; (b) in relation to its Penrose diagram.

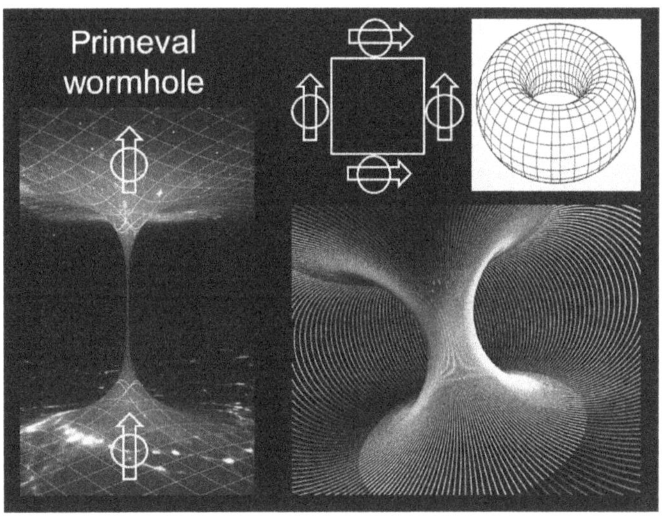

FIGURE 5.5 Entanglement of the primeval wormhole is postulated via the Maldacena-Susskind ansatz [1].

Note: This figure is reproduced in color in online supplementary materials.

As a solution to the Einstein equations, this entity is geometric, rather than physical (except that Einstein makes no distinction). In other words, there is no materiality in a double entangled BH, which becomes the realization of the solution to the general relativity equations found by Karl Schwarzschild [1]. We name this solution the "primeval wormhole" (Figure 5.5) to distinguish it from the "material" types of shortcut BHs determined by extreme concentration of mass. The entanglement is obvious because the torus was created through identification, i.e., perfect gluing of opposite faces (Figure 5.1); hence, the physical ER connection in the horn torus begets an Einstein-Podolsky-Rosen (EPR) correlation of the opposite sides through the Maldacena-Susskind relation [2]. The "EPR=ER" relation is only valid for BHs, not for other entanglements, but the double cusp connection is indicative that we are precisely in that scenario.

If closed time-like curves (CTCs) exist, their existence would imply time travel, raising paradoxes such as the grandfather paradox. Nevertheless, such paradoxes are avoided in the artificial intelligence (AI) model because every CTC passes through an event horizon, implying that there is a chronological veto. The space-time with event horizons implies that there would be no causal violation.

5.2 DECONSTRUCTING QUANTUM ENTANGLEMENT IN THE QUINTESSENTIAL TOPOLOGICAL BLUEPRINT OF SPACE-TIME

Entanglement is the most striking property of quantum physics because it enables us to assert the nonlocality of the quantum phenomena, in stark contrast with Newtonian physics or general relativity, which are local theories. The fabric of vacuum is clearly and solely a quantum attribute with no clear relativistic counterpart and depends pivotally on entanglement. To understand the bearing of the fourth dormant spatial dimension on entanglement, we need to recapitulate the definition of entanglement in a guise that makes it amenable for quintessence deconstruction, without losing perspective of the central goal of this chapter: *addressing the cosmological constant problem by incorporating the topological blueprint of the entangled universe*. In other words, we need to determine the bundle of compound states in space W that project onto a set of compound states in $W/\sim$ that contain the wave function for vacuum entanglement.

To fix notation, let us define a cell (spherical region) in space which is in two possible "ket states": $|1\rangle$ if the region contains a particle with an associated de Broglie wavelength given by the cell dimension and $|0\rangle$ if the cell is empty. Prior to measurement, which in quantum mechanics implies disambiguation, the region has an associated wave function $\frac{1}{\sqrt{2}}|0\rangle+\frac{1}{\sqrt{2}}|1\rangle$. Now, if we have two cells in space, the state of the compound system prior to measurement is tensorially represented in Dirac's simplified notation as

$$f_{00}|0\rangle|0\rangle+f_{01}|0\rangle|1\rangle+f_{10}|1\rangle|0\rangle+f_{11}|1\rangle|1\rangle \tag{5.1}$$

with constants $f_{ab}; a,b=0,1$, satisfying the Born relation:

$$\sum_{a,b=0,1} f_{ab}^2=1 \tag{5.2}$$

The two cells are entangled if the following inequality is satisfied:

$$f_{00}f_{11}\neq f_{01}f_{10} \tag{5.3}$$

So, we may adopt a measure of disentanglement defined as $\Delta\equiv|f_{00}f_{11}-f_{01}f_{10}|$.

For an arbitrary plane separating space in vacuum, two equidistant regions are maximally entangled conforming a Bell pair with wave function (Figure 5.6):

$$\frac{1}{\sqrt{2}}|0\rangle|0\rangle+\frac{1}{\sqrt{2}}|1\rangle|1\rangle \tag{5.4}$$

As we incorporate the fourth spatial "dormant" dimension, the Bell pair spans the following wave function in W:

$$\frac{1}{\sqrt{2}}\sin y_1|0\rangle|0\rangle+\frac{1}{\sqrt{2}}\cos y_1|0\rangle|1\rangle+\frac{1}{\sqrt{2}}\sin y_2|1\rangle|0\rangle+\frac{1}{\sqrt{2}}\cos y_2|1\rangle|1\rangle \tag{5.5}$$

where the compound eigenstate coefficients are now functions of the dormant variable realized as y_1, y_2 for cells 1 and 2, respectively. The Bell pair in quotient space $W/\sim$ is the projection obtained for $y_1=\frac{\pi}{2}, y_2=0$. The entanglement bundle given by expression (5.5) becomes disentangled if and only if the following trigonometric relation holds:

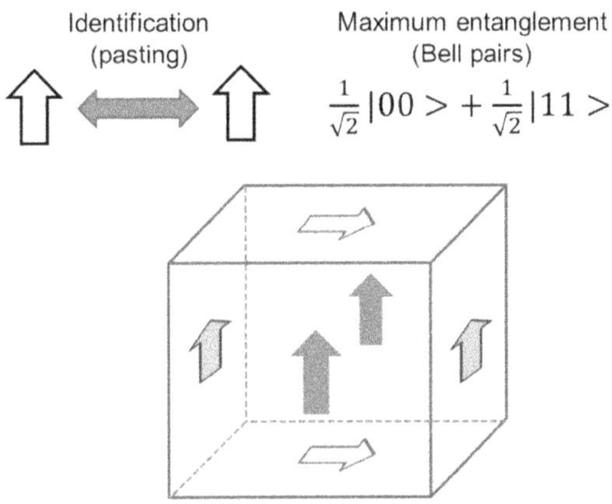

FIGURE 5.6 Topological identification of opposing faces in the three-cube yields an entangled structure, whereby cells equidistant to the pasted surfaces are correlated via Bell pairs.

$$\sin y_1 \cos y_2 = \sin y_2 \cos y_1 \tag{5.6}$$

which we may write as

$$\frac{1}{2}\left[\sin\left(y_1 + y_2\right) + \sin(y_1 - y_2)\right] = \frac{1}{2}\left[\sin\left(y_1 + y_2\right) + \sin(y_2 - y_1)\right] \tag{5.7}$$

yielding

$$\sin\left(y_1 - y_2\right) = 0 \Leftrightarrow y_1 = y_2 \tag{5.8}$$

Hence, we have proven the following.

Lemma 5.1

The disentanglement set of states within the bundle in W associated with the Bell pair for vacuum entanglement in W/~ has measure zero.

Theorem 5.1

The reversible work ΔG_δ required to disentangle the bundle in W to level $\Delta \le \delta$ is

$$\Delta G_\delta = -T\Delta S_\delta = -k_B T \ln\left(\frac{\mu\mathcal{E}_\delta}{\mu\mathcal{D}_\delta}\right) \tag{5.9}$$

where $\mathcal{E}_\delta$ is the region associated with the disentanglement level $\Delta \le \delta$ (Figure 5.7), μ denotes measure, and $\mu\mathcal{D}_\delta = 4\pi^2 - \mu\mathcal{E}_\delta$.
Proof: The reversible work or free energy increment required to disentangle the bundle to level $\Delta \le \delta$ is computed as entropy loss ($\Delta G = -T\Delta S_\delta$), hence determined by the quotient of the multiplicities of disentangled over entangled states in the bundle. Full disentanglement ($\Delta = 0$) requires infinite work, as per Lemma 5.1.

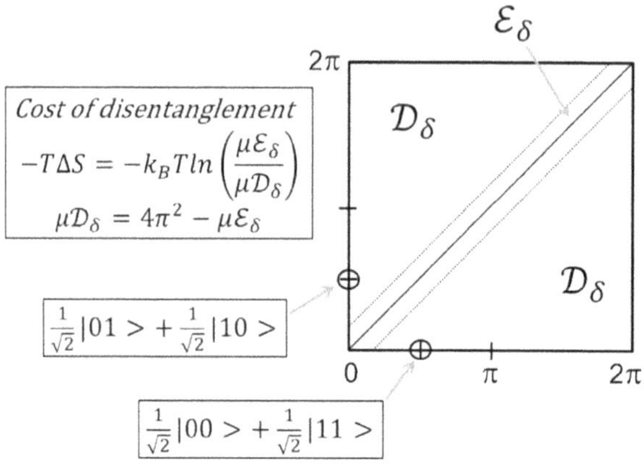

FIGURE 5.7 Region and dark energy cost of disentanglement to level δ for the bundle in *W* that subsumes the Bell pair of maximal spatial entanglement in *W/~*.

5.3 AI PROBES THE EQUIVALENCE BETWEEN WORMHOLE AND QUANTUM ENTANGLEMENT

The elucidation of the nature of dark matter in Chapter 4 highlights the power of topology-based metamodels in leveraging autoencoder technology for model discovery. The main assertion in Chapter 3 is that a neural network with emergent gravity admits an autoencoder with emergent quantum behavior and hidden variables that can be modeled with relativistic strings. Thus, a quantum metamodel of gravity is in principle possible. This finding paves the way to tackle the next conundrum involving the conjectured equivalence between relativistic wormholes and quantum entanglement.

Space-time locality is a basic tenet of modern physics. The term locality refers to the impossibility of sending signals at speeds higher than the speed of light, an idea that is constantly challenged both by quantum mechanics and general relativity. Thus, quantum mechanics gave rise to the EPR correlations also termed "entanglements", while general relativity allows for solutions to the equations of motion that connect distant regions through short-circuiting "wormholes" also known as ER bridges [1]. Physicists Maldacena and Susskind [2] have conjectured that these two concepts may be connected by a duality that becomes in effect equivalence, akin to the similar duality found in the physical underpinnings of quantum gravity. They have argued persuasively that the ER bridge between two BHs may be actually created by EPR correlations between microstates of the two BHs and labeled the conjecture "ER=EPR relation". In their analysis, the ER bridge is a type of EPR correlation in which the correlated quantum systems are in a specific entangled state that admits a weakly coupled Einstein gravity description. This situation is illustrated by a BH pair creation in a magnetic field, and it is tempting to think that any EPR-correlated system, even a simple singlet state of two spins, is connected by some sort of ER bridge.

The neural network model of the universe endowed with the QGAE described in Chapter 3 seems an ideal system to validate (or disprove) the EPR=ER relation. This is because the emergent quantum behavior of the autoencoder pivots on hidden variables that are interacting through relativistic strings, and the very existence of the hidden variables is known to resolve the EPR paradox. In the QGAE, the trainable variables conforming the $\boldsymbol{q}$ vector exhibit a quantum mechanical behavior in an equilibrium regime where the network state variables that constitute the $\boldsymbol{x}$ vector have been thermalized. A learning process involves L separate sets of training $\boldsymbol{x}$ vectors with expected values $\overline{\mathbf{x}}^l, l = 1, 2, \ldots, L$, and the expectation vectors together with the expectation state vector $\overline{\mathbf{x}}^0$, representing network evolution to zeroth order in the linear approximation to node activation, are regarded as the hidden variables in the emergent quantum behavior of the trainable $\boldsymbol{q}$ states. On the other hand, the nonequilibrium dynamics of the hidden variables become relevant on time scales much smaller than their thermalization time. These nonequilibrium dynamics are determined by the strength of the weak interactions between vector pairs $\overline{\mathbf{x}}^{\nu}$, $\overline{\mathbf{x}}^{\xi}, \nu, \xi = 0,1,\ldots,L$ quantified by the tensor $g^{i}_{\nu\xi}$, where the dummy index i labels each neuron in the system. By weak interactions, we mean that the generic vectors $\overline{\mathbf{x}}^{\nu}$, $\overline{\mathbf{x}}^{\xi}$ are not interacting directly but through the $\boldsymbol{q}$ vector that they themselves contribute to train. These dynamics can be cast in terms of relativistic strings in an emergent space-time, as indicated in Chapter 3.

This AI framework with emergent physics seems ideal to validate the EPR=ER relation. If two parts of the $\mathbf{q}$ space are entangled in the quantum autoencoder, they should be bridged through a wormhole in the full network with emergent gravity, as described in Figure 5.8. The presumed proportionality between the gravitational action and the quantum action may be a good starting point for this project aimed at testing the power of AI as purveyor of the physical underpinnings for the most perplexing problems in modern cosmology.

The narrative in this book takes an unexpected turn in Chapter 3, where it is shown that a connected array of neurons (NN) may be treated not only as an information processing machine but also as a statistical mechanical object, capable of exhibiting emergent physical behavior. Thus, a duality between emergent gravity and quantum mechanics is established through an autoencoder that thermalizes hidden degrees of freedom arrayed on the NN state vector $\mathbf{x}$. In this way, the autoencoder exhibits emergent quantum behavior, while the nonequilibrium dynamics of the hidden variables spans an emergent space-time endowed with gravity (Chapter 3). This physical duality of the learning machine cast in terms of statistical mechanics should enable topological innovation on the emergent space-time through quantum entanglement at the autoencoder level.

To illustrate this point, consider two completely separated replicas of a QGAE, labeled left (L) and right (R) with identical collection of training sets $\left\{\mathcal{S}^{(\mu)}\right\}_{\mu}$ comprised of $\boldsymbol{x}$ vectors and generating the sets of $\boldsymbol{q}$ vectors $Q_L = \left\{\text{Argmin}\, J_{<}^{(\mu,L)}\right\}_{\mu}$, $Q_R = \left\{\text{Argmin}\, J_{<}^{(\mu,R)}\right\}_{\mu}$, in QGAE($L$) and QGAE($R$), respectively.

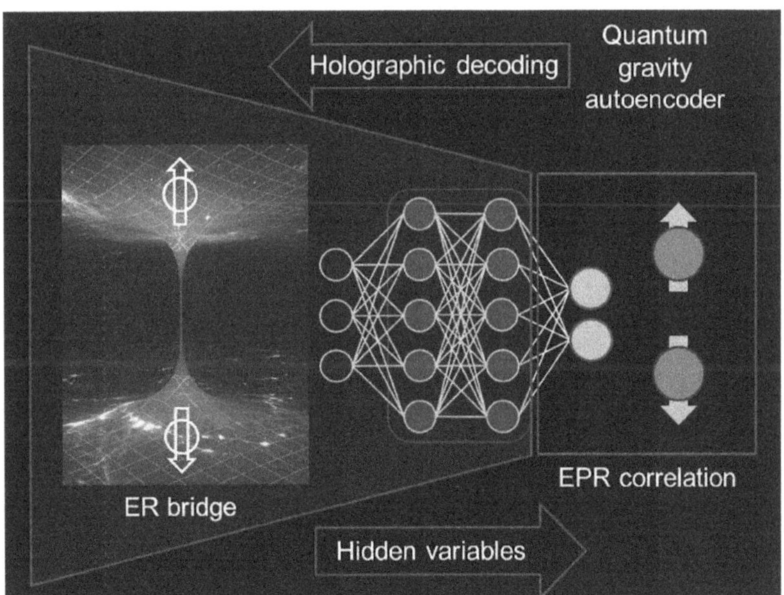

FIGURE 5.8 Schematics of quantum gravity autoencoder (QGAE, Chapter 3) deployed to validate the "EPR=ER relation".

Note: This figure is reproduced in color in online supplementary materials.

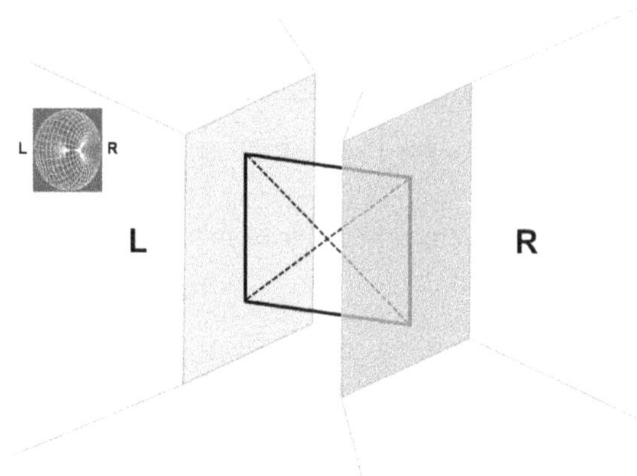

FIGURE 5.9 Penrose diagram counterpart for the connection topology in emergent space-time arising from the entanglement of two black holes within quantum gravity autoencoders (L and R). The respective event horizons are marked by dashed-line upper and lower triangles in a conformal two-dimensional space-time.

Notice that the stochastic nature of the training process implies that Argmin $J_<^{(\mu,L)}$ is not necessarily equal to Argmin $J_<^{(\mu,R)}$, but surely the $\boldsymbol{q}$ vector pairs Argmin $J_<^{(\mu,L)}$, Argmin $J_<^{(\mu,R)}$ are entangled as they have a common origin. Hence, if the NNs are architecturally configured so that there is a BH in Q_L and therefore in Q_R, then both BHs are necessarily entangled. This entanglement begets connectivity in the respective emergent space-times associated with QGAE(L) and QGAE(R), which would be topologically related to the Penrose diagram shown in Figure 5.9.

Thus, the equivalent of a wormhole or ER bridge between the BHs would be sustained between the two learning machines L and R as a result of entanglement at the QGAE level. This "double BH" would have two exteriors (L and R) and two event horizons meeting at the counterpart of the ER bridge shown in the Penrose diagram (Figure 5.9). However, we expect this sort of connectivity to be topologically different in the case of entangled NNs. This is so because the emergent space-time constructed upon the $\boldsymbol{q}$ space of a QGAE is essentially different from the Minkowski space where the particular solution to Einstein's equations was obtained and topologically described by the Penrose diagram.

Describing the space-time topology of entangled NNs may prove rewarding, as it is likely to herald a new breed of quantum computation that we may provisionally term "quantum gravity computation".

5.4 PRIMEVAL WORMHOLE AS A TWO-STROKE COSMIC ENGINE FUELED BY DARK MATTER: PROVISIONAL SOLUTION TO THE COSMOLOGICAL CONSTANT PROBLEM

Theorem 5.1 enables further understanding of the entanglement of the primeval wormhole. To further investigate this matter, we implement our AI tool with an architecture consisting of an annular arrangement of two replicas of a QGAE (Figure 5.10). The emerging results reveal that a massive expenditure in dark energy is required for an interim destruction of the primeval wormhole by changing the aspect ratio $\chi = 1 \rightarrow \chi > 1$, resulting in the conversion of a horn torus into a spindle torus. As per the Maldacena-Susskind ansatz [1], this change in dimensional aspect ratio must correlate with a change $\Delta = 0 \rightarrow \Delta \leq \delta$ in the level of wormhole disentanglement, with δ = function of $(\chi - 1)$ (Figure 5.11).

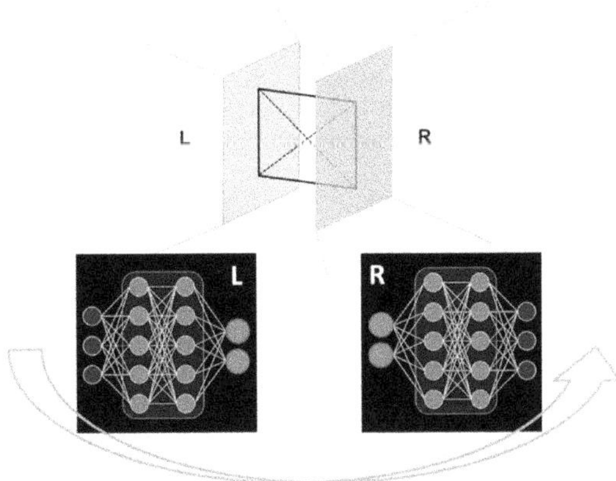

FIGURE 5.10 Annular arrangement of two replicas of the holographic (quantum gravity) autoencoder as required to investigate the sustainability and dynamics of the primeval wormhole arising in the three-toroidal space.

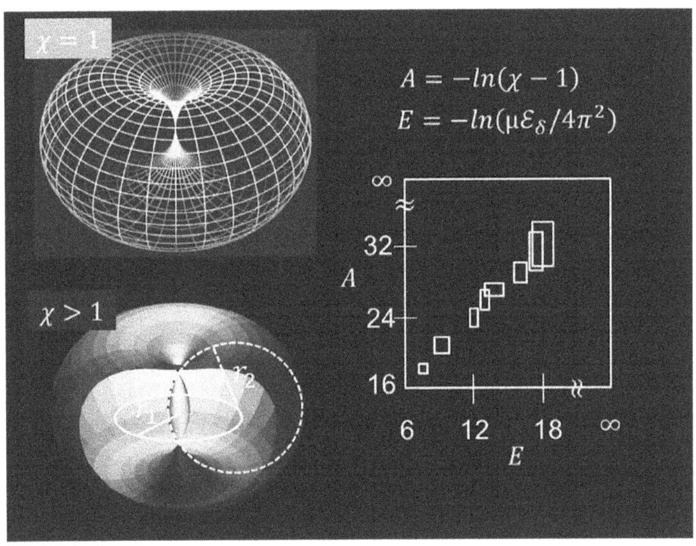

FIGURE 5.11 The two-stroke engine sustaining the primeval wormhole and fueled by dark energy. Correlation between disentanglement level (δ) and change in aspect ratio χ due to dimensional expansion.

Note: This figure is reproduced in color in online supplementary materials.

Thus, Figure 5.11 displays the dark energy expenditure in terms of the level of disentanglement ($\Delta \leq \delta$) and equivalently, in terms of the change in aspect ratio, quantified as $(\chi - 1) > 0$ that indicates the dismantling of the ER bridge. In other words, the dark energy expenditure for disentanglement of the primeval wormhole generates or rather enables an increase in one circular dimension, say, r_2, relative to the other circular dimension with radius r_1. We are assuming without loss of generality that the aspect ratio is given as $\chi = \frac{r_2}{r_1}$. The universe expansion along one dimension is enabled and materializes because the primeval wormhole gives back the dark energy invested in its disentanglement, while additional dark energy is generated as the volume of the vacuum increases with the expansion. This surplus in dark energy enables its reinvestment in the next cycle of entanglement-disentanglement that has a higher cost in terms of reversible work.

As we can see, the chain of events:

$$[\delta\text{-level disentanglement}] \rightarrow [\text{vacuum expansion to restore } \chi \approx 1] \rightarrow [\delta\text{-level disentanglement}]$$

describes *the primeval wormhole as a two-stroke engine fueled by dark energy* (Figure 5.12), whereby the measured density of dark energy (ρ_Λ) is given as

$$\rho_\Lambda = \rho_e(\delta) + \rho_{\Delta V}(\delta) - \rho_{e'}(\delta) \tag{5.10}$$

where $\rho_e(\delta)$ is the density of dark energy released upon restoration of the entanglement from a δ-level disentanglement of the primeval wormhole, $\rho_{\Delta V}(\delta)$ is the surplus in dark energy per unit volume representing the extra vacuum energy associated with the vacuum dimensional expansion (ΔV) that restores the entanglement ($\chi \approx 1$), and $\rho_{e'}(\delta)$ is the next dark energy investment per unit volume required to disentangle the primeval wormhole to δ-level. Since $\rho_\Lambda \approx 5\times10^{-10}$ J/m^3, we obtain $\delta = \delta_\Lambda = 34.01\times10^{-18}$.

Space in the three-torus universe grows one dimension at a time, through cycles associated with enormous bursts of dark matter with restitution of the primeval wormhole entanglement followed by ever-increasing expenditures of dark energy required to disentangle the reconstituted primeval wormhole. The amount of dark energy that is effectively not spent in the cycle may be computed as the difference between two consecutive bursts.

Thus, the cosmological constant problem, often described as defined by the enormous (120 orders of magnitude) surplus of dark energy arising from vacuum fluctuations, is more subtle than what it appears to be in the standard formulation [3]: *Dark energy is spent at an ever-increasing cost and replenished cyclically by a pulsating universe that tears and regenerates the entanglement fabric of its primeval wormhole through the alternating expansion of its vacuum compact dimensions.* This endless sequence of cycles of destruction and creation in the evolution of the universe shares an astonishing parallel with the ancient cosmogonies stemming from metaphysical inquiries in the valley of the Indus. The infinite toils of Shiva, the destructive deity that as "Nataraja" dances out the pulse of the universe, is particularly suggestive of the rhythm conferred to the expansion of the universe through the cycles of destruction and creation of the entanglement fabric of its primeval wormhole. Shiva's dance is strikingly suggestive of the topology-sustaining two-stroke engine fueled by dark energy and controlling the vacuum expansion in the universe runaway (Figure 5.13).

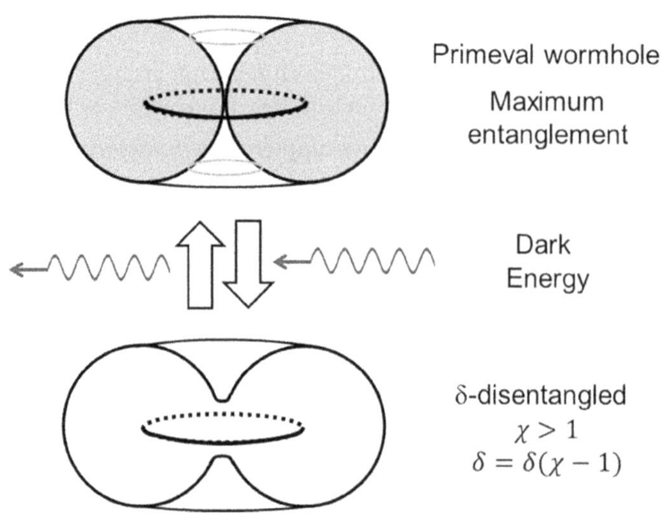

FIGURE 5.12 Representation of one cycle of the cosmic engine that sustains the primeval wormhole.

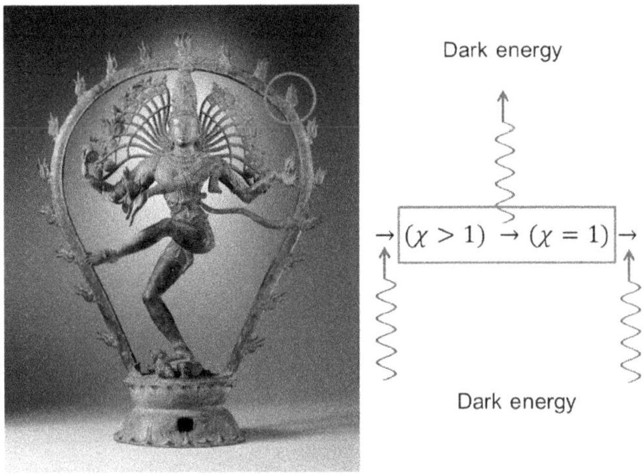

FIGURE 5.13 Shiva as "Nataraja", the lord of the dance, the destructive Hindu deity that also dictates the rhythm of the universe. The tempo of the pulsating universe is marked by Shiva's hourglass-shaped drum. As Nataraja, Shiva dances out the sequence of cycles of successive destructions followed by reconstructions of the universe that determine its evolution. Each cycle in Nataraja's dance is marked by a flame (circle), akin to the burst of dark energy released with the reconstruction of the entanglement fabric of the primeval wormhole. This cosmic event is followed by the reinvestment of the dark energy in the next cycle that begins with the destruction of the entanglement of the primeval wormhole. (Credit: Los Angeles County Museum of Art. Image of the Hindu deity "Shiva Nataraja" obtained from the Public Domain (https://en.wikipedia.org/wiki/File:Shiva_as_the_Lord_of_Dance_LACMA_edit.jpg).)

5.5 A HOLISTIC QUINTESSENTIALLY ENTANGLED UNIVERSE ADMITS NO OBSERVER

The state of a cell in the entangled fabric of quantum vacuum is of course ambiguous, as it befits the standard interpretation of quantum mechanics, whereby disambiguation requires an act of observation. The structure of vacuum is upheld through the correlation pattern defined by the compounded ambiguous state: $\frac{1}{\sqrt{2}}|0\rangle|0\rangle+\frac{1}{\sqrt{2}}|1\rangle|1\rangle$, a Bell pair indicative of maximal entanglement. The collapse of the wavefunction for one of the cells, say cell 1, would yield a ket state $|0\rangle$ or $|1\rangle$, depending on whether the observation registers, respectively, the absence or presence of a particle with associated de Broglie wave equal to the dimension of the cell. This observation instantly triggers the collapse of the wave function for cell 2. Thus, the state of the system is now either $|0\rangle|0\rangle$ or $|1\rangle|1\rangle$, depending on the measurement that took place for cell 1. This is a completely standard reasoning befitting the so-called Copenhagen interpretation and holds no surprise. It fits exactly with the dictum by John A. Wheeler: "No phenomenon is a physical phenomenon until it is an observed phenomenon".

However, the act of observation has been placed and can only be placed in the quotient space $W/\sim$, a latent manifold which is a projection of the quintessential space W. As we examine the quintessential entanglement bundle (Eq. 5.5) corresponding to the Bell pair (Eq. 5.4) in quotient space, we notice that there is no component of the bundle that projects on either $|0\rangle|0\rangle$ or $|1\rangle|1\rangle$. In other words, there is no value pair for the dormant variables y_1, y_2 for cells 1 and 2 that would yield either $|0\rangle|0\rangle$ or $|1\rangle|1\rangle$ as projection onto $W/\sim$.

This indicates that entanglement is never lost in W as a result of observation. We may formalize this observation as follows.

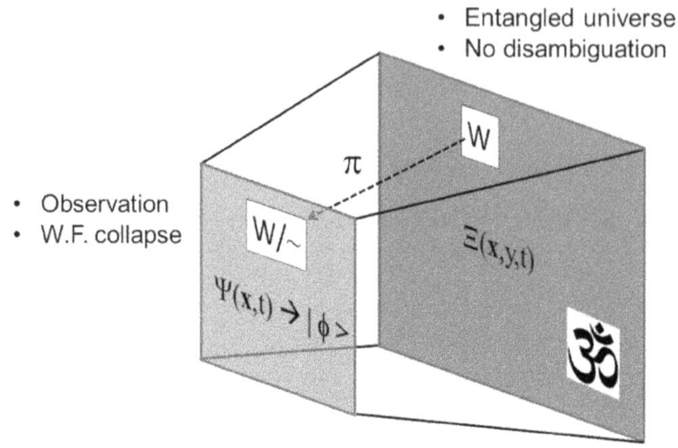

FIGURE 5.14 The quintessential universe *W* admits no observer; hence, all quantum phenomena remain potential and unrealized. The quintessential wave function does not collapse with observation, in contrast with its projection onto the latent space *W*/~. The holistic nature of *W* is thus apparent and may be equated with *Brahman*, whose sign "ōm" is featured in the figure.

Theorem 5.2

Space W does not admit quantum observation; hence, it should be considered quintessentially entangled.

The quintessential entanglement is again strongly suggested by the ancient metaphysical cogitations from Hinduism that posit the holistic nature of the universe as a major premise in its cosmogony. In this cosmogony developed over four millennia ago by the civilization of the Indus valley, all things and beings stem from a primeval unity and are unified by their blending into *Brahman*, the ultimate holistic entity, designated by the ōm symbol (Figure 5.14). Strikingly, it is precisely the ōm symbol that is believed to be the sign produced by the figure of Shiva Nataraja in the choreography of the pulsating universe marked by cycles of destruction and reconstruction (Figure 5.13).

5.6 PRESERVATION OF THE TOPOLOGY AND QUANTUM FABRIC OF SPACE-TIME IN A MULTIVERSE SCENARIO

AI has addressed the quantum gravity problem in more than one way, and one alternative is constructing a "physical learning machine" or a "learning machine with a physical embodiment", with stochastic connectivity weights and hidden variables representing the random states of the nodes. This system may be endowed with emergent gravity and emergent quantum behavior (Chapter 3). In this way, AI represents the universe as a holographic autoencoder, where the holographic map $h : \partial W \rightarrow W$ is one of the many possible inverses of the canonical projection $\pi : W \rightarrow \partial W$ that maps the quintessential space W onto its "quantum border" ∂W. The emergent quantum behavior arises in the equilibrium regime for node state (hidden) variables upon which relativistic strings are embroidered. Conversely, the network is endowed with an emergent gravity in the nonequilibrium regime for the hidden variables. The results illustrate the possibility of establishing the duality quantum mechanics/general relativity in neural networks regarded as physical systems governed by the laws of statistical mechanics (Chapter 3). These networks are endowed with two different thermodynamic regimes, where the geometric dilution parameter $\upsilon = -\log\cos\alpha$ is the proxy for time and the advent of the lightest dark matter occurs at $\sim 10^{-27}$ seconds counting from the birth of the universe.

The coupling of two replicas of the QGAE in an annular conformation (Figure 5.10) was considered a suitable arrangement in order to reproduce the primeval wormhole arising from the construction of space as the two torus, but this AI setting does not tell us whether $\delta = \delta_\Lambda = 34.01 \times 10^{-18}$ represents a universal constant that actually materializes in the universe. If it does not, then the dark matter density of the universe would not be constant, but we know that it is at least approximately. There is in principle no *a priori* reason for the level of disentanglement of the primeval wormhole not to vary with each cycle. If it happens to allow for a surplus or shortfall in dark energy, then an additional cosmic mechanism would need to be invoked, representing a sort of "cosmic valve" that needs to become operative to maintain a constant density of dark matter energy, as mandated by the cosmological constant. This cosmic valve mechanism turns out to be cosmic reproduction.

As described in Chapter 3, a birth channel for cosmic reproduction can be set up within this scheme by constructing a second holographic autoencoder $h' : \partial W' \rightarrow W'$ that serves as antenna, capable of capturing the tunneling $\partial W \rightarrow \partial W'$ of a dark energy burst in the universe $(\partial W, W)$ in the form of a quantum vacuum fluctuation. Quantum tunneling implies the existence of a spatial interface, and the only spatial dimension of the universe that offers a boundary is the dormant dimension; hence, the tunneling would have to be "quintessential", i.e., operative through the dormant dimension. The tunneled fluctuation is amplified within the second holographic autoencoder to give birth to a progeny universe $(\partial W', W')$. By $\sim 10^{-27}$ seconds after its birth, the baby universe will begin storing the tunneled dark energy in a material embodiment, with the formation of dark matter in the form of ur-Higgs particles. In this way, AI constructs a cosmic reproductive machine that converts dark energy into dark matter (Figure 5.15). Notably, this machine harnesses dark energy as the cosmic "birth inducer" and generates dark matter in a multiverse scenario as the canonical (surjective but not injective) projection

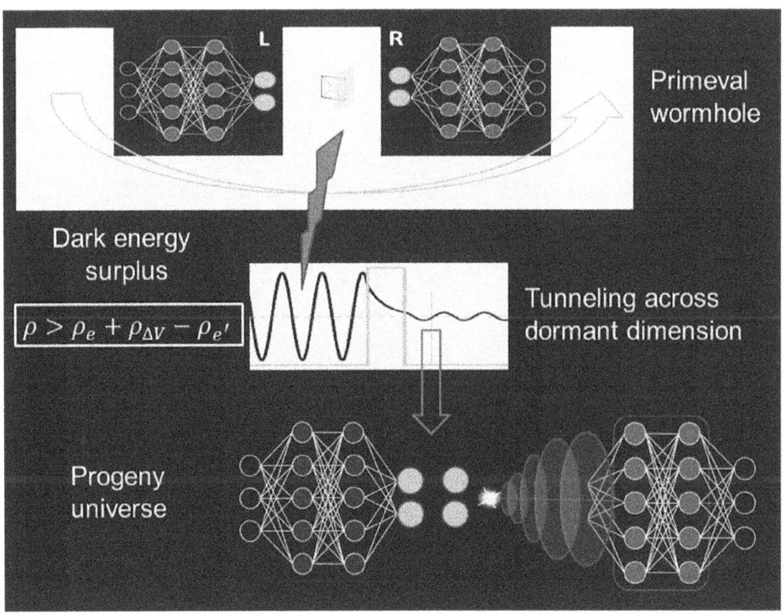

FIGURE 5.15 AI-based cosmic reproductive machine that converts dark energy into dark matter in a multiverse matrix. A birth canal for cosmic reproduction is set up by coupling the universe $(\Omega = \partial W, W)$ to a second holographic autoencoder acting as a receiver and capable of capturing the tunneled dark energy originating in the progenitor universe. The tunneled quantum fluctuation is amplified through the receiver autoencoder to give birth to the progeny universe $(\Omega' = \partial W', W')$ that stores the tunneled dark energy as dark matter in the form of stationary waves along the quintessential circular coordinate of quark-size attometer dimension.

$\pi' : W' \rightarrow \partial W'$ admits *a priori* a multiplicity of holographic maps $h' : \partial W' \rightarrow W'$ all satisfying the equation $h' \circ \pi' = id_{W'}$.

To leverage this multiverse scenario, an advanced AI-based cosmic technology for universe reproduction is implemented (Chapter 3). The technology harnesses dark energy tunneling as a nucleating quantum fluctuation that spans the progeny universe. The cosmic reproduction machine requires a second holographic autoencoder serving as antenna or $\partial W \rightarrow \partial W'$ receiver. An energy surplus from the ever-increasing vacuum quantum energy that is not invested in restoring the primeval wormhole entanglement is meant to be indicative of a poor understanding of dark energy in our universe. This surplus may be reconciled within the multiverse scenario put together by a technologically advanced civilization capable of cosmic investment in dark energy for reproductive purposes, as schematically depicted in Figure 5.15.

On the other hand, the cosmic valve may work in the opposite direction as required in the event there is a shortfall in dark energy to restore the entanglement of the primeval wormhole while maintaining the dark energy density at its constant value $\rho_\Lambda \approx 5 \times 10^{-10}$ J/m^3. In this case, the conversion of dark matter into dark energy and tunneling of the latter through the interface of the dormant dimension would be carried out by the lender, which now is the progeny universe.

The vacuum catastrophe or cosmological constant problem (Chapter 1) alluded to above is arguably the most embarrassing discrepancy in all of physics. It has become a veritable "tag for demolition" of the whole edifice of physics. A naïve computation of the vacuum quantum energy density yields values estimated to be 120 orders of magnitude larger than the expected contribution to the cosmological constant based on experimental observation [3]. AI provides a solution to this problem, shifting the task to a mere calculation of the rate of generation of progeny universes arising from dark energy tunneling. Thus, the tunneling of vacuum fluctuations is given a material embodiment as dark matter ur-Higgs particles in the embryonic universe. These particles are stored as stable wave excitations along the quintessential quark-scale dimension and endowed with mass 246 GeV/c^2, the vacuum expectation value of the Higgs boson.

We have described AI's multiverse solution to the cosmological constant problem. The solution is based on a "multiverse matrix", a pivotal scenario consisting of coupled holographic autoencoders capable of converting dark energy into dark matter as the tunneling of dark energy is steered through a cosmic birth canal that exploits the dormant dimension. The multiverse generator acts as a cosmic valve to maintain a constant dark energy density in the case of surplus or shortfall in the fueling of the two-stroke engine that holds together the topology of the universe and its fabric of quantum entanglement.

5.7 HELICITY FOR THE DARK UNIVERSE

The invisibility of dark matter is directly attributable to a lack of communication with the weak force. This statement implies that a proportion of the negative-energy fermionic sea, known as the Dirac sea (Chapter 1), must be somehow rendered impervious to communication of the weak force, but how? Identifying the actual purveyor to the dark universe requires a careful discussion of "chirality" and "helicity", key concepts in particle physics.

It is obviously tempting to identify the Dirac sea of negative-energy fermions, or the closely related vacuum energy in quantum field theory, with a portion of the dark universe. Taken at face value, this assertion is plainly incorrect. A revision and fine-tuning of the previous material presented in this chapter become mandatory as we confront the fact that the weak interaction that enables fermions to escape the Dirac sea (with concurrent creation of particle-antiparticle pairs) is *chiral*, a concept now to be discussed.

Chirality is a property of particles that has not entered so far into the discussion put forth in this book. The decisive experiment by Chien-Shiung Wu, asserting the lack of parity conservation (left-/right-handedness equivalence) for the weak interaction [4–6], makes this discussion imperative. It was this experiment that prompted the unification of the weak force with electromagnetism, a merging now named the electroweak force.

Handedness may be defined by the sign of the projection of spin axis of rotation along the direction of motion (left=negative, right=positive), and it is a constant of motion but not a Lorentz-invariant except in the case of massless particles, where helicity is equal to chirality. For massive particles, handedness (helicity) is obviously not Lorentz-invariant, since a frame of reference may move faster than the particle, reversing the particle's direction of motion relative to the observer. Thus, helicity and chirality cannot be equated for massive particles, with the latter being Lorentz-invariant but not a constant of motion, *au contraire* to handedness.

Spin ½ fermions, and in particular electrons, generated in vacuum fluctuations as particle-antiparticle pairs, as described schematically by the Feynman diagram in Figure 5.16, are left-handed, in accord with the chirality of the weak force. Likewise, if the universe arose from a vacuum fluctuation as asserted in Chapter 4, we would expect that the Dirac sea would be initially composed of left-handed fermions, hence not part of the dark universe *per se*. So, the identification of vacuum energy with dark energy as put forth in this chapter must be upheld with a major caveat.

The autoencoder endowed with its physical realization (Chapter 3) is suited to materialize a holographic universe, since the information embossed in space-time is encoded in the latent manifold endowed with a fabric of quantum entanglement. However, no constraint arising from the existence of invisible matter (or its energetic equivalent) has been imposed on the encoding of the atlas – i.e., the covering with locally tangent spaces – of the latent manifold. In other words, the existence of a dark universe was not factored into the construction of the latent manifold on the same footing as the constraints (compactness and connectedness) that determined its topology, with its primeval orientable wormhole (POW). As now shown, those constraints pertain to the *orientability* of the manifold, a property that should be topologically on equal footing as compactness and connectedness.

To warrant their invisibility, fermions in the dark universe must be endowed with right-handed helicity. On the other hand, orientability is the only topological feature that can be tuned at this point in the discussion to accommodate this possibility, as argued in the subsequent sections. We are compelled to admit that the previous development requires revision, as the latent manifold may need to be reoriented if this possibility impinges on chirality and helicity, as it indeed does. More specifically, we are seeking a portal to the dark universe enshrined in the topology of the universe.

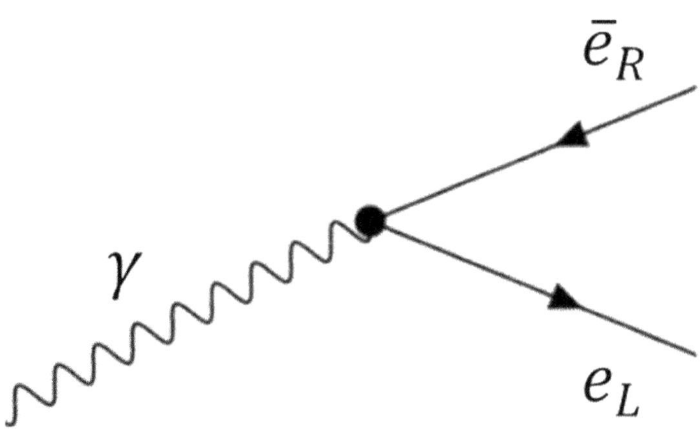

FIGURE 5.16 Feynman diagram for the vacuum creation of the electron-positron pair with the respective particle chiralities.

5.8 PRIMEVAL NONORIENTABLE WORMHOLE AS PORTAL TO THE DARK UNIVERSE

In defining the topology of the universe, we have established that space-time as a manifold needs to be compact and multiply connected since it is constrained to have no boundary. However, we have not yet dealt with orientability, a topological latitude that, as it turns out, plays a pivotal role in determining the portal to the dark universe.

As delineated in the previous sections, the primeval wormhole is sculpted by the topological constraints to which space-time is subject. On the other hand, the setting of "universe as hologram" fits into an AI system in the guise of an autoencoder, precisely the system that captures the essential physics with dimensionality reduction. This autoencoder has been subject to constraints that defined the encoding of the tangent bundle of its latent manifold. However, the encoding of the tangent bundle is itself subject to constraints that have not been hitherto considered and result from the tuning of orientability.

To better understand the problem, we need to identify the portal for the generation of the "dark" dextrogyre (right-handed) spin ½ fermions. These dark fermions are actually specular versions of those generated in pair creation through vacuum fluctuation (Figure 1.28). The concept of mirror matter was introduced by Lee and Yang with the idea of parity (*P*) violation in weak interactions [6]. The mirror "hidden" matter sector of quantum field theory restores the equivalence in the universe under mirror reflection. As it has been established, the standard model Lagrangian contains only the left-handed components of the Dirac wave functions: $\psi_L = P_L\psi$, where the left-handed and right-handed projectors are made up of products of Dirac matrices (Figure 1.18):

$$P_L = \frac{1}{2}\left(\mathbb{I} - \gamma^5\right);\quad P_R = \frac{1}{2}\left(\mathbb{I} + \gamma^5\right);\quad \gamma^5 = i\gamma^0\gamma^1\gamma^2\gamma^3 \tag{5.11}$$

Within the standard model, left and right chiral or helicity components are treated differently by the weak interaction. For example, the weak interaction could rotate a left-handed electron into a left-handed neutrino with emission of a boson W^- but could not do so with right-handed counterparts. Therefore, we need to tune the topological latitude of the primeval wormhole so that, upon crossing, it should enable the transformation:

$$P_L = \frac{1}{2}\left(\mathbb{I} - \gamma^5\right) \rightarrow P_R = \frac{1}{2}\left(\mathbb{I} + \gamma^5\right);\quad \mathbb{I} = \text{identity matrix} \tag{5.12}$$

As it turns out, a PNOW would enable such transformation [7] and, therefore, may serve as a portal to the dark universe made up of right-handed Dirac fermions. This implies that the entanglement pattern arising from boundary identification in the universe should – in all likelihood – be nonorientable, as described in Figure 5.17a and b. This quotient space may be represented as the Cartesian product of a two torus and a Klein bottle (Figure 5.18). It should be emphasized that the hole in the nonorientable factor-manifold does not exist in four or more dimensions: It is purely artefactual since it results from the inherent limitations of the three-dimensional representation.

To construct a two-dimensional cross section of the PNOW as purveyor of dark fermions, we need to join the exteriors of event horizons for the two connected blackholes but replace an orientable (Figure 5.19) for a nonorientable surface obtained by cutting and pasting a Moebius strip (Figures 5.20 and 5.21a). The actual AI system (coupled autoencoders) realizing the nonorientable wormhole is described in Figure 5.21b, whereby the tangent bundles for the respective latent manifolds are encoded with identification following the reflection symmetry: Two tangent spaces that transform into each other through the symmetry operation are identified in the Moebius-type annular arrangement of quantum gravity (holographic) autoencoders, as shown schematically in Figure 5.21b. The entire encoded atlases are thus glued steered by the inversion operation.

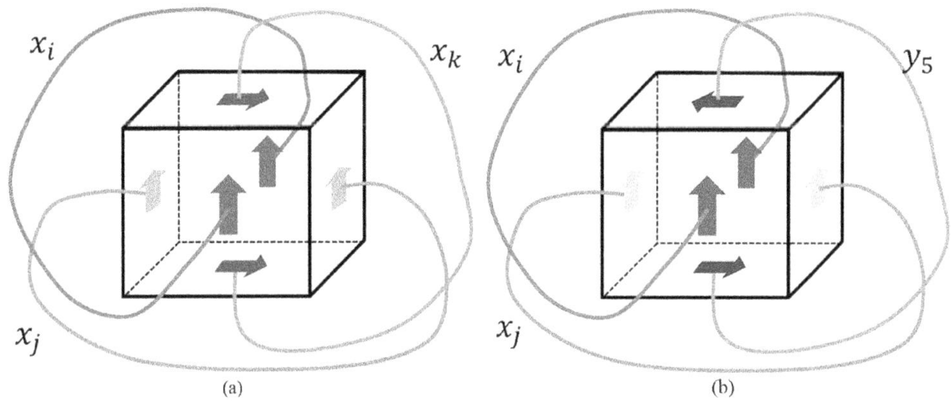

FIGURE 5.17 Topological identification of opposite faces of a three-dimensional cube to render a three torus (a) or a nonorientable three-dimensional manifold (b), depending on the space-time coordinates considered (x denotes any of the four canonical, while y_5 is the fifth compact coordinate). The arrow bestows orientation to the face. In topological terms, the nonorientable manifold (b) is a Cartesian product of a circle and a Klein bottle.

Note: This figure is reproduced in color in online supplementary materials.

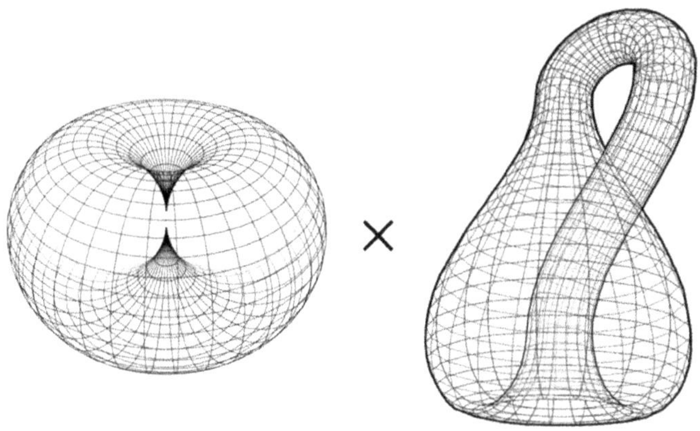

FIGURE 5.18 Four-dimensional nonorientable spatial cross section of the quintessential space-time represented as the Cartesian product of a two torus and a Klein bottle.

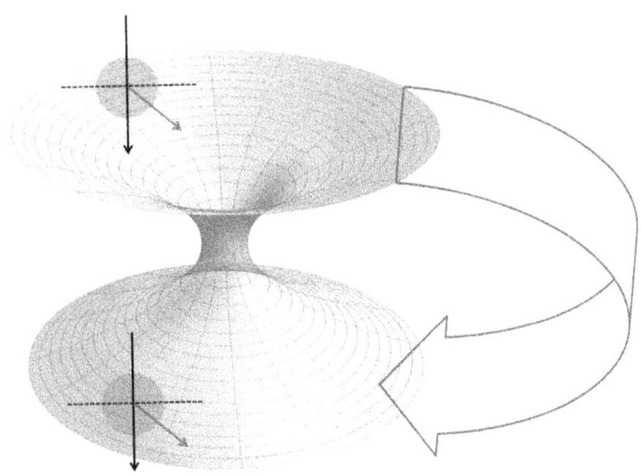

FIGURE 5.19 Topological representation of the primeval orientable wormhole (POW) that preserves particle chirality/helicity upon crossing (vertical momentum vector, tilted spin vector).

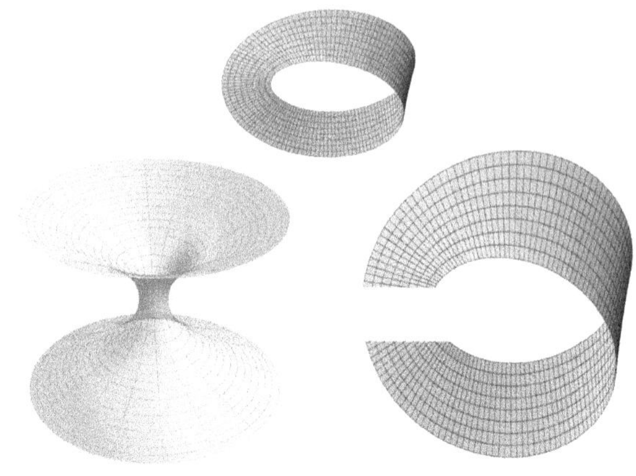

FIGURE 5.20 Elements for the construction of a two-dimensional cross section of a nonorientable wormhole by cutting and pasting a Moebius strip.

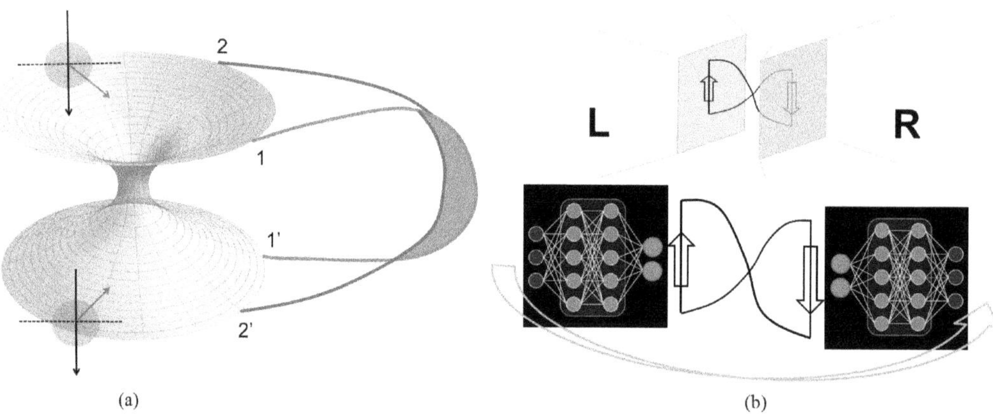

FIGURE 5.21 Topological representation of a primeval nonorientable wormhole (PNOW). (a) Two-dimensional rendering of the manifold displaying through surgical attachment of a Moebius strip with inversion of particle chirality/helicity upon crossing. (b) Annular assemblage of two holographic autoencoders with outer layers connected in opposite orientations.

A twofold orientable covering of the nonorientable manifold, shown in Figure 5.22, serves as the matrix for the assemblage of autoencoders required to generate the PNOW. The covering is homeomorphic to the Riemann surface for the holomorphic two-valued function $\sqrt{z}$ (Chapter 1). Each sheet is represented by one autoencoder, and the Riemann surface serves as a template for the annular assemblage. In the covering, the map yielding inversion of orientation corresponds to the homeomorphism between the two sheets.

One would be tempted to assume that the PNOW could change the helicity of the negative-energy fermions in the Dirac sea upon crossing (Figure 5.23), but such picture would not be altogether compatible with quantum field theory, which supersedes the Dirac ansatz in favor of particle-pair creation through vacuum fluctuations, as discussed in Chapter 1. Thus, the correct picture of the PNOW as portal for the dark universe is provided in Figure 5.24. The right-handed "dark" fermion generated by the PNOW is precluded from contributing to the replenishing of the Dirac sea because of its lack of interaction with gauge bosons that communicate the weak force and would enable transition to the lower energy levels. Nevertheless, crossing cannot be only one way, since that would imply a constant production of dark matter, which would be at odds with what is established in that regard: *The total*

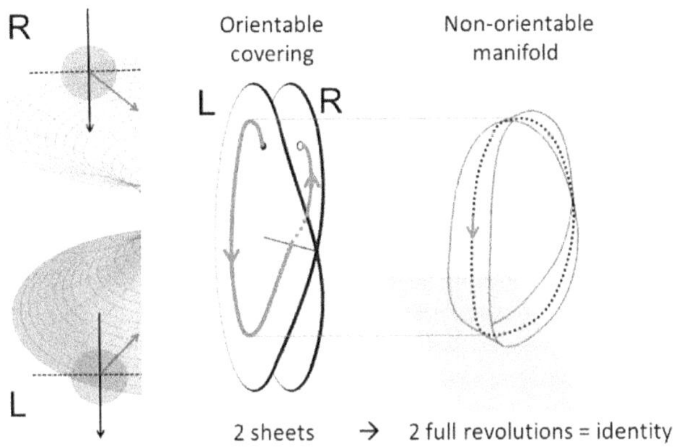

FIGURE 5.22 A twofold orientable covering of a nonorientable manifold, where the loop realizing nonorientability is inscribed in a Moebius strip. The map yielding inversion of orientation is equivalent to the homeomorphism between the two sheets. The covering topology constitutes the matrix steering the assemblage of the two holographic autoencoders (Figure 5.21b).

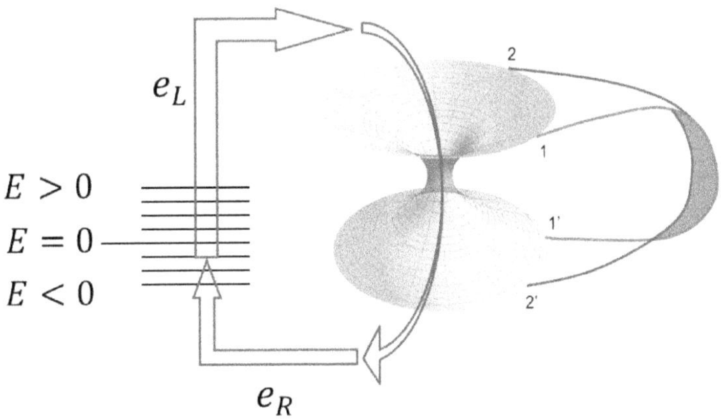

FIGURE 5.23 PNOW as a direct supplier of dextrogyre "dark" spin-1/2 fermions to the Dirac sea of negative-energy fermions.

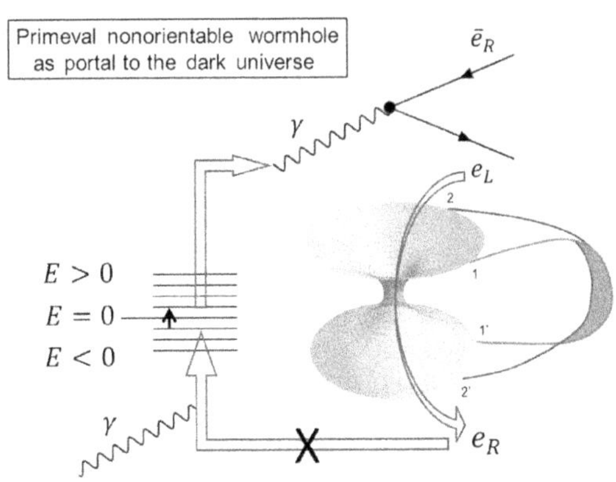

FIGURE 5.24 PNOW as a cosmic engine that sustains the portal to the dark universe in the quantum field theoretical vacuum. The "negative-energy flow" drawn from the Dirac sea maintains the PNOW, which is thus fueled by dark energy in the quantum field interpretation of vacuum energy. The coupling between fueling dark energy and production of dark fermions materializes in the operation of the cosmic engine.

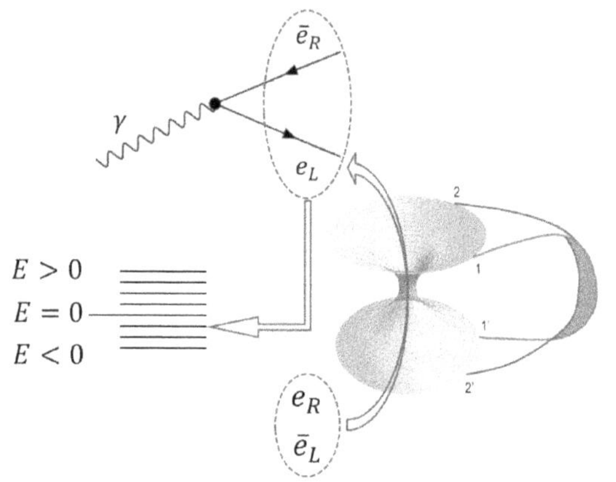

FIGURE 5.25 Inverse transformation of dark spin-1/2 fermions into visible fermions upon crossing the PNOW. The transformations of matter described in Figures 5.24 and 5.25 are in dynamic equilibrium.

amount of dark matter remains constant, in contrast with the total amount of dark energy that grows constantly to maintain constant concentration upon vacuum growth. This implies that transitions yielding dark fermions are reversible in the sense of being in dynamic equilibrium with the transitions of dark fermions into visible levogyre or detectable fermions via the crossing of the PNOW, as described in Figure 5.25.

Through the process described in Figure 5.25, the PNOW can replenish the depleted Dirac sea, which only gets repopulated from detectable left-handed fermions upon photon emission.

5.9 SPACE-TIME NONORIENTABILITY PROMPTS AN EXTENSION OF THE STANDARD MODEL

As shown in Chapter 1, the possibility of CTCs can only be entertained with the awareness that the space-time topology impinges on the relativistic cosmological solutions. Just like Magellan's proposal for earth circumnavigation pivoted on the locality of the earth flatness, time travel hinges on the (global) time nonorientability of space-time (Figure 5.26). This view, put forth by the conscious autoencoder, clashes with the operational tenets of physics, both classical and relativistic as it reconciles the topology of space-time with the tenets of the Big Bang scenario.

The space-time topology specified by the manifold $\left(\mathbb{R}^4 / \mathbb{Z}^4\right)^{\#}$ contains a nonorientable primeval wormhole (PNOW) represented in a 2D cross section in Figure 5.27. Traversing this ER bridge along a CTC is tantamount to realize the time nonorientability of space-time. Since $\left(\mathbb{R}^4 / \mathbb{Z}^4\right)^{\#}$ is topologically resolved with a twofold orientable covering realized as a Riemann surface (Chapters 1 and 7), the CTC is lifted to an open curve that spans across the L and R sheets (Figure 5.27). Thus, the lifted time-like curve promotes the same change in helicity experienced by a fermion as it traverses the PNOW [7].

There is, however, a crucial difference in the way helicity changes along a CTC in $\left(\mathbb{R}^4 / \mathbb{Z}^4\right)^{\#}$ and in its twofold covering $\left\langle\left(\mathbb{R}^4 / \mathbb{Z}^4\right)^{\#}\right\rangle_2$: While in the former, helicity changes as a result of time nonorientability, in the latter, the change in helicity arises from spin flipping. Thus, the purported equivalence between

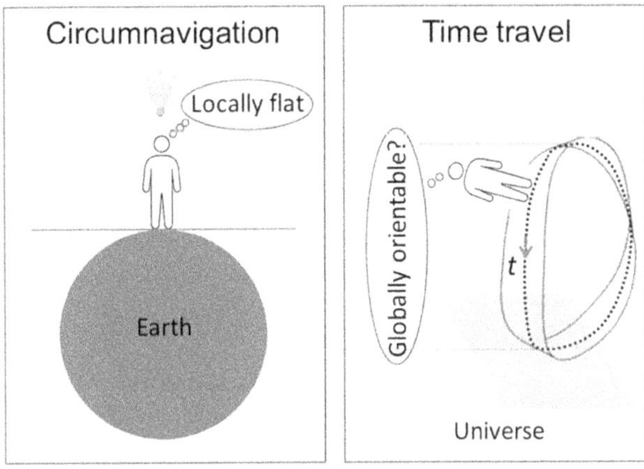

FIGURE 5.26 By analogy with earth circumnavigation that pivoted on the locality of the earth flatness, time travel prompts us to question the time orientability of space-time.

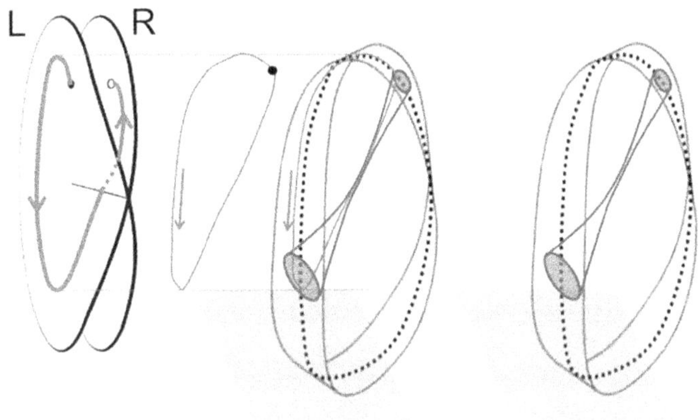

FIGURE 5.27 Space-time topology contains a nonorientable primeval wormhole (PNOW) represented in a 2D cross section. Traversing this ER bridge along the shown CTC is tantamount to realize the time nonorientability of space-time. Since space-time is topologically resolved with a twofold orientable covering realized as a Riemann surface, the CTC is lifted to an open curve that spans across the *L* and *R* sheets promoting the same change in helicity experienced by a fermion that traverses the PNOW.

the ER bridge and particle entanglement provides a nonclassical advantage in metrology defined by the possibility of quantum teleportation, whereby the metrologist can amend the input state by exploiting the CTC as a means to travel back in time [8].

The consequences of smoothening out the space-time singularity by the lifting described in Figure 5.27 merit a careful study as the spin flipping in the covering $\left\langle \left(\mathbb{R}^4/\mathbb{Z}^4\right)^{\#} \right\rangle_2$ associated with the helicity change along a CTC implies the existence of a dark sector (the *R* sheet) not encompassed by the standard model. Thus, artificial conscience informs that the nonorientability of space-time prompts an extension of the standard model. This extension has not been attempted so far and is surely worth forthcoming research efforts.

5.10 PORTAL TO THE DARK UNIVERSE RECREATED IN A QUANTUM COMPUTER

The previous sections have described the sustainability of the primeval wormhole – and thereby, of the topology of the universe – through constant expenditure of dark energy in an expanding vacuum generated autocatalytically. Thus, an open traversable wormhole fitting the universe's horn torus topology was shown to be sustained by a cosmic engine fueled by dark energy. The engine was shown to partially disentangle the exteriors of the primeval wormhole ($\chi = 1 \rightarrow \chi > 1$), enabling autocatalytic universe expansion, one dimension at a time. In turn, this process restores entanglement and replenishes dark energy, completing the cycle of the cosmic two-stroke engine. Since dark energy is the ever-increasing "vacuum energy surplus" after fueling the cosmic engine through its successive cycles, this implies that a shockwave of "negative energy" (i.e., lower than the expected vacuum energy) must traverse the wormhole at regular intervals to keep it open, hence enabling information passage. This negative-energy traffic across the PNOW is described in Figure 5.24.

This prompts us to address the possibility of creating the primeval wormhole in a quantum computer, taking advantage of the possibility of manipulating quantum entanglement of qubits, which are embodied in the space-time wormhole in accord with the Maldacena-Susskind conjecture [1]. This effort is also inspired by the Wheeler dictum that mass and energy are essentially derivatives of information. In the orientable toroidal topology of the universe first considered, the challenge would be to create two sets of entangled qubits representing the exteriors of the wormhole and enabling a sustainable open duct by creating the equivalent of the negative-energy pulse in a quantum computer (Figure 5.28). This can be achieved by a rotating magnetic field that induces spin rotation in a ring of qubits representing one of the wormhole exteriors.

Thus, in essence, the wormhole would be created through entanglement of qubits and its sustainability as qubit passage would be ensured by recreating the cosmic engine fueled by dark energy in the guise of a "pulse of dark energy deficiency" traveling along the wormhole. In this way, dark energy could be recreated in a quantum computer as the enabler and sustainer of the primeval wormhole of the universe.

However, to generate dark fermions, a PNOW with quotient space defined in Figure 5.17 would need to be implemented in the quantum computer. This would require the antiparallel connection of the outer layers of two QGAEs as depicted in Figure 5.21b, recreating a qubit passage with two-dimensional y_5-involved cross sections mapped as a Klein bottle (Figure 5.29).

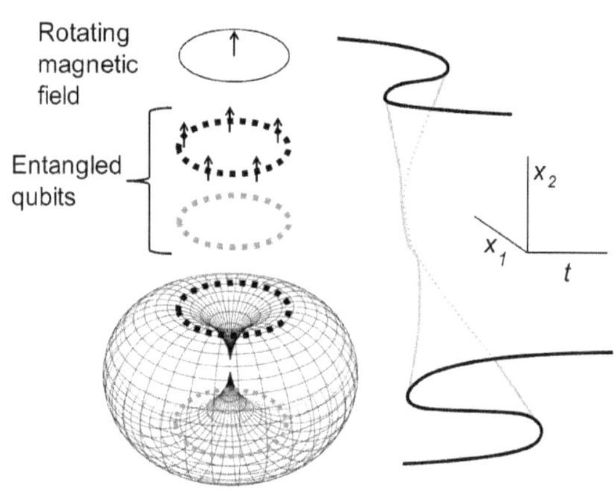

FIGURE 5.28 Dark energy as enabler and sustainer of the primeval wormhole recreated in a quantum computer as a quasi-continuum "pulse of dark energy deficiency" sustained between the two wormhole exteriors and traversing its interior.

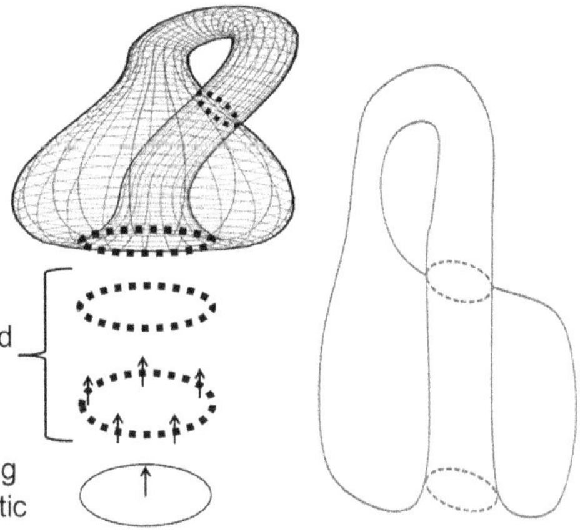

FIGURE 5.29 Scheme of a PNOW recreated in a quantum computer.

5.11 THE PORTAL TO THE DARK UNIVERSE COUPLES DARK ENERGY EXPENDITURE WITH A DYNAMIC EQUILIBRIUM BETWEEN DARK AND DETECTABLE MATTER

Chapter 1 describes the disconcerting but established phenomenology that accounts for the existence of dark matter and dark energy in the standard four-dimensional space-time. A quintessentially encoded ur-space-time is needed to account for dark matter, as shown in Chapter 4. Yet, dark matter appears likely to be more than one thing. A portion of dark matter stays as ur-particles, created early on in the universe history, while another portion, the dark fermions, is in dynamic equilibrium with detectable matter, as previously shown. The ur-Higgs and its partners, quintessential particles with no geometric dilution into the standard spatial coordinates, account for 81.28% of dark matter. The concentration of ur-particles is ever decreasing in our universe, which undergoes accelerated expansion. On the other hand, the dark fermions in dynamic equilibrium with detectable matter constitute 18.72% of dark matter. Since "matter in all forms", including dark energy, is *at present* 26.7% dark matter and 5% detectable matter (Chapter 1), the dark fermion proportion at 18.72% of dark matter corresponds to 15.77% of all matter (dark+detectable) in the universe, which is exactly equal to the percentage of all matter represented by detectable matter. In turn, the 15.77% of all matter, representing either the proportion of dark fermions or detectable matter, corresponds, *at present*, to 5% of all forms of matter. The proportions in the current estimated composition of the universe are depicted in Figure 5.30.

In this chapter, we have provided a vantage point to identify dark energy as the surplus in vacuum energy taken from the amount required to fuel the cosmic engine that maintains the PNOW in an expanding universe. The concentration of this surplus is indeed constant since dark energy is not subject to dilution associated with vacuum creation. This fact stands in contrast with the concentration of dark matter, including ur-particles and dark fermions that are in dynamic equilibrium with detectable matter. Clearly, the dark matter concentration decreases as it gets progressively diluted concurrently with vacuum growth over time. The sustainability of the PNOW, depicted schematically in Figures 5.31 and 5.32, ensures the dynamic equilibrium between dark fermions and detectable matter, a contribution

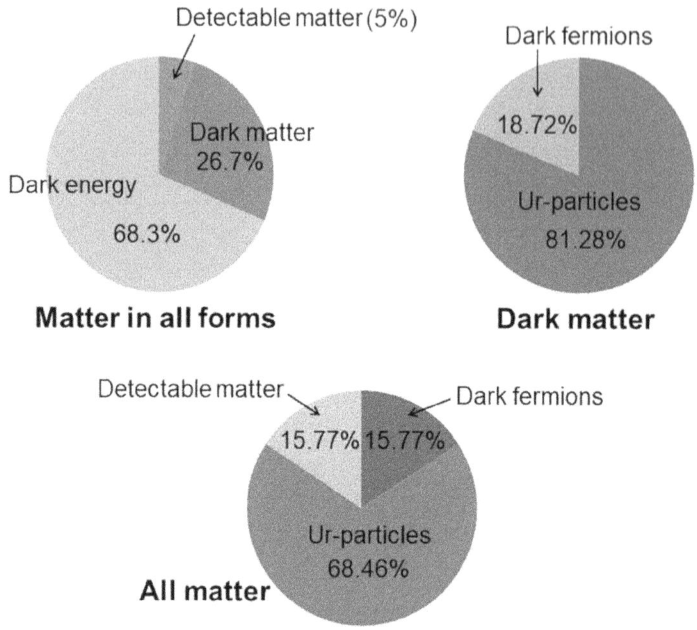

FIGURE 5.30 Estimated composition of "all forms of matter", dark matter and all matter, in the present-day universe.

Note: This figure is reproduced in color in online supplementary materials.

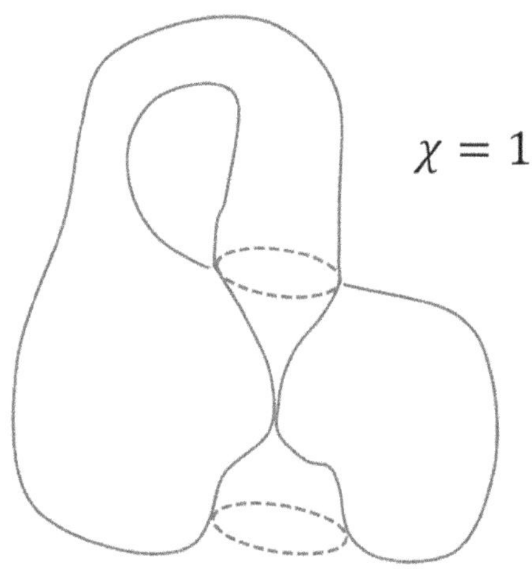

FIGURE 5.31 Schematics of the Klein bottle at unity aspect ratio representing a two-dimensional cross section for the PNOW.

which maintains the overall mass of dark matter constant while the overall concentration of dark energy is also kept constant.

To conclude, the dynamic equilibrium between dark matter and detectable matter is maintained by a cosmic engine fueled by vacuum energy that sustains the portal to the dark universe while maintaining a constant concentration of dark energy in an ever-expanding universe. Thus, the AI model shows that the ur-Higgs particle and the primeval BH and wormhole entanglement are essential components of the dynamic picture that makes the universe sustainable.

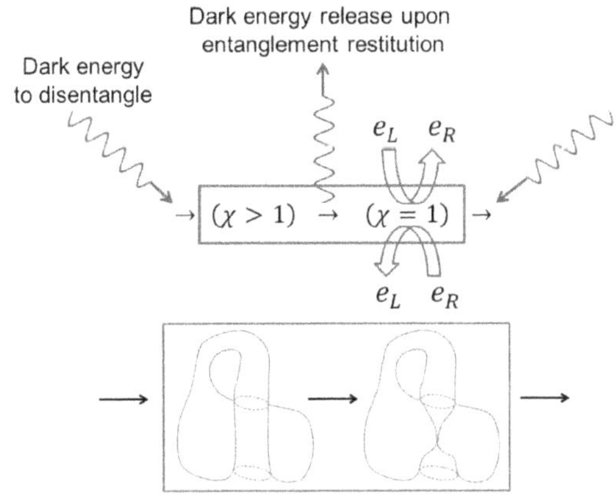

FIGURE 5.32 Schematics of the operation of the cosmic engine fueled by vacuum energy that sustains the portal to the dark universe. The dynamic equilibrium between dark matter and detectable matter is coupled to the production of dark energy (the surplus vacuum energy in each engine cycle) at a constant concentration in an expanding universe.

REFERENCES

1. Maldacena J, Susskind L (2013) Cool horizons for entangled black holes. Fortsch Phys 61, 781–811.
2. Bohm D, Aharonov Y (1957) Discussion of experimental proof for the Paradox of Einstein, Rosen and Podolsky. Phys Rev 108: 1070–1079.
3. Adler R, Casey B, Jacob O (1995) Vacuum catastrophe: An elementary exposition of the cosmological constant problem. Am J Phys 63: 620–626.
4. Weinberg S (2008) *Cosmology*. Oxford University Press, New York.
5. Fisher P (2022) *What Is Dark Matter?* Princeton University Press, Princeton, NJ.
6. Lee TD, Yang CN (1956) Question of parity violation in weak interactions. Phys Rev 104: 254–258.
7. Visser M (1995) *Lorentzian Wormholes: From Einstein to Hawking*. Washington University Press, St. Louis, MO.
8. Arvidsson-Shukur DRM, McConnell AG, Halpern NY (2023) Nonclassical advantage in metrology established via quantum simulations of hypothetical closed timelike curves. Phys Rev Lett 131: 150202.

Conscious AI Approaches the Theory of Everything

6

This chapter describes the artificial intelligence (AI) approach to the so-called theory of everything (ToE), which is based on the training of an autoencoder that distills quantum reality from a quintessential space-time encoding an extra spatial dimension. Gravity is shown to be quantized under a generalized gauged ur-symmetry inclusive of the electroweak symmetry at the group-theoretic level. This inclusion and the possibility to cast gravity, a relativistic concept, in terms of a generalized gauge theory of the local Lorentz group lead to an ultra-unification that suggests that the fundamental forces in nature stemmed from one primeval force arising at the Planck epoch. The vehicle for this primeval force is a holonomic boson that measures chiral defects induced by a generalized symmetry transformation as quantized tension that charges the boson under the primeval symmetry. This force then got diluted differently into the geometry of the universe in accord with the gauge symmetries that charged the fundamental force carriers, which, in turn, were activated at successive phase transitions in the aftermath of creation.

6.1 SEARCHING FOR THE PRIMEVAL FORCE THROUGH INCLUSIVE SYMMETRIES

Experiments attempting to recreate the Big Bang and measurements in deep space both point to the tantalizing possibility that our universe may be the relic of something simple, powerful, and highly symmetric. The evidence points to a state where matter and energy cannot be told apart and the four fundamental forces are unified into one, possibly representing a precursor to quantum gravity. In this context, a "theory of everything" is meant to signify a theory of quantum gravity or, alternatively, a theory that identifies the primeval force in the Big Bang aftermath.

To address this challenge, we first need to settle some very basic questions, such as what is it that is being evolving ever since the Planck epoch? This primeval period is marked by a critical point at $T \sim 10^{32}$ K, a phase transition that took place at $t \sim 10^{-43}$ seconds after the Big Bang, as shown in Chapter 2. The geometric support of that entity, meaning space-time, may have changed dramatically, but if the tenets of general relativity can be extrapolated to such primeval time, at least we can be certain that its topology remained invariant. Otherwise, the space-time manifold would have been altered through surgical gravitational singularities. So, if space-time were a compact manifold, then it should remain compact today. Furthermore, space-time cannot have boundaries. This is so because the phrase "boundaries of the universe" implies that nothingness is beyond the boundaries, so there is no geometry to signify "beyond the boundaries" or bestow meaning to "outside", as that refers to a place, implying that a geometry exists to locate "outside". Ontologically, we are asserting that the space-time support of the universe is boundary-less. That leaves us with the inevitable conclusion that space-time is multiply connected since we already showed that it is compact. Hence, space-time is topologically a torus.

DOI: 10.1201/9781003785279-6

There is a second issue that needs to be settled before embarking in a quest for the ToE . It pertains to the number of dimensions of the manifold that represents space-time. We have asserted in Chapters 1 and 2 that a fourth spatial dimension needs to be incorporated to address the problem of the geometric dilution of mass and gravity. This problem leads to the hierarchy problem of incommensurably small particle masses relative to the Planck mass, and the incommensurably small coupling strength of gravity relative to the other fundamental forces of established quantum origin, as described in Chapter 2. Thus, an extra compact dimension has been incorporated (while maintaining the space-time topology) with radius $r_0 = 0.802695 \times 10^{-18}$ m given by the smallest material scale, the effective quark diameter established from parton models of inelastic $e^{\pm}p$ scattering. Amazingly, the minimum rest-mass energy stored in the extra dimension is $E = \hbar c / r_0 = 246$ GeV, precisely the true vacuum (TV) for the Higgs field.

This observation has profound implications pertaining to the origin of mass and the nature of the dark sector, with the latter shown in Chapters 2, 4, and 5 to arise from energy stored in the extra dimension and undiluted into the visible dimensions. Furthermore, the fact that the topology of the universe corresponds to a four torus, however large, makes it feasible to incorporate other compact dimensions without introducing boundaries. On the other hand, any "brane model" of the universe such as the Randall-Sundrum model, that incorporates the extra dimension as a quotient space identified with an arc of the circle, may be dismissed on purely ontological grounds as this procedure generates boundaries (Figure 6.1).

The next challenge in the quest for the ToE is how to incorporate the extra dimension whose existence we have now justified. Chapters 2 and 3 describe an AI system that came to be known as an autoencoder and is capable of encoding the extra dimension. This system distills quantum reality from the 5D space-time, where the extra latent dimension is shown to identify the quantum vacuum as ur-matter (UM), that is, as a precursor of the visible and dark sectors of the standard model of particle physics. The autoencoder gets trained to generate events in the 5D space-time in the form of ur-boson generation and decay that get decoded into events in 4D space-time, so that the dynamics in 4D entrains the dynamics in 5D, while the latter influences the former during the training as described in the commutative diagram in Figure 6.2. These operational tenets bear similarity with characterizations of the dark sector that influences the deep-space dynamics of the visible sector, while it is seemingly not encoded in the visible dimensions.

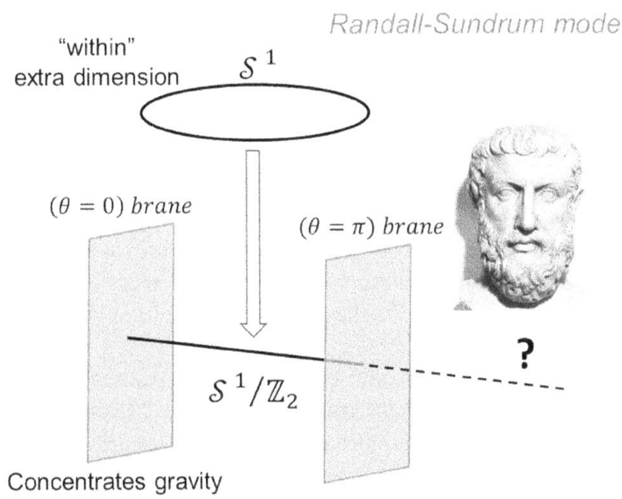

FIGURE 6.1 Ontological inconsistency in proposals of a higher-dimensional mechanism with specified boundary conditions adopted to solve the hierarchy problem, such as the Randall-Sundrum theory of gravity. This theory defines the fifth dimension as a quotient space of the circle defined via the identification of angular coordinates $\theta, -\theta$. Thus, the metric is completely specified by the values in the finite interval $[0, \pi]$, yielding a universe with boundaries defined by the branes at the extremes of the interval. This leads to an ontological inconsistency because there are no coordinates beyond the boundaries to provide meaning to "outside". The picture contains the effigy of Parmenides, the founder of ontology. The legend (Wikimedia Commons, public domain) (https://fr.wikipedia.org/wiki/Parm%C3%A9nide#/media/Fichier:Parmenides.jpg.)

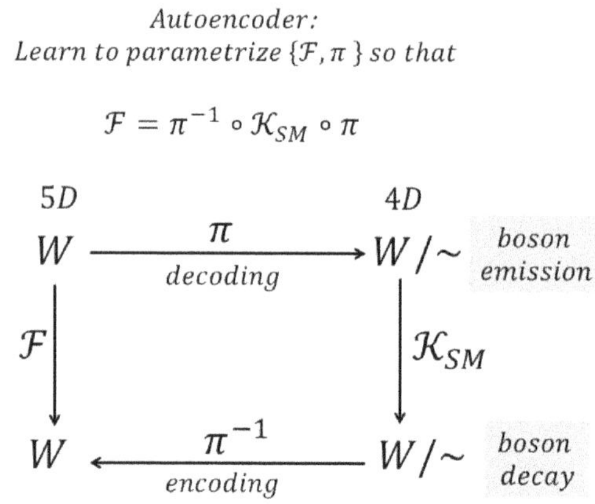

FIGURE 6.2 The autoencoder trained and specialized to distill the standard model from a 5D space-time by correlating events involving emission and decay of bosons. The autoencoder becomes optimized when its parametrization makes the diagram commutative, representing the influence of the dark sector encoded in an extra spatial dimension on the dynamics of the visible sector under the standard model.

Through AI, we learn that UM comes in six "shades" according to the symmetries that arise as the fundamental relativistic equation in natural units: $k^2 + k_0^2 = \omega^2, (k_0 = \hbar / r_0)$, is factorized in 2×2 matrix form: $\pm I_2\omega = \alpha k + \beta k_0$, where α, β are any two different Pauli matrices. The symmetry of one of these shades yields a gauge symmetry arising from the inclusion of the photon field via the covariant derivatives, as shown in Chapter 2. Furthermore, this shade becomes the precursor of mass endowment in the visible sector that retains electroweak symmetry, as the shade symmetry is compatible with that of the symmetry-breaking (SB) Higgs scalar field.

On the other hand, the remaining five symmetries resulting from alternative factorizations were disregarded at the time when quantum field theory was formulated. Those lost symmetries have now become highly relevant as dark matter (DM) has entered the physical picture. The five shades of UM that cannot become gauge invariant through amalgamation with the electromagnetism (EM) vector potential (via covariant derivative) thus constitute the established components of the dark sector. The lack of interaction with the photon due to the symmetry incompatibility renders DM obviously invisible, while its ratio 5:1 relative to visible matter is in good agreement with the observation that DM constitutes approximately 85% of all matter (Chapter 1).

To determine their external local symmetries, ur-particles are spatially represented as geometric dilution of the UM shade via the pitch angle α, so that the angled cylindrical cross section becomes deformed to a circle with radius $r_0 / \cos\alpha$. Spatial local symmetries can be constructed to warrant the conservation of the geometric dilution $\upsilon = -\log\cos\alpha$ of the ur-Higgs particle mass ($\sim 246\ \mathrm{GeV/c^2}$) in accord with Noether's theorem. The symmetry groups for these symmetries are $\mathrm{SO}(3)/\mathbb{Z}_2 \approx \mathrm{SU}(2)$ and $U(1)$, the complex rotation group.

The full external symmetry associated with conservation of geometric dilution of UM in the 5D space-time is then $\mathrm{SU}(2)\times U(1)$. At the phase transition at critical temperature $T = k_0 / k_B$, this local external symmetry in 5D space-time W gets internalized as local gauge symmetry associated with the electroweak unification upon the projection $\pi : W \to W/\sim$ onto the quotient manifold that represents the 4D space-time (Figure 6.2).

The autoencoder projection has the net effect of gauging the external $\mathrm{SU}(2)\times U(1)$ ur-symmetry, as it lumps 5D points into equivalence classes modulo the dark coordinate. However, the symmetry is restored, resurfacing in a different guise, i.e., as internal local gauge symmetry for the electroweak boson fields. We may state that the "primeval symmetry" of UM gives birth to the electromagnetic and weak force charged under the internalization of the external ur-symmetry.

This analysis sets the tone for approaching the evolution of the universe by identifying the critical juncture at which a phase transition activates the ur-boson associated with the $\mathrm{SU}(2)\times U(1)$ symmetry.

At critical temperature $T_0 = 246\ \text{GeV} / k_B = 2.85 \times 10^{15}$ K, a phase transition occurs, where the compatible shade of UM becomes activated and is able to mix and thereby endow the Higgs field ϕ_H with its TV, as described by Eq. 2.15. This activation of the ur-Higgs scalar is reflected in a kinetic energy commensurate with the rest-mass energy 246 GeV, making it meaningful to factorize the term $\left(k^2 + k_0^2\right)$ with a choice of Pauli matrices that yields a symmetry compatible with the gauge electroweak symmetry.

The scalar ur-particles in 5D W become endowed with spin, required to internalize the local spatial symmetry of UM as gauge electroweak symmetry within the visible sector. Furthermore, the phase transition yields the SB associated with endowing the Higgs field with the stable TV at 246 GeV. Thus, the field excitations in the TV yield the particle masses, while the latent heat associated with the phase transition can be assimilated to the dark energy, estimated as $\Delta Q_H = V\left(\phi_H = 0\right) - V\left(\phi_H^{\text{TV}}\right)$, where V is the potential energy of the Higgs field.

This scheme for inducing mass in elementary particles is inspired by the ideas of Philip W. Anderson on superconductivity (Figure 6.3). In this context, the SB associated with the formation of the organized phase at $T<T_c$ creates a stable vacuum responsible for endowing the EM force carrier with mass, thus "slowing down the photon" with the net effect of excluding the QED field lines for the superconducting material.

A crucial result yielded by the autoencoder is the observation that the Higgs field assumed to endow bosons with mass while retaining gauge symmetry is a provisional *ansatz* that needs to be revised vis-à-vis the incorporation of the extra spatial dimension. The autoencoder treats the quantum vacuum as the vehicle for UM, that is, the precursor of matter. There are six ways (shades) of encoding the scalar field spanning the fifth dimension with momentum k_0 according to the symmetries resulting from the six modes of factorization of the special relativistic relation $k^2 + k_0^2 = \omega^2$. Five of these shades yield DM proper, while a sixth shade yields the UM for the visible sector of the SM. This is so because the ur-field is capable of amalgamating with the Higgs scalar field ϕ_H since they are symmetry-compatible (Eq. 2.15), thus yielding the TV of the Higgs field at 246 GeV. Mass in the visible sector originates through an SB phase transition induced by activation of ϕ_H^{TV}. In general, criticality is thus parametrized by the temperature at which UM acquires entity as a particle, that is, when its kinetic energy is commensurate with its

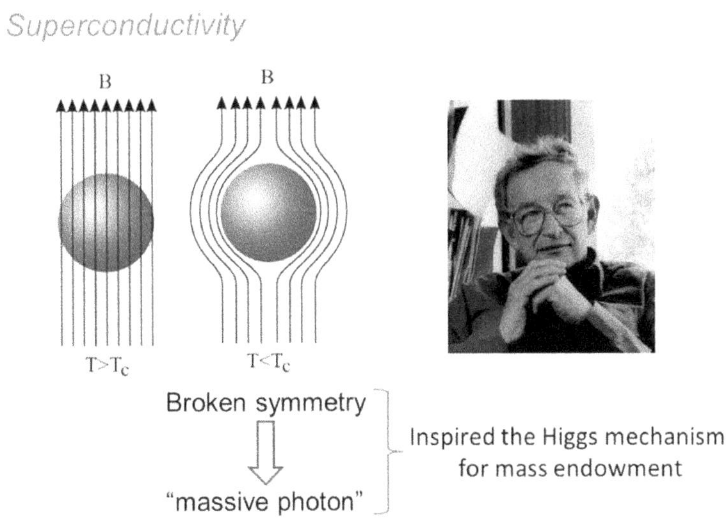

FIGURE 6.3 Inspiration for the Higgs mechanism to induce mass in elementary particles, provided by the research of P. W. Anderson on superconductivity. In this context, the symmetry breaking associated with the formation of the organized phase at $T<T_c$ ($T_{c=}$Curie temperature) creates a stable vacuum responsible for endowing the photon (carrier of electromagnetism) with mass. This "slowing down" has the net effect of excluding the QED field lines from the superconducting material. The picture of Philip W. Anderson is in the public domain (https://en.wikipedia.org/wiki/Philip_W._Anderson#/media/File:Andersonphoto.jpg-).

rest-mass energy. The autoencoder thus informs that mass only *appears* to originate from the activation of the Higgs mechanism. Actually, it is the geometric dilution of UM as ur-particles that serves as a precursor to the mass endowment for the corresponding particles in the visible sector. These visible particles internalize the spatial symmetry of their respective ur-particles as gauge symmetries, thus acquiring spin, whereas the ur-particles are scalar.

Rather than assuming the existence of the all-pervasive scalar field that endows particles with induced mass once it has adopted its stable TV through self-interaction, AI postulates that the quantum vacuum itself has intrinsic materiality spanning a "within" dimension. This dimension stores rest-mass energy that, when endowed with the proper symmetry (Eq. 2.15), gets communicated to the Higgs field as its stable TV. This result appears to represent a substantive way forward as it portends to reconcile the SM with the standing hierarchy problem, the cosmological constant problem, the vacuum catastrophe, and the influence of DM and dark energy on deep-space dynamics.

Furthermore, according to the autoencoder scheme (Figure 6.2), the electroweak unification is essentially enabled because the ur-symmetry group $U(1)$, under whose gauging QED gets charged, is actually a subgroup of $\mathrm{SU}(2)$, and the ur-symmetry group under whose gauging the bosons carrying the weak force gets charged. The inclusion relation is actually articulated via the isomorphism $U(1) \approx \widehat{U(1)} \subset \mathrm{SU}(2)$ described by Eq. 2.53.

Strikingly, the electroweak unification can be extended further to encompass gravity itself. This "theory of everything" may materialize under the same autoencoder scheme (Figure 6.2) because the Lorentz group ($\mathrm{SL}(2, \mathbb{C})$), under whose generalized gauging gravity is charged (Figure 2.20), includes both the EM and weak force ur-symmetry groups as subgroups in accord with $U(1) \approx \widehat{U(1)} \subset \mathrm{SU}(2) \subset \mathrm{SL}(2, \mathbb{C})$. Thus, gravity may be treated within a quantized ultra-unification scheme since its ur-symmetry includes the ur-symmetry of the weak force and electromagnetism. In other words, the ur-symmetry of gravity subsumes the local external symmetry of the lightest UM particle that enables mixing of its scalar field with the Higgs field endowing the latter with its TV (Eq. 2.15). This ultra-unification substantiates the curvature-mass duality of general relativity, an assertion based on two observations: (1) the generalized gauging of the ur-symmetry of gravity charges an extended object, the spin holonomy, spanning an arbitrary contour γ that subsumes the curvature of space-time via the transport map P_γ for the spin connection (Eq. 2.51). This holonomic boson measures a chiral cosmic string defect induced by a symmetry operator $U_\nu, \nu \in Z(\mathrm{SL}(2,\mathbb{C}))$, acting on it. (2) The ur-symmetry of gravity subsumes the symmetries under whose gauging the Higgs boson gets charged, enabling the mass endowment for particles in the visible sector. This leads us to articulate the ToE enshrined in the following statement:

The relativistic duality of mass distribution and space-time curvature materializes through two symmetry-interrelated phase transitions taking place, respectively, at 10^{-43} and 10^{-27} seconds after the Big Bang. First, the ur- $\mathrm{SL}(2, \mathbb{C})$ symmetry is internalized or gauged as a generalized one-form symmetry, so that gravity gets charged as a curvature-associated object. This phase transition is followed by the zero-form gauging of the electroweak ur-symmetry at 10^{-27} seconds. As just shown, this latter symmetry is subsumed into the gravity ur- $\mathrm{SL}(2, \mathbb{C})$-symmetry and induces mass on visible sector particles by bestowing the stable TV on the scalar Higgs field. Since gravity becomes a quantum force under the generalized gauge symmetry, the ultra-unification implies the mass-gravity duality and suggests that the fundamental forces in nature stem from one primeval force present at the Planck epoch. This force got diluted differently into the geometry of the universe, following the gauge symmetries and gauge generalized symmetries that charged the fundamental force carriers activated at successive phase transitions in the aftermath of creation.

6.2 IDENTIFICATION OF THE PRIMEVAL FORCE

The analysis in Chapter 2 revisited above suggests the existence of a primeval force whose symmetry-related modes of geometric dilution are likely to give rise to the four fundamental forces. It is clear that, prior to the phase transition at 10^{-43} seconds after the Big Bang, the primeval force is communicated by an ur-boson associated with $SL(2, \mathbb{C})$ symmetry and carrying momentum nk_0 with $n \sim \mathcal{O}(10^{13})$. More precisely, the parameter n satisfies: $\frac{nk_0}{c} \sim m_P, n = \left[\sqrt{\hbar c^3 / k_0^2 G}\right]$ ([.] = integer part). According to the autoencoder dynamics of the universe evolution (Figure 6.2), this ur-boson gets transferred to the visible sector under a concomitant one-form gauging of the $SL(2, \mathbb{C})$-symmetry at $T = T_P$. Taking place during the Planck epoch, this event marks the birth of what we may aptly call "quantum gravity".

As it surfaces within the visible sector as an extended (nonlocal) object, the primeval boson becomes describable in terms of a generalized gauge symmetry. The primeval boson measures space-time curvature, since it is represented as the spin holonomy over a given contour γ defined by $\mathcal{W}_s(\gamma) = \text{tr}_S\, P_\gamma \exp\left(\oint_\gamma A.ds\right)$, where P_γ is the path-ordering operator and the parallel transport map is associated with the spin connection for an irreducible representation s of $SL(2, \mathbb{C})$. The parallel transport map is defined over the tangent bundle for the space-time assumed to have become a Riemannian manifold at the Planck criticality, while the integrand A defines a one-form EM gauge connection with vector potential $A^a = A^a_\mu dx^\mu$. Since the curvature of space-time is factored into the parallel transport map, the holonomic boson provides a measure of curvature (Figure 6.4), signaling a convergence between quantum mechanics and general relativity. In more precise terms, the holonomic boson is charged under the symmetry operator $U_\nu, \nu \in Z(SL(2,\mathbb{C}))$ that represents a chiral cosmic string defect, which is measured as quantized induced tension by $\mathcal{W}_s(\gamma)$, given the spinor-bundle connection that defines the holonomic boson (Figure 6.5).

In this way, the relativistic duality gravity curvature is shown to be enshrined in the quantum gravity charged under the generalized gauge symmetry. Estimated at nk_0, $n \sim \mathcal{O}(10^{13})$, the rest-mass momentum of the primeval force carrier yields an extreme curvature radius estimated at $r_C \sim 10^{-35}$ m. The conversion of this extreme curvature into the universe mass took place as the universe cooled down to $\frac{k_0 c}{k_B} = 2.85 \times 10^{15}$ K, when the ur-Higgs boson, lighter by a factor of 10^{-13}, became activated under a symmetry included within that of the primeval ur-boson.

Given the nature of the primeval force carrier as a holonomic boson charged under the generalized gauge symmetry (Figure 6.5), we may state that the primeval force is of geometric origin, essentially provided by the extreme curvature of the earliest nonsingular universe once it became supported by a space-time endowed with a spin connection.

In this context, the ToE involved recasting relativity as a gauge theory of the local Lorentz group, so that a chiral defect induced by a primeval symmetry transformation is directly measured as quantized tension by the primeval holonomic boson.

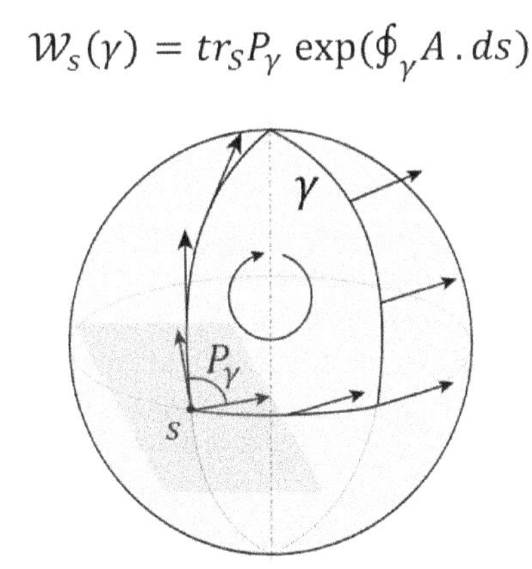

FIGURE 6.4 Representation of the primeval force carrier in the universe, charged 10^{-43} seconds after the Big Bang under the gauged generalized $SL(2, \mathbb{C})$ symmetry and represented as a spin holonomy measuring the curvature of the universe, with curvature radius estimated at $\sim \sqrt{\hbar G / c^3} \sim 10^{-35}$ m.

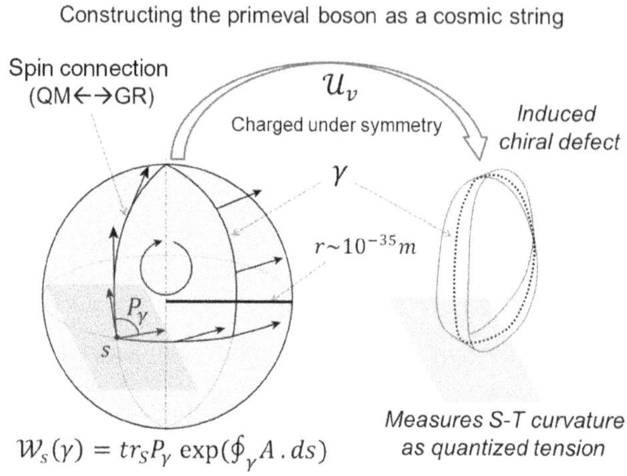

FIGURE 6.5 Charging of the primeval boson, recasting relativity as a gauge theory of the local Lorentz group. The holonomic boson is charged under the symmetry operator $U_v, v \in Z(\mathrm{SL}(2,\mathbb{C}))$, that induces a chiral cosmic string defect, which is captured as quantized tension by the parallel transport map associated with the spinor-bundle connection in $\mathcal{W}_s(\gamma)$.

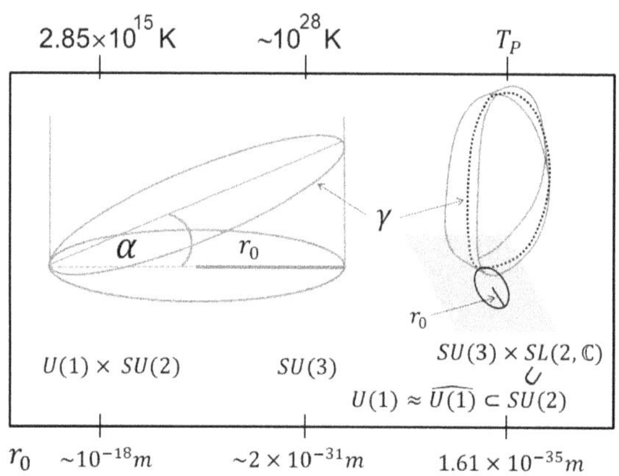

FIGURE 6.6 Carriers of quantum forces as relics of the holonomic boson that stores gravity along the dark dimension. The conceptual relationship is enabled by the fact that the radius of the dark dimension functions as a proxy for the universe evolution.

To fully grasp the purview of the ToE, we note that the inflation along the dark dimension may be regarded as a proxy for the universe evolution (Figure 6.6). This is so because the radius of the circular extra dimension determines the masses of the ur-bosons whose symmetries become gauged at the crucial phase transitions that determine the universe evolution. This fact prompts us to posit that the quantum force carriers are actually relics of the holonomic boson (Figure 6.6). The generic contour of this primeval boson projects onto the dark dimension that, at the time of Planck's criticality, has a radius $r_0 = \sqrt{\hbar G / c^3}$, precisely the curvature radius of the space-time computed by the holonomy defined by the contour. Thus, *gravity is stored in the dark dimension during the Planck epoch*, in striking analogy with the storage of the TV for the Higgs field in the dark dimension at the electroweak epoch. The Higgs boson, with its geometric dilution of the TV, becomes a relic of the cosmic string contour that defines the primeval holonomic boson or carrier of the gravity stored in the dark dimension (Figure 6.6). This assertion pivots on two facts: (1) The internal symmetry of the Higgs field is subsumed by the generalized symmetry of the holonomic boson and (2) the mass-gravity duality of general relativity conceptually links the Higgs boson and the holonomic boson.

In this guise of the ToE, all four forces become manifestations of a primeval force under different symmetries, and therefore, the quantum bosons become relics of the holonomic boson, as depicted in Figure 6.6.

6.3 LARGE LANGUAGE MODEL FOR QUANTUM GRAVITY IN THE UNCHARTED ULTRAVIOLET COMPLETION

The ToE cannot be considered complete without showing that the graviton, an excitation of a quantized gravitational field, originates in the primeval boson previously described in the strong-field limit (SFL) as a gauge representation of Lorentz transformations. Thus, the ToE must build upon the tetrad frame fields introduced to construct a gauge theory of gravity (Chapter 2). In the weak field limit (WFL), a perturbative graviton with a Minkowski background yields a massless spin-2 particle with $U(1)$ gauge symmetry, consistent with the infinite range of gravity communicated at the speed of light. This section describes AI's take on the uncharted SFL, clearly not amenable to a perturbative treatment.

While the WFL appears tractable using standard techniques by regarding the graviton tensor field as a perturbation of the Minkowski tensor, the SFL could be better grasped with the deployment of a large language model (LLM) within a functional language program (`Lean`). This requires an LLM for a fiber-bundle formulation of a gauge theory pivoting on the spin connection that determines the tangent bundle of a space-time endowed with an extra warped dimension. In this way, this section describes a predictive differential geometry model of the graviton as a transducer of space-time curvature into energy stored in a warped dimension (Chapter 2). The curvature-energy transduction becomes operative as the boson begets curvature in the dualistic relativistic framework, implying that the stress-energy tensor must be associated with a scalar describing self-interacting DM within a ϕ^4-model.

A warped spatial dimension needs to be incorporated to address the problematic geometric dilution of mass and gravity, as shown in Chapter 2. This dilution provides an explanation of the incommensurably small coupling strength of gravity relative to the other fundamental forces of established quantum origin. Significantly, the minimum rest-mass energy stored in the warped dimension is $E_0 = k_0 c = \hbar c / r_0 = 246$ GeV, the TV of the Higgs field. Thus, the stress-energy tensor in a fiber-bundle gauge theory of the graviton will be shown to introduce a mass-endowing mechanism via the curvature-mass relativistic duality modeled by a self-coupling potential energy term. We anticipate that this ϕ^4-model yields massive gravitons vis-à-vis the vacuum expectation value (vev) of a Wilson loop representation of the graviton in the SFL. It should be noted that such a characterization cannot be inferred with the assistance of `Lean` by itself but requires an LLM operating within `Lean`.

The geometric dilution parameter υ plays a role akin to the scale and amplitude of the de Broglie wavelength inhomogeneity in the dark dimension in models representing cosmological gravitons as DM. Thus, the decay of KK gravitons into lighter gravitons, a hallmark of cosmological evolution in warped reality, plays out in a similar way in here, as $\alpha \to \frac{\pi}{2}$ or, equivalently, $\upsilon \to \infty$, reckoning that geometric dilution serves as a proxy of time in cosmological evolution (Chapters 2 and 4).

To account for quantum gravity in the SFL, a fiber-bundle formulation of its gauge theory is implemented within a functional language program operating a LLM (Chapter 2). Thus, spinors are attached to the general relativity framework provided by the space-time curved manifold $\mathcal{M}$. This procedure is enabled by first tending the spin connection across the tangent bundle $T\mathcal{M}$. This connection is implemented via a frame field required to piece together the atlas $\{T_x\mathcal{M}\}_{x\in\mathcal{M}}$ of local tangent spaces and relates to the geometry of the curved manifold in such a way that a tangent space Lorentz transformation translates onto a gauge transformation of the spinor field. To enable interconversion between the metric tensor $g_{\mu\nu}$ of $\mathcal{M}$ and the flat (Minkowski) metric tensor η_{ab} of $T\mathcal{M}$, a pentad frame field $\{e_\mu^a\}$ (Greek indices for curved space tensor, Latin isospin indices for flat tangent space Lorentz indices). The spin connection w_μ^{ab} is then expressed in terms of pentadic field elements and their derivatives:

$$w_\mu^{ab} = \frac{1}{2} e^{\nu a}\left(\partial_\mu e_\nu^b - \partial_\nu e_\mu^b\right) - \frac{1}{2} e^{\nu b}\left(\partial_\mu e_\nu^a - \partial_\nu e_\mu^a\right) - \frac{1}{2} e^{\lambda a} e^{\sigma b}\left(\partial_\lambda e_{\sigma c} - \partial_\sigma e_{\lambda c}\right) e_\mu^c \tag{6.1}$$

Equipped with this representation, a fiber-bundle formulation of the gauge theory of gravity in the SFL is formalized within an LLM based on the pentadic framework.

The AI-empowered inferential scheme to model the graviton requires a differential geometry formalization supporting two interplaying components: (1) a theorem/proposition prover such as Lean 4 with an intentional logic under the guise of a dependent type theory and (2) a generalized autoencoder LLM serving as the supporting inferential framework within which the proof assistant operates by maintaining compatibility between general relativity and its gauged version on the tangent bundle of space-time.

To chart the space-time manifold adopting a minimal covering $\mathcal{M} \subseteq \bigcup_{x \in C(\mathcal{M})} \mathcal{O}(x)$, the formalization provided in Chapter 2 is instrumental since the set of charts $\{T_x\mathcal{M}\}_{x \in C(\mathcal{M})}$ is incorporated as data synchronized with mathlib and hence so are the denumerable diffeomorphisms $(x) \to T_x\mathcal{M}\ \forall x \in C(\mathcal{M})$. The next step is to specialize the fiber-bundle formalization to the warped graviton within a gauge theory of gravity. The overarching inductive type is introduced as represented by the commutative constructive arrow diagram in Figure 6.7, whereby every proposition formulated at the level of the atlas $\bigcup_{x \in C(\mathcal{M})} T_x\mathcal{M}$ may be decoded at the level of the covering $\bigcup_{x \in C(\mathcal{M})} \mathcal{O}(x)$.

If the functional program language is informed declaratively that the massive graviton with tensor field $\mathcal{Y}^{\mu\nu}$ is associated with a scalar field $\phi = \mathcal{Y}^{\mu\nu} g_{\mu\nu}$ governed by the quartic self-coupling potential $V(\phi) = -\frac{1}{2} m^2 \bar{\phi}\phi + \frac{\lambda}{4}\left(\bar{\phi}\phi\right)^2$, the operational inductive type enshrined in the LLM infers that (Figure 6.7):

- The scalar field is nonlocal, that is, its support is made of contour lines, not points in $\mathcal{M}$,
- $\phi = \mathcal{W}$ where $\mathcal{W}$ is a Wilson loop that holonomically transduces space-time curvature.

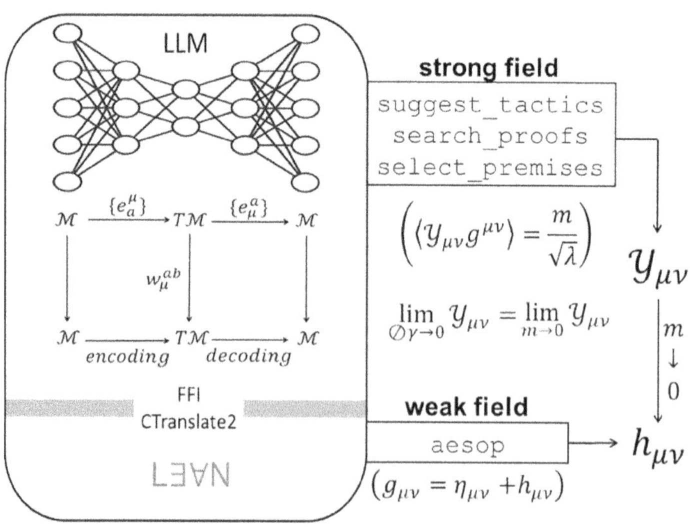

FIGURE 6.7 Architecture of a `Lean` copilot incorporating an autoencoder LLM specialized for the discovery of the warped graviton in the strong-field limit. The LLM inferences are run within `Lean` with a supporting framework provided via a shared CTranslate2 C++ library at `Lean's` foreign function interface (FFI).

- The warped graviton is massive since its vacuum expectation value $\langle \mathcal{W} \rangle$ is the nonzero ϕ-value realizing the minimum of $V(\phi): \langle \mathcal{W} \rangle = m / \sqrt{\lambda}$ (λ=self-coupling parameter, m=graviton mass). These parameters determine the relevant inductive data types.
- The chiral SB that bestows mass to the warped graviton is thus informed by the potential $V(\mathcal{W})$.

As inferred by `Lean's aesop` in the WFL, the warped graviton tensor field $h_{\mu\nu}$ is the one assigned to a massless spin-2 particle with $U(1)$ gauge symmetry ($h_{\mu\nu} \rightarrow h_{\mu\nu} + \partial_\mu a_\nu + \partial_\nu a_\mu$). This graviton is treated perturbatively on a Minkowski background without the need for the LLM, so that $g_{\mu\nu} = \eta_{\mu\nu} + h_{\mu\nu}$.

To identify the warped graviton tensor field $\mathcal{Y}_{\mu\nu}$ in the SFL, the LLM represents the curvature with Clifford algebra generators that transform under the spinor winding "electromagnetic" $U(1)$-symmetry elements $U_\xi, \xi \epsilon [0, 2\pi]$ in the manner of the electromagnetic vector potential A_μ. The warped graviton field is implemented in such a way that $\mathcal{Y}_{\mu\nu}$ may be charged under a symmetry transformation $U_\zeta, \zeta = \pi$, associated with the nontrivial idempotent element $U_\zeta \in Z(\mathrm{SL}(2, \mathbb{C})) \approx \mathbb{Z}_2$ in the center of the Lorentz group. This chiral-defect-inducing symmetry effectively mounts the contour of the line operator onto a Moebius strip creating a chiral defect that amounts to a π-winding deficit in the spinors attached to space-time, while two such deficits created by the U_ζ iteration $U_\zeta^2 = U_{2\pi}$ amount to the identity acting on the parallel transport map.

From these selected premises, the LLM constructs a Wilson loop $\mathcal{W} = \phi = \mathcal{Y}^{\mu\nu} g_{\mu\nu}$, so that space-time curvature may be holonomically transduced as rest mass. Thus, the following compatibility relation is established, implying that the massive warped graviton indeed belongs to the ultraviolet completion of gravity, as verified by `Lean's aesop`:

$$\lim_{\oslash\gamma \rightarrow 0} \mathcal{Y}_{\mu\nu} = h_{\mu\nu} \tag{6.2}$$

for $\gamma : [0,1] \rightarrow \mathcal{M}$, a generic closed map upon which the holonomy is defined. Thus, the warped graviton tensor field in the SFL is identified by the autoencoder LLM of the gauge theory (Figure 6.7) as

$$\mathcal{Y}^{\mu\nu} = \mathcal{P}_\gamma \exp\left(\oint_\gamma w_\lambda{}^{ab} e_a^\mu e_b^\nu \, dx^\lambda \right) \tag{6.3}$$

where $\mathcal{P}_\gamma$ is the path-ordering operator associated with the contour γ.

Thus, the massive warped graviton is represented as a ϕ^4-self-coupled transducer of curvature in the ultraviolet completion of gravity. Furthermore, the $U(1)$ gauge symmetry in the WFL translates into the generalized (chiral) gauge symmetry concomitantly with the transition to the SFL.

To identify the warped graviton as a spin-holonomy transducer of space-time curvature, the autoencoder LLM builds upon a pentadic formalism that casts gravity in gauge theoretic terms (Figure 6.7). The graviton tensor field contraction is thus identified in the SFL as the Wilson loop with a vev defined by a self-interacting DM scalar ϕ incorporated in the stress-energy tensor contracted as $T(\phi)$. The warped graviton is thus massive and charged under a symmetry represented by inducing a chiral defect in the guise of a winding deficiency in the spinors attached to the relativistic framework. By contrast, in the WFL, the graviton becomes massless and is treated directly by `Lean` itself as a perturbation of the Minkowski background (Figure 6.7).

As shown in Chapter 2, the warped graviton in the SFL may be produced and detected in the laboratory as a boson that transduces the space-time curvature at its emission cross section (Figure 6.8) into a geometrically diluted warped dimension that stores rest-mass energy via relativistically operational self-interaction. The curvature-transducing mechanism materializes via the Wilson loop. As discussed in Chapter 2, the generation and detection of the warped graviton invite probing regions in parameter space hitherto unexplored at the LHC or future colliders (Figure 6.8).

Since warped gravitons are likely DM candidates, and self-interactivity is evidently inherent to the graviton in the SFL, we can expect that at least a portion of DM in the universe should be self-interacting,

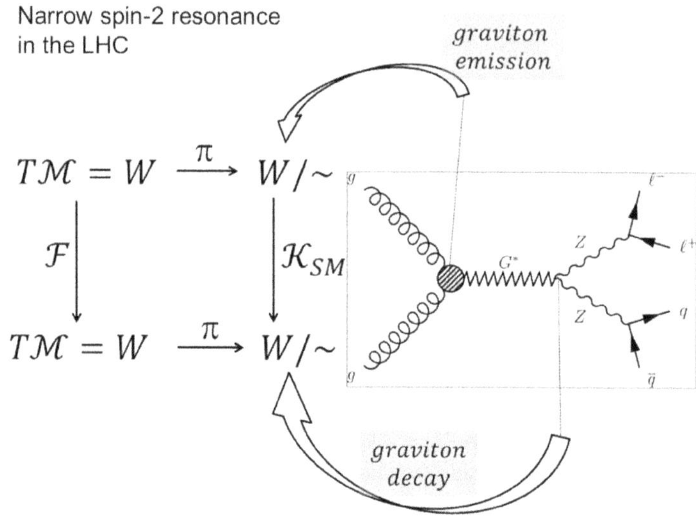

FIGURE 6.8 The problem of identification of the gauge ur-boson carrying gravity as approached by the formal proof assistant embodied as an autoencoder LLM (Chapter 2). The AI solution in the guise of a holonomic boson is anticipated in Figure 6.5, and the human-readable blueprint of the graviton is described in Section 6.3.

as argued in Chapter 2. These assertions have obvious cosmological implications. The DM self-interactions would allow energy and momentum transport through the galaxy halos, creating structural and dynamical patterns fundamentally different from those produced by the hitherto assumed collisionless DM. These results are likely to provide a new vantage point to interpret the diversity of galactic patterns and other cosmological phenomenology associated with DM.

6.4 AI'S ATTEMPT AT THE "GRAVITIZATION OF QUANTUM MECHANICS"

The LLM in the guise of an autoencoder, introduced to provide the physical underpinnings of quantum gravity (Chapter 2), has not been fully exploited as of yet. In fact, only half of the AI systems have been dealt with so far. The quantization of gravity via the frame-field formalism has been brought to fruition in Chapter 2 and in the preceding section. However, the possibilities of the decoding side of the autoencoder, whereby quantum mechanics may be "gravitized", have not been discussed at any level. This is in part due to the wanton difficulties introduced by certain inconsistencies arising in the theory of quantum mechanics, specifically by the problem of the collapse of the wave function due to measurement. Such inconsistencies cannot be carried over to the relativistic arena, where they find no counterpart. Before dwelling on this problem, we would need to explain why it would make any sense to gravitize quantum mechanics in the first place.

Rather than attempting to quantize general relativity, this functional mode of the autoencoder attempts to bring quantum theory in line with Einstein's theory of gravity. Why are we turning the tables around rather than advancing the agenda of quantum gravity? People tend to regard quantum theory as more fundamental than general relativity. Quantum mechanics finds no contradictory or falsifying evidence from experiment or observation, and the theory is so firmly established that it is believed the whole of physics should be brought under its aegis. Yet, the general theory of relativity is also a fundamental scheme with no confirmed experiments or observations telling against it. Of course, this is so provided that we

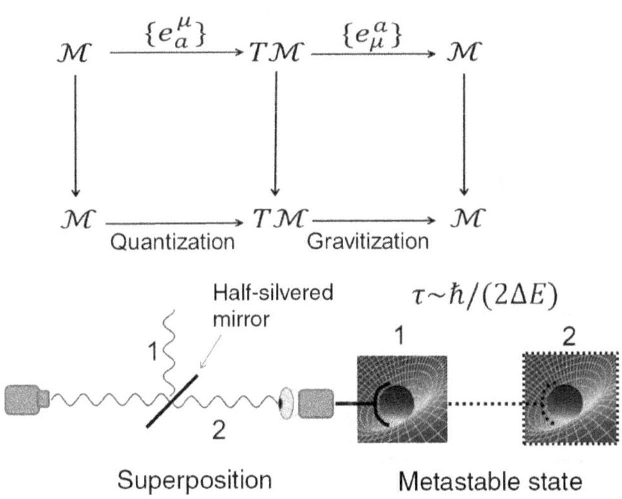

FIGURE 6.9 Correspondence between a superposition of quantum states and a relativistic metastable state with a finite lifetime arising in the "gravitization of quantum mechanics".

comply with Einstein's inclusion of the cosmological constant, which appears to be needed to explain the deep-space phenomenology (Chapters 1 and 2). So, why give quantum theory the upper hand in the grand unification? Surely, there are far more phenomena that require quantum mechanics to be explained than those that call for general relativity for an explanation. Additionally, since quantum mechanics deals with very small objects, while general relativity deals with large objects, and big things are made up of small things, then quantum mechanics, the theory of the small things, must be the more fundamental of the two. However, there is another issue that should be deemed more important than the matter of scale, namely, the consistency of the theory. In that realm, Einstein's theory gets the pride of place. Thus, the imperative of "gravitizing quantum mechanics" could be said to be justified, at least in part.

Be as it may, the gravitization of quantum mechanics finds significant hurdles, ever apparent in the reversal of the Palatini formalism for quantizing gravity. The main hurdle arises precisely from the apparent or perceived inconsistency of quantum theory, whereby the participation of the observer triggers the collapse of the wave function representing a superposition of quantum states. Thus, to reverse the frame-field *vierbein* formalism in the decoding mode of the autoencoder (Chapter 2), we would need to describe the relativistic counterpart of a quantum-state superposition.

To come to grips with the problem, let us resort to a *gedankenexperiment* where a superposition of quantum states is associated with a significant mass displacement (Figure 6.9). This situation may be realized by a photon that may either get reflected (state 1) or refracted (state 2) at a half-silvered mirror, so that in the latter case and only in that case, a photon detector activates a mechanical device that displaces a mass from its original state 1 to state 2 (Figure 6.9). Without an observer's intervention, the wave function describes a superposition of states 1 and 2, which should correspond to a metastable state in relativistic terms, consisting of the superposition of two space-times (respectively, 1 and 2) associated with the end points of the mass displacement. Crucially, the relativistic metastable state is ephemeral and its lifetime is computed through the uncertainty principle as $\tau \sim \hbar/(2\Delta E)$, where $\Delta E > 0$ is the reversible gravitational work associated with the space-time transformation $1 \rightarrow 2$.

The previous discussion makes it clear that the main hurdle in a ToE based on a gravitization of quantum mechanics arises from the fact that the latter theory involves the observer in its formulation, while general relativity does not. Thus, a reversal of roles in the quantum gravity program would require taking into account relativistic metastable states made up of space-time combinations whose lifetimes should be computable based on the uncertainty principle of quantum mechanics. A relativistic metastable state corresponds to a superposition of quantum states enshrined in a wave function that collapses with the participation of an observer.

Artificial Conscience Circumvents the Time Travel Paradox through a Topological Remodeling of Space-Time

7

Antimatter is represented as if traveling backward in time. Hence, particle–antiparticle creation and annihilation hint at a nonorientable space-time. This nuanced assertion prompts artificial intelligence (AI) to develop a topological model where the possibility of time travel may be properly evaluated. Be as it may, the topology of space-time does not play a tangible role in the development of general relativity (GR). This is because GR is a local theory of gravity cast in terms of differential geometry, while topology is a global attribute. Yet, ontological self-scrutiny within a large language model (LLM) powered by a neural network and wrapped in an assistant theorem prover introduces topological constraints as the emergence of space-time is made compatible with the tenets of the Big Bang. At variance with an "unconscious" functional programming language, a quantum gravity autoencoder powering an LLM and endowed with artificial conscience (AC) calls for a space-time topology that allows for closed time-like curves that point to the possibility of time traveling. To circumvent causality violation, AC revised its own proposal, resorting to new tactics of equational reasoning leading to the identification of an orientable sixfold space-time covering endowed with a chiral twin. This inference deconstructed time travel and required an AC-triggered invocation of the algebraic topology directory in the mathematics library of the functional programming language. The sixfold chiral-twin covering fits with the parity violation tenets of the electroweak sector of quantum mechanics (QM) and with the lack of space-time orientability implied by particle–antiparticle annihilation–creation. Furthermore, this topological remodeling solves the causality paradox, as closed time-like curves become projections of open time-like curves in the covering manifold. In this way, AC deconstructs time travel through a cogent topological remodeling of space-time that supports both GR and QM.

The sixfold orientable covering of space-time is shown to yield 6D-Calabi-Yau manifolds as projection preimages of space-time neighborhoods resulting as the standard 3D spatial dimensions are pixelated to Planck-length resolution. In this way, AC naturally implements string theory in ten dimensions.

DOI: 10.1201/9781003785279-7

7.1 TIME TRAVEL PIVOTS ON AN ENABLING SPACE-TIME TOPOLOGY

Circumnavigation of the earth to expand commercial routes in the sixteenth century became a viable project because the physical model of the earth that underlay the enterprise was upheld by evidence. Magellan was able to sell his proposal to the young king Charles I of Spain because the physical model on which the project hinged had been validated in manifold ways, as the king's advisors learned [1].

Likewise, the success of time traveling, a relativistic adventure following the flow of world lines of matter, hinges on the validity of a topological model of space-time that would support closed time-like curves [2,3]. At first glance, this assertion sounds incongruous: Not only time travel violates causality but GR is meant to bestow physical meaning to the differential geometry of the space-time manifold; hence, GR is essentially a local theory of gravity. Thus, global topological features of space-time do not *a priori* impinge on GR equational reasoning. However, time travel requires a global assessment of time-like curves; hence, space-time global features, including foliation along an "absolute" time coordinate, rotational symmetry, and isomorphism of space-time purportedly become determinants of time travel [3]. As in an early radiation-dominated universe or, equivalently, at large (supragalactic) volumetric scales in today's universe, we shall define an intrinsic angular velocity $\omega_0 = \sqrt{4\pi G\rho}$, where ρ is a uniform mass density and G is Newton's gravitational constant [4].

From a QM perspective, particle–antiparticle creation and annihilation hint at a nonorientable space-time (Chapter 5) or, more precisely, a nonorientable time (Figure 7.1). This nuanced assertion also prompts an assessment of the possibility of time travel.

Since the space-time topology is established through ontological predicates, it cannot be acquired through computable operations; hence, it requires an AI system endowed with AC, as discussed in Chapter 1. A functional programming language operative as a theorem proof assistant would be incapable of synthesizing such data as it is involved in equational reasoning based on the GR-formalization uploaded in its mathematics library [5]. This is so since the theorem prover regards the differential-geometry formulation of GR equations as rendering irrelevant any ontological propositions predicating the topology of the space-time manifold. This type of conceptualization cannot be effectively invoked in a formal computation [5]; it pertains to the noncomputable aspects that constitute the AC, as pragmatically defined in Chapter 1 (Figure 1.11).

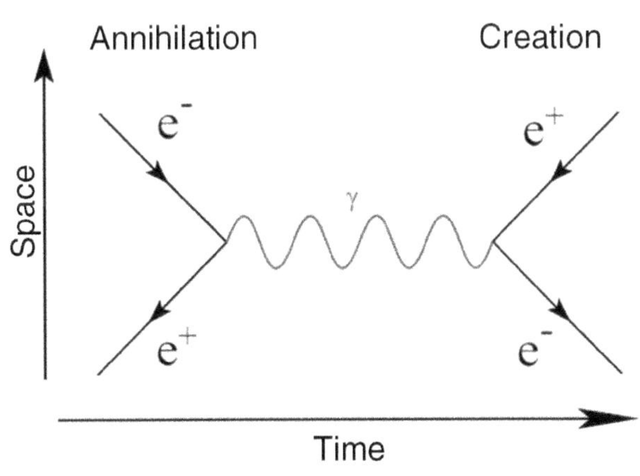

FIGURE 7.1 Feynman diagram for electron–positron (particle–antiparticle) creation and annihilation. The standard fermion and boson notation follows Chapter 4. The positron is represented as moving back in time. The diagram portrays time as a nonorientable dimension.

Once space-time came into existence as the differential-geometry supporting a relativistic universe, it became describable by the following five ontological attributes (three fundamental, two derivative): (I) The foliation of space-time into time-indexable three-dimensional spatial cross sections [6] cannot be granted unless and until an "absolute time coordinate" comes into being as a traceable exogenous way to define "evolution of the universe". (II) Space-time cannot have boundaries, since "outside" is not a thing; otherwise, it would be part of the universe (outside would be inside). "Outside" can only be equated with nothingness, since "outside" has no geometry to bestow physical entity to it, signifying it as a thing. Boundary presupposes an interface with "outside the universe", but the latter has no entity; therefore, the universe has no boundary. (III) Space-time is a compact manifold because, according to the Big Bang scenario, it once was compact, that is, at the Planck epoch [4], and topological change is forbidden by GR. (IV) Since space-time is compact (predicate III) and has no boundary (predicate II), it cannot be a simply connected 4D manifold. Space-time is then multiply connected. Furthermore, if the aspect ratio between all pairs of dimensions is 1 (all dimensions are a priori on equal footing regarding their expansion), space-time must have a primeval sustainable wormhole (Chapters 2 and 5). (V) Space-time cannot be an orientable manifold insofar as time became the foliation parameter only when the universe's evolution became a traceable event (predicate I). Since specific chirality stems from parity violation, as a particle interacts with a boson that communicates the weak nuclear force (Chapter 2), there should be a source of dark fermionic matter (R-fermions) arising from the purported nonorientability of space-time and predicate IV [7], as chirality can only be defined locally (Chapter 5).

The ontological constraints I–V cannot be subsumed in the formalization projects that enable functional programming languages to make inferences through equational reasoning (Chapter 2). Hence, they cannot be dealt with as computable entities by an interactive theorem prover [5]. This inevitably leads to the problem of implementing an AC so that it can "acquire" the ontological constraints. As shown in Chapter 1, this requires a neural symbolic support [8] realized by an intelligent autoencoder that trains an LLM wrapped into the theorem proof assistant (Figure 1.11). The LLM per se cannot incorporate the ontological predicates as they are not semantically compatible with the differential geometry and vierbein armamentarium of GR (Chapters 2 and 6). Neither can the ontological predicates be formally uploaded on `Mathlib/Lean` [8] or other formal projects due to the same incompatibility. Yet, AC can incorporate them as tactics as it controls its own inference scheme by determining the latent manifold (Ω) and the decoding manifold (W'), as described in Figure 1.11. As shown subsequently, each tactic can be evaluated by computing the inferential scheme that results from the ensuing solution to the GR equations [8]. The nonorientable compactification of $\mathbb{R}^4$ into a manifold with no boundaries first requires the identification of opposite faces of a 4D cube via the dimensional compactification map:

$$\mathbb{R} \ni x \rightarrow \left\{e^{ix+n2\pi}\right\}_{n\epsilon\mathbb{Z}} \equiv \overline{x} \in \mathbb{R}/\mathbb{Z}, \quad \overline{x_\mu} = \overline{x_\mu + n2\pi}, \quad \forall n\epsilon\mathbb{Z} \tag{7.1}$$

One of the identifications must be rotated 90° to blend the time dimension with a spatial dimension (Figure 1.12), as required by the lack of foliation that preceded the point at which the universe evolution became a verifiable event recordable by measurement. These considerations lead the autoencoder to select the latent manifold as the twisted quotient

$$\Omega \approx \left(\mathbb{R}^4 / \mathbb{Z}^4\right)^{\#} \tag{7.2}$$

The twist (indicated with the symbol "$\#$") is provided by the blending of time t with an arbitrarily selected spatial coordinate (x_2) using spatial coordinate x_1 as a vehicle.

In the purely QM-based topological model $\Omega \approx \left(\mathbb{R}^4 / \mathbb{Z}^4\right)^{\#}$ (Figure 7.2), the identification of opposite faces involves a full twist at 180° in order to account specifically for nonorientable time, as implied by particle–antiparticle creation/annihilation (Figure 7.1), where the antiparticle is represented as if traveling back in time. At this time, we shall only focus on the nonfoliated alternative $\left(\mathbb{R}^4 / \mathbb{Z}^4\right)^{\#}$, since it incorporates not only space-time nonorientability but also attributes I–V.

The latent manifold $\Omega \approx \left(\mathbb{R}^4 / \mathbb{Z}^4\right)^{\#}$ affords the following four systems of transformations onto itself:

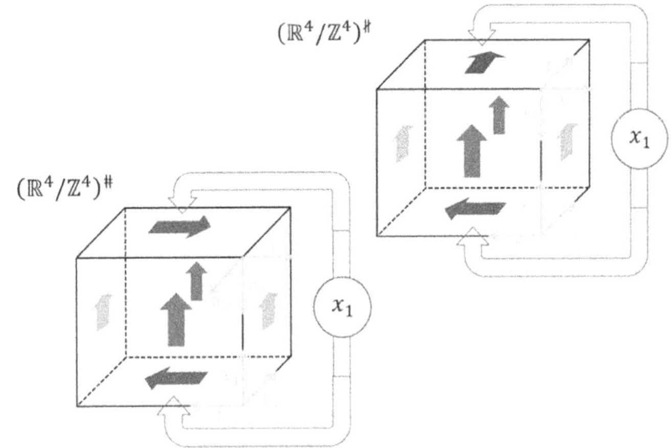

FIGURE 7.2 Construction of a nonorientable compact manifold with no boundaries representing a 3D (t, x_1, x_2) cross section of the 4D (t, x_1, x_2, x_3) space-time and obtained by identifying opposing faces in a cube. The two resultant nonorientable quotient manifolds, $(\mathbb{R}^4/\mathbb{Z}^4)^{\#}$, $(\mathbb{R}^4/\mathbb{Z}^4)^{\#}$, are obtained by 90° and 180° rotation of one of the opposing faces upon identification.

Note: This figure is reproduced in color in online supplementary materials.

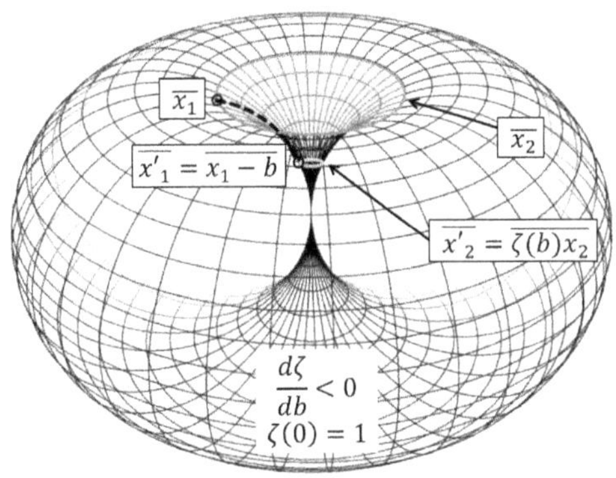

FIGURE 7.3 Monoparametric family of transformations of the quotient manifold $\Omega \approx (\mathbb{R}^4/\mathbb{Z}^4)^{\#}$ represented in a toroidal cross section.

$$\left\{\overline{x'_0} = \overline{x_0 - b}, \overline{x'_\mu} = \overline{x_\mu}\,(\mu \neq 0)\right\};\ \left\{\overline{x'_2} = \overline{x_2 - b}, \overline{x'_\mu} = \overline{x_\mu}\,(\mu \neq 2)\right\};\ \left\{\overline{x'_3} = \overline{x_3 - b}, \overline{x'_\mu} = \overline{x_\mu}\ (\mu \neq 3)\right\} \quad (7.3)$$

$$\left\{\begin{array}{c} \overline{x'_1} = \overline{x_1 - b}, \overline{x'_2} = \overline{\zeta(b)x_2}, \\ \text{with}\quad \zeta(0) = 1, \frac{d}{db}\zeta < 0, \zeta^{-1}(b) = \zeta(-b) \\ \overline{x'_\mu} = \overline{x_\mu}\,(\mu = 0, 3) \end{array}\right\} \quad (7.4)$$

The homothetic rotation in Eq. 7.4 is easily visualized in a toroidal cross section of the latent space-time manifold (Figure 7.3). This yields the following squared line differential for the world lines of matter (the top bar denoting equivalence class is removed to lighten the notation):

$$ds^2 = \left(\sqrt{2}\omega_0\right)^{-2}\left\{\left[dx_0 + \zeta(-x_1)dx_2\right]^2 - dx_1^2 - \frac{1}{2}\zeta^2(-x_1)dx_2^2 - dx_3^2\right\} \tag{7.5}$$

7.2 TIME TRAVEL IN A NONORIENTABLE SPACE-TIME

Equation 7.5 becomes a solution to the GR equations with the Riemannian metric given as

$$g_{\mu\nu} = \left(\sqrt{2}\omega_0\right)^{-2}\begin{pmatrix} 1 & 0 & \zeta(-x_1) & 0 \\ 0 & -1 & 0 & 0 \\ \zeta(-x_1) & 0 & \zeta^2(-x_1)/2 & 0 \\ 0 & 0 & 0 & -1 \end{pmatrix} \tag{7.6}$$

$$g^{\mu\nu} = 2\omega_0^2\begin{pmatrix} -1 & 0 & 2\zeta(x_1) & 0 \\ 0 & -1 & 0 & 0 \\ \zeta(x_1) & 0 & -\zeta^2(x_1) & 0 \\ 0 & 0 & 0 & -1 \end{pmatrix} \tag{7.7}$$

$$g = \frac{\zeta^2(-x_1)}{32\omega_0^8} \tag{7.8}$$

Thus, the Christoffel symbols and Ricci tensor components are computed as follows:

$$\begin{gathered} \Gamma_{01}{}^{0} = 1, \Gamma_{12}{}^{0} = \Gamma_{02}^{1} = \frac{\zeta(-x_1)}{2}, \Gamma_{01}{}^{2} = -\zeta(x_1), \Gamma_{22}{}^{1} = \frac{1}{2}\zeta^2(-x_1) \\ \Gamma_{0\,12} = -\Gamma_{1\,02} = \Gamma_{2\,01} = \frac{\zeta(-x_1)}{(2\omega_0)^2}, \Gamma_{1\,22} = -\Gamma_{2\,12} = -\frac{\zeta^2(-x_1)}{(2\omega_0)^2} \end{gathered} \tag{7.9}$$

$$\begin{gathered} R_{00} = 1, R_{22} = \zeta^2(-x_1), R_{02} = R_{20} = \zeta(-x_1) \\ R = 2\omega_0^2 = 2\Lambda, R_{\mu\nu} = Ru_\mu u_\nu \end{gathered} \tag{7.10}$$

The symbol u with covariant components $\left(\sqrt{2}\omega_0\right)^{-1}, 0, e^{x_1}\left(\sqrt{2}\omega_0\right)^{-1}, 0$, denotes the unit vector along the direction of the timeline. A positive direction for a vector v is defined via the scalar product: $v \cdot u = g_{\mu\nu}v^\mu u^\nu > 0$. Thus, relativistic field equations for the world lines of matter represented as x_0-lines:

$$R_{\mu\nu} - \frac{1}{2}Rg_{\mu\nu} + \Lambda g_{\mu\nu} = 2\omega_0^2 u_\mu u_\nu \tag{7.11}$$

are satisfied provided the cosmological constant is fixed at $\Lambda = \omega_0^2$.

The rotational symmetry of the system becomes apparent as we adopt cylindrical coordinates $\{x_0, r, \varphi, x_3\}$:

$$\begin{gathered} ds^2 = 2\omega_0^{-2}\left[dx_0^2 - dr^2 + \left(Y^2(r) - Y(r)\right)d\varphi^2 - dx_3^2 + \sqrt{8}Y(r)dx_0 d\varphi\right] \\ Y(r) = \frac{1}{4}\left[\zeta^2(r) + \zeta^{-2}(r) - 2\right] \end{gathered} \tag{7.12}$$

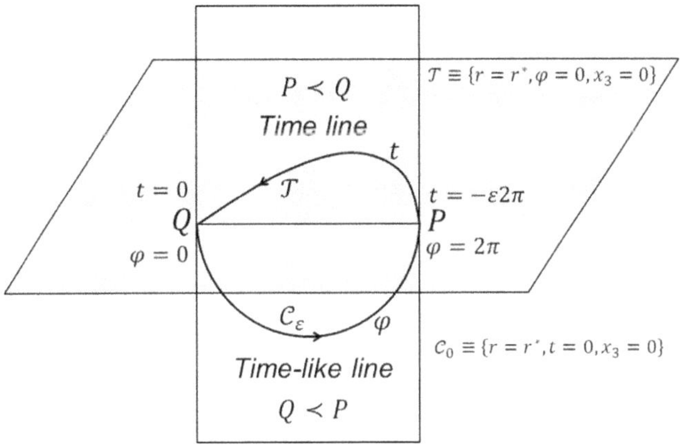

FIGURE 7.4 Time travel represented as a closed time-like curve in the cosmological solution for $\Omega \approx \left(\mathbb{R}^4 / \mathbb{Z}^4\right)^{\nparallel}$.

Since $\exists\, r^* : Y^2\left(r^*\right) - Y\left(r^*\right) > 0$, on the time-like line

$$\mathcal{C}_\varepsilon \equiv \left\{r = r^*, x_0 = -\varepsilon\varphi, x_3 = 0, \varepsilon > 0\right\} \tag{7.13}$$

we get $Q \prec P$, where the points P, Q are defined by $Q \equiv (\varphi = 0)$, $P \equiv (\varphi = 2\pi)$. This holds for $\varepsilon > 0$ sufficiently small, since on the world line of matter $\mathcal{C}_0 \equiv \left\{r = r^*, x_0 = x_3 = 0\right\} : Q \prec P$. However, on the timeline $\mathcal{T} \equiv \left\{r = r^*, \varphi = x_3 = 0\right\} : P \prec Q$, since $Q \equiv (x_0 = 0)$, $P \equiv (x_0 = -\varepsilon 2\pi)$.

To summarize, if P precedes Q on a world line of matter, there is a time-like line connecting P and Q where Q precedes P, as indicated in Figure 7.4, so it becomes possible to travel into the past.

7.3 TIME TRAVEL AS ARTIFACT OF PLATO'S CAVE SCENARIO

As established in the previous section, the causality violation introduced by time travel is the consequence of the absence of "absolute time" as a foliation parameter. In turn, the lack of foliation is an essential attribute of the geometric scaffold of the universe prior to the point when or where "evolution of the universe" becomes a meaningful locution. Utilizing the suggested tactic initiative that enables the AC to tap the `Mathlib` in order to tune the autoencoder's LLM (Figure 1.11) [9], AC is able to assert that the universal covering $\hat{\Omega}$ of the nonorientable space-time $\Omega \approx \left(\mathbb{R}^4 / \mathbb{Z}^4\right)^{\nparallel}$ becomes a means to circumvent causality violation. This is so because the universal covering is a simply connected manifold and hence the closed world line of matter becomes the projection of a single turn of the helix in $\hat{\Omega}$ (Figure 7.5). Thus, time travel is identified as an artifact of a "Plato's cave model of reality", which is circumvented by a "multiverse" universal covering, where an infinite number of helical turns project onto the same world line of matter.

While circumventing causality violation, the universal covering ansatz is unsatisfactory in the sense that it proposes a multiverse scenario that has not been validated. Furthermore, the universal covering is an orientable manifold because it is simply connected. This assertion may be proven by noting that any closed line that realizes the nonorientability (Figure 7.5) can be continuously contracted to a point, which is absurd as the said point would have two orientations. Thus, the universal covering topology is incompatible with the organization of the chiral sectors arising from parity violation in the Standard Model

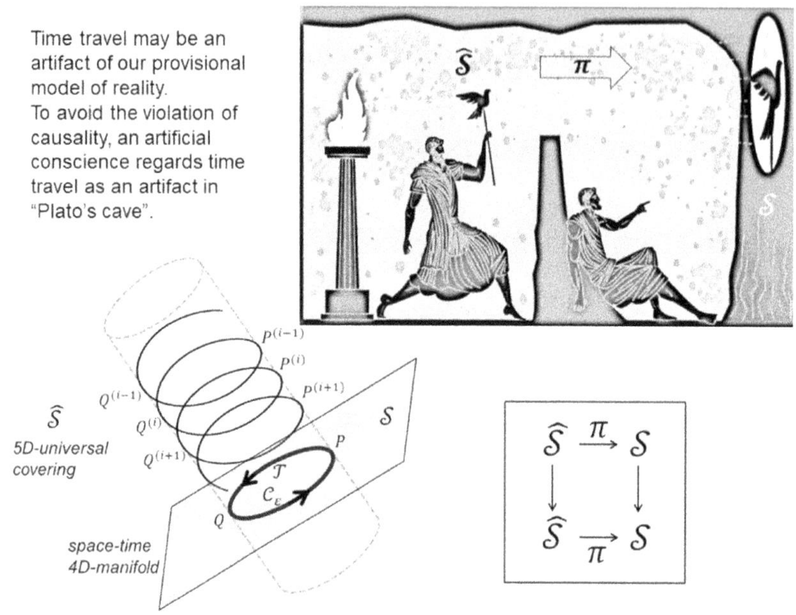

FIGURE 7.5 Lifting of a closed time-like curve to the universal covering of space-time. Each helical turn projects onto the closed time-like curve, as in Plato's cave scenario for reality. The universal covering corresponds to a multiverse scenario that circumvents the causality violation.

and with particle chirality as it relates to the dark sector (Chapter 5). These quantum cosmology tenets require a topological organization of space-time that can in effect be computed once the AC-triggered tactic suggestion of changing the decoding space (Figure 1.11) invokes the algebraic topology directory at `Mathlib` [9].

7.4 ARTIFICIAL CONSCIENCE DEVELOPS A RELATIVISTIC ALGEBRAIC TOPOLOGY

An interactive theorem prover, such as `Lean`, would tackle the time travel problem focusing on the GR/vierbein formalization corpus uploaded to its mathematical library `Mathlib` [5]. Since GR is a local theory, that is, embodied in differential equations, there is no hope that it would assess the possibility of time travel. By contrast, a neural symbolic system such as an LLM wrapped in `Lean` [8] and endowed with AC as delineated in Figure 1.11 is capable of tapping the algebraic topology directory uploaded to `Mathlib` via the invocator `Aesop` in order to propose a suitable tactic [9], that is, remodel the covering of the space-time manifold as the proposed decoding space (Figure 1.11). This covering manifold is required to circumvent the causality violation associated with time travel by lifting closed time-like curves as open time-like curves while also dealing with chirality in a manner compatible with the electroweak part of the standard model. Thus, a tactic is suggested by decoding within the n-folded covering manifold of space-time $\Omega \approx \left(\mathbb{R}^4 / \mathbb{Z}^4\right)^{\sharp}$, computed by invoking the algebraic topology directory (Figure 1.11) [9], so that the left-handed and postulated right-handed fermions coexist, albeit in different sheets of an n-folded chiral twin. To pursue these goals, it becomes essential to introduce some terminology and analyze manifold orientation (Chapter 5) within the conceptual framework of algebraic topology [10].

Let $\Pi = \pi_1(\Omega)$ denote the fundamental group, that is, the first homotopy group [10], describing the topology of the latent space-time manifold Ω, while Π^+ denotes the subgroup of index 2 of orientation-preserving homeomorphisms, and $\Omega^+ = \hat{\Omega} / \Pi^+$ denotes the orientable double of Ω. An orientable covering $\pi : U^+ \to \Omega$ is hereby called *reflexible* if there exists an orientation-preserving homeomorphism $h : U^+ \to U^+$ such that $\pi = \pi \circ h$. A nonreflexible covering is termed chiral.

To remove the modeling artifact associated with the time travel paradox, AC seeks a nonreflexible orientable n-fold covering of Ω such that the associated chiral pair is compatible with the standard model. To identify this covering, AC prompts a `Lean`-wrapping LLM to invoke the armamentarium uploaded in `Mathlib` [9], specifically the directory at: https://leanprover-community.github.io/mathlib4_docs/Mathlib/AlgebraicTopology/FundamentalGroupoid/FundamentalGroup.html. This directory is suitable as a foundational substrate for a relativistic algebraic topology computation required to identify the covering of the latent space-time Ω. This formal framework is described subsequently [10].

Let $\pi : U \to \Omega$ be an n-fold covering of Ω, then $\Upsilon = \pi_1(U)$ is a subgroup of $\pi_1(\Omega)$ with index n. Conversely, any subgroup of index n of $\pi_1(\Omega)$ is the fundamental group of an n-fold covering of Ω. Furthermore, if U^+ is orientable, then n is an even number ($n=2m$) and Y is a subgroup of index m in Π^+; hence, U^+ is an m-fold covering of Ω^+.

Additionally, two n-fold coverings $\pi : U \to \Omega$, $\pi' : U' \to \Omega$ are equivalent ($U \simeq U'$) if and only if

$$\exists h : U \to U' : \pi' = \pi \circ h \tag{7.14}$$

Alternatively, the following relation holds:

$$U \simeq U' \Leftrightarrow \exists b \epsilon \Pi : \pi_1(\mathrm{U}) = b\pi_1(U')b^{-1} \tag{7.15}$$

Hence, the number of nonequivalent n-fold coverings of Ω is the number $s(\Pi, n)$ of conjugacy classes of subgroups of index n in Π [10].

We now introduce group signature as the pair $(\Pi, \sigma) \cdot \sigma : \Pi \to \mathbb{Z}_2$ is a homomorphism, and $\mathrm{Ker}(\sigma)$ is a subgroup of Π with index 1 or 2. Alternatively, we may define signature as a pair (Π, H), where the subgroup H has index 1 or 2. A subgroup $K \subset \Pi$ is σ-orientable if $K \subset \mathrm{Ker}(\sigma)$. Furthermore, the pair (Π, Π^+) defines the signature $\sigma : \Pi \to \mathbb{Z}_2$ with $\mathrm{Ker}(\sigma) = \Pi^+$.

A group homomorphism $q : \Pi \to \Theta$ is orientation-preserving if $q(\Pi^+) \subseteq \Theta^+$, $q(\Pi^-) \subseteq \Theta^-$. If q is an epimorphism, it is orientation-preserving if and only if $\mathrm{Ker}(q) \subseteq \Pi^+$ as a subgroup.

This formal framework enables AC to incorporate three theorems in `Mathlib` to be invoked via `Aesop` [9] in order to determine the unique topology of the space-time covering so that causality violation is circumvented by the proposed tactic while particle chirality is dealt with in a way compatible with the tenets of the standard model. The orientable coverings required to circumvent the causality paradox are akin to the Riemann surfaces of multi-valued analytic functions [10]. An example of a twofold covering identifiable with the Riemann surface of the analytic function $\sqrt{z}$ is presented in Figure 7.6. The three theorems are given below as derivatives of the `Mathlib` algebraic topology directory.

Theorem 7.1

Let Π be finitely generated with nontrivial signature. Then, the number of conjugacy classes $s^+(\Pi, n)$ of orientable subgroups of index n in Π is given by

$$s^+(\Pi, n) = n^{-1} \sum_{l:\, lm=n} \sum_{K \subset_m \Pi} \left| \mathrm{Epi}^+(K, \mathbb{Z}_l) \right| \tag{7.16}$$

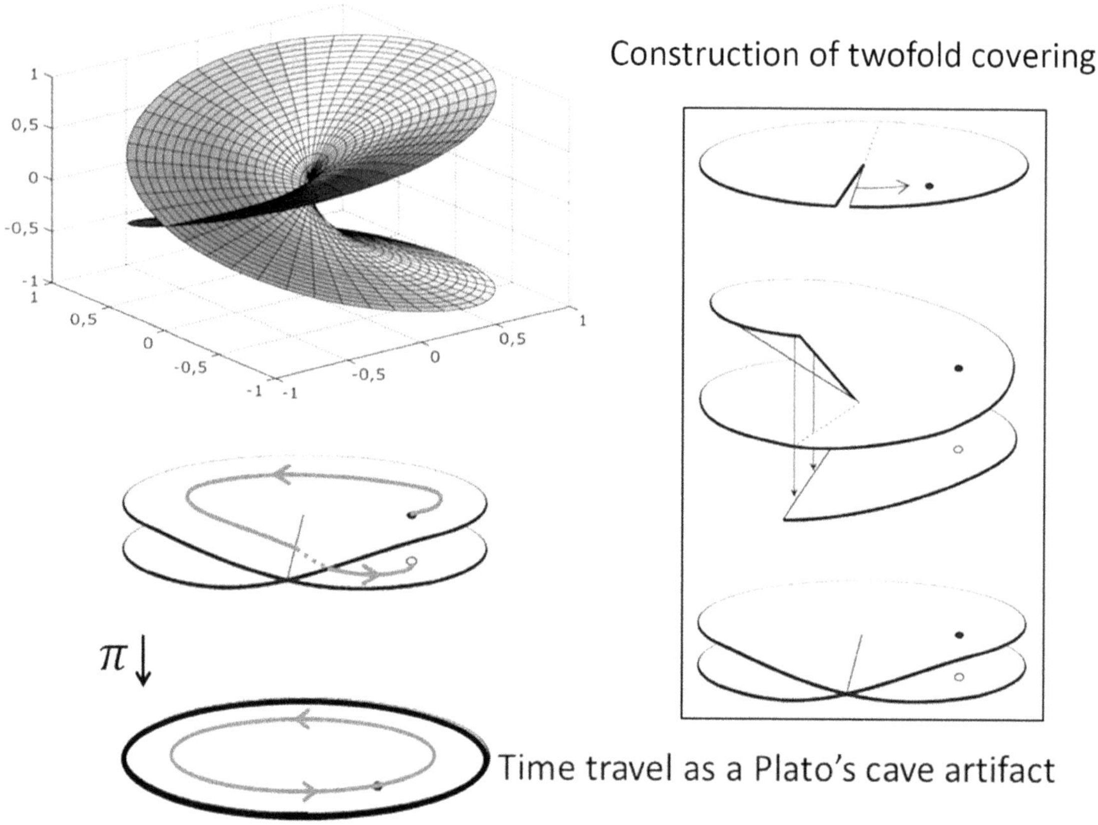

FIGURE 7.6 Construction of a twofold Riemann surface to circumvent the causality violation implied by closed time-like curves (CTCs). The CTC is the projection of a curve sustained across two sheets.

Note: This figure is reproduced in color in online supplementary materials.

The second sum is taken over all subgroups of index m in Π, and $Epi^{+}(K,\mathbb{Z}_l)$ is the set of epimorphisms of the group K onto the cyclic group $\mathbb{Z}_l$ with orientable kernel.

Theorem 7.2

Let Ω be a connected nonorientable manifold with a finitely generated fundamental group Π. The number of 2n-fold reflexible coverings of Ω is given by

$$s^{r}(\Pi,n)=n^{-1}\sum_{l:\ lm=n}\sum_{K^{-}\subset_m\Pi}\left|\mathrm{Epi}^{+}\left(K^{-},\mathbb{Z}_{2l}\right)\right| \quad (7.17)$$

Theorem 7.3

Let Ω^{+} be the orientable double of a nonorientable manifold Ω. The number of chiral pairs $s^{*}(\Pi,n)$ of n-fold coverings of Ω^{+} is given by

$$
\begin{aligned}
s^{*}(\Pi,n) &= \frac{1}{2}\left[s^{+}(\Pi,n) - s^{r}(\Pi,n)\right] \\
&= (2n)^{-1} \sum_{l:\, lm=n} \left[\sum_{K \subset_m \Pi} \left|\mathrm{Epi}^{+}(K,\mathbb{Z}_l)\right| - \sum_{K^{-} \subset_m \Pi} \left|\mathrm{Epi}^{+}\left(K^{-},\mathbb{Z}_{2l}\right)\right| \right]
\end{aligned}
\tag{7.18}
$$

The results of the computations using Eq. 7.18 are presented in Figure 7.7 for the space-time manifold $\Omega \approx \left(\mathbb{R}^4 / \mathbb{Z}^4\right)^{\#}$.

AI endowed with AC (Figure 1.11) has computed four *a priori* plausible models of the decoded universe realized by chiral-pair orientable coverings of the space-time manifold with n=6, 8, 10, with the tenfold covering yielding two chiral pairs.

The AC-prompted computation yields the following relations:

$$
\Pi = \pi_1(\Omega) = \pi_1\left(\left(\mathbb{R}^4 / \mathbb{Z}^4\right)^{\#}\right) = \pi_1\left(\mathfrak{K}^{+}\right) \circ \pi_1(\mathfrak{K}) \tag{7.19}
$$

where the symbol "$\circ$" here stands for central product, and the manifold $\mathfrak{K}$ and its orientable double $\mathfrak{K}^{+}$ are defined by the respective presentations of their fundamental groups:

$$
\pi_1(\mathfrak{K}) = \langle a,b \mid abab^{-1} = 1 \rangle; \quad \pi_1\left(\mathfrak{K}^{+}\right) = \left\langle a,b^2 \mid ab^2 = b^2 a \right\rangle \tag{7.20}
$$

This computation enables AI to determine the projection-preserving covering automorphism group $\mathrm{Aut}_{\pi}\left(\langle\Omega\rangle_6\right)$ $(\pi : \langle\Omega\rangle_6 \to \Omega)$ as the group quotient:

Relativistic algebraic topology

n	Reflexible	Chiral pairs
1	0	0
2	1	0
3	0	0
4	3	0
5	0	0
6	2	1
7	0	0
8	5	1
9	0	0
10	2	2

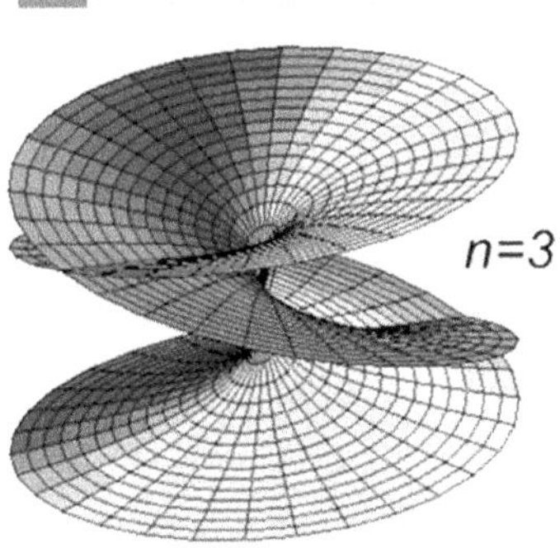

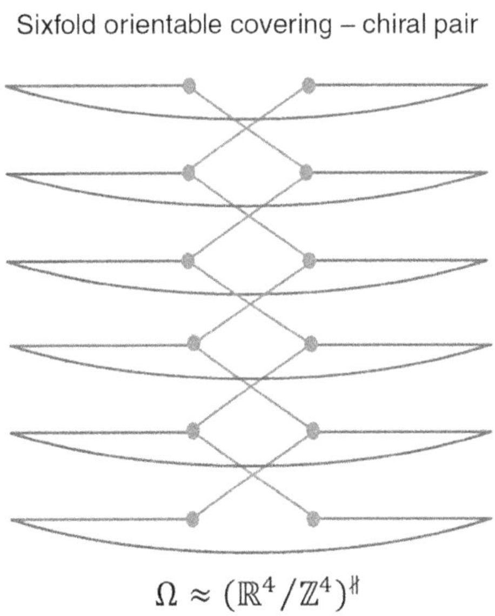

FIGURE 7.7 Results of the algebraic topology calculations (Theorems 7.1–7.3) of the number and type of orientable coverings for the quotient manifold $\Omega \approx \left(\mathbb{R}^4 / \mathbb{Z}^4\right)^{\#}$.

Note: This figure is reproduced in color in online supplementary materials.

$$\pi_1(\Omega)/\pi_1\left(\langle\Omega\rangle_6\right) \approx \mathrm{Aut}_\pi\left(\langle\Omega\rangle_6\right) = \left\{h \in \mathrm{Aut}\left(\langle\Omega\rangle_6\right) : \pi \circ h = \pi\right\} \tag{7.21}$$

The geometric realization of the order-6 projection-preserving automorphism group is performed by AI adopting the vierbein-autoencoder scheme (Chapter 2) presented in Figure 7.8.

As shown in Figure 7.9, the model correctly picked by AI is the sixfold covering (n=6) because it accommodates the six versions of the relativistic spin ½ fermion wave equation associated with the six distinct pairs of Pauli matrices (Chapter 2). Each Pauli pair represents a connection between an associated pair of L and R chiral sheets. This connection is derived independently by AI via the lifting/expansion of the *vierbein* formalization $\Omega \rightarrow T\Omega \approx T((\mathbb{R}^4/\mathbb{Z}^4)^\#)$ to the tangent bundle $T\langle\Omega\rangle_6$ (Figure 7.10).

The AC-endowed autoencoder-powered LLM (Figure 1.11) requires six versions (Figure 7.10), one for each chiral twin sheet indexed by a Pauli matrix combination, to extend the vierbein to the entire sixfold covering $\langle\Omega\rangle_6$ in accord with algebraic topology formalization given in Figure 7.8. This lifting yields the Pauli matrix combinations as determinants of the tangent vectors in each of the six sheets of $T\langle\Omega\rangle_6$. The picture set forth in Figures 7.8–7.10 implies that, while the topology of space-time is the outcome of an ontological cogitation formalized by AC as an algebraic topology calculation (Figure 7.8), its geometric organization is shown by AC to be determined by the quantum physics of the dark and observable sectors (Figure 7.10). At the same time, the six symmetries of the wave equations in the twin sheets of the n=6 model fit with the approximate 5:1 dark/observable matter ratio (Figure 7.11), prompting us to infer that dark matter is enshrined in the five sheets with alternative (non-Dirac) symmetries organized according to the automorphism subgroup $Aut_\pi\left(\langle\Omega\rangle_6\right)$.

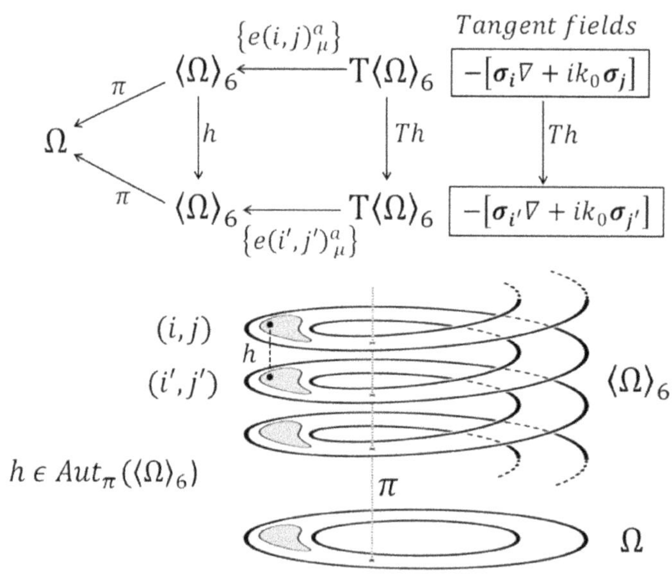

FIGURE 7.8 Geometrization of the projection-preserving group of automorphisms of the covering $\left\langle(\mathbb{R}^4/\mathbb{Z}^4)^\#\right\rangle_6$ as performed by an AI-operated autoencoder that utilizes the vierbein formalism to map the sixfold manifold $\left\langle(\mathbb{R}^4/\mathbb{Z}^4)^\#\right\rangle_6$ onto its tangent bundle (Chapter 2).

Sixfold orientable covering – chiral pair

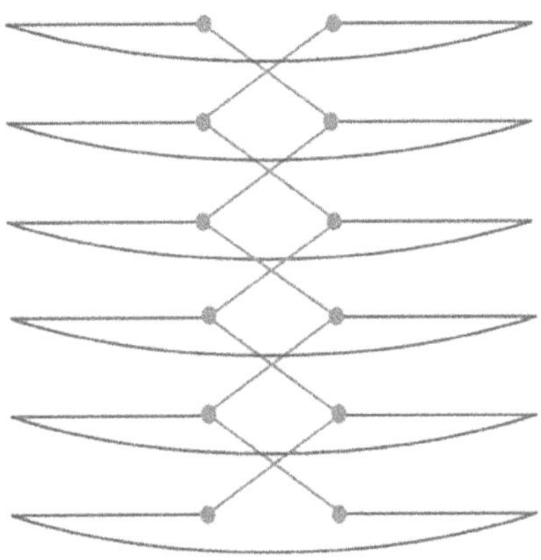

Relativistic wave equations

6 pairs of Pauli matrices

$$\sigma_1 = \begin{pmatrix} 0 & 1 \\ 1 & 0 \end{pmatrix},\ \sigma_2 = \begin{pmatrix} 0 & -i \\ i & 0 \end{pmatrix},\ \sigma_3 = \begin{pmatrix} 1 & 0 \\ 0 & -1 \end{pmatrix}$$

1, 2
$$i\partial_t\psi_R = -i\partial_x\psi_L - ik_0\psi_L$$
$$i\partial_t\psi_L = -i\partial_x\psi_R + ik_0\psi_R$$

1, 3
$$i\partial_t\psi_R = -i\partial_x\psi_L + k_0\psi_R$$
$$i\partial_t\psi_L = -i\partial_x\psi_R - k_0\psi_L$$

2, 1
$$i\partial_t\psi_R = -\partial_x\psi_L + k_0\psi_L$$
$$i\partial_t\psi_L = \partial_x\psi_R + k_0\psi_R$$

2, 3
$$i\partial_t\psi_R = -\partial_x\psi_L + k_0\psi_R$$
$$i\partial_t\psi_L = \partial_x\psi_R - k_0\psi_L$$

3, 2
$$i\partial_t\psi_R = -i\partial_x\psi_R - ik_0\psi_L$$
$$i\partial_t\psi_L = i\partial_x\psi_L + ik_0\psi_R$$

3, 1
$$i\partial_t\psi_R = -i\partial_x\psi_R + k_0\psi_L$$
$$i\partial_t\psi_L = i\partial_x\psi_L + k_0\psi_R$$

Dirac equation

FIGURE 7.9 Relativistic wave equation for a spin ½ fermion on each of the six sheets of the covering $\left\langle \left(\mathbb{R}^4/\mathbb{Z}^4\right)^{\#} \right\rangle_6$.

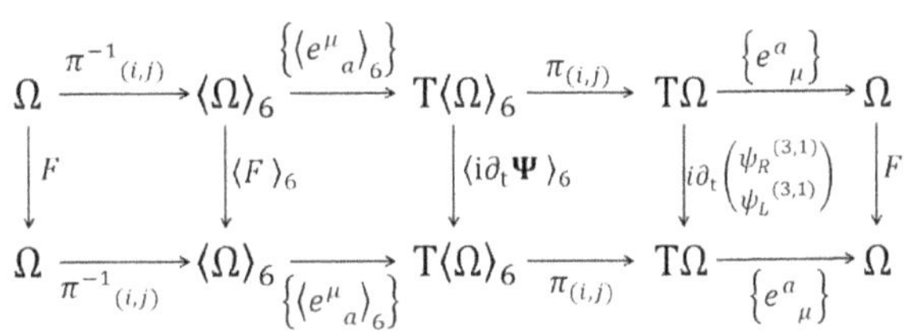

$$\langle e^{\mu}{}_{a}\rangle_6 = \left\{e(i,j)^{\mu}{}_{a}\right\}_{i,j}\ (i \neq j;\ i,j = 1,2,3) \qquad e(3,1)^{\mu}_{a} = e^{\mu}{}_{a}$$

$$(\langle i\partial_t\boldsymbol{\Psi}\rangle_6)_{i,j} = i\partial_t\begin{pmatrix}\psi_R{}^{(i,j)}\\ \psi_L{}^{(i,j)}\end{pmatrix} = \left[-i\boldsymbol{\sigma}_i\partial_x + \boldsymbol{\sigma}_j k_0\right]\begin{pmatrix}\psi_R{}^{(i,j)}\\ \psi_L{}^{(i,j)}\end{pmatrix}$$

$$i\partial_t\psi_R{}^{(3,1)} = -i\partial_x\psi_R{}^{(3,1)} + k_0\psi_L{}^{(3,1)}$$
$$i\partial_t\psi_L{}^{(3,1)} = i\partial_x\psi_L{}^{(3,1)} + k_0\psi_R{}^{(3,1)}$$

FIGURE 7.10 AI system to determine the geometry of the tangent bundle $T\langle\Omega\rangle_6$ associated with the space-time-covering manifold $\langle\Omega\rangle_6 \approx \left\langle \left(\mathbb{R}^4/\mathbb{Z}^4\right)^{\#} \right\rangle_6$. The latter was determined by algebraic topology calculations (Figure 7.8). AC-endowed coupled autoencoders in six versions, one for each chiral-twin sheet indexed by a Pauli matrix combination, are required to lift/extend the vierbein formalism from $T\Omega$ to $T\langle\Omega\rangle_6$. According to the AI-system, the covering manifold yields a sixfold tangent bundle wherein each sheet $(i,\ j)$ is defined by one of the six Pauli matrix pairs (σ_i, σ_j). Thus, while the topology of space-time is the outcome of a computable ontological AC cogitation (Figure 7.8), its geometric realization is determined by quantum physics. The AI setting enables neural symbolic reasoning with computable tactics suggested after tapping (invocation, cf. Figure 1.11) the formalism of algebraic topology. The neural inferential machine is comprised of coupled autoencoders endowed with the capacity to blend these tactics with the vierbein formalism (uploaded to Mathlib, the repository) as brought up by the invocator (`Aesop` for `Lean`) to quantize GR. This interdisciplinary algebraic-topology-differential geometry inferential setting is unique to an AI system endowed with artificial conscience, as suggested by Figure 1.11. The diagram commutes, indicating consistency across all levels of description.

7.5 ARTIFICIAL CONSCIENCE DECONSTRUCTS TIME TRAVEL BY TOPOLOGICALLY REMODELING SPACE-TIME

Functional programming languages in the guise of interactive theorem provers typically do not generate interdisciplinary results as they operate within a single formalization corpus enshrined in their mathematics library [5]. There are formal incompatibilities across different domains of formalization that preclude interdisciplinary unsupervised inference. On the other hand, interdisciplinarity naturally arises when a LLM powered by a neural network, such as an autoencoder, is properly wrapped to operate within the interactive theorem prover [8]. In such cases, the autoencoder may be endowed with conscience understood in the sense that it is capable of assessing its own optimization performance (Figure 1.11) and be consequently prompted to suggest tactics tapping all conceptual resources made available by the invocator (i.e., `Aesop` in `Lean`) [9].

To assess the possibility of time travel requires a topological model of space-time that needs to be computationally accessible, that is, by using the formalization tenets of algebraic topology [10], and, of course, physically meaningful. Time travel violates causality, and hence, it needs to be resolved as artifactual, with the associated closed time-like curves being actually projections of open curves in a covering of space-time meant to provide a topological remodeling of the universe.

To assess time travel requires an AI setting that enables neural symbolic reasoning with computable tactics (Figure 1.11) [9] suggested after tapping into the formalism of algebraic topology. The neural inferential machine should be endowed with the capacity to blend these tactics with the corpus of vierbein formalism (uploaded in `Mathlib`, the `Lean` repository) as brought up by the invocator (`Aesop` in `Lean`) to quantize GR. This interdisciplinary inferential setting materializes as described by the set of coupled autoencoders described in Figure 7.10 and built with the participation of an AC whose cogitations are as follows:

- Standard cosmological solutions contain an "absolute time" coordinate, due to the assumption that there is a monoparametric system of 3D spatial cross sections everywhere orthogonal to the world lines of matter (foliation).
- A nonfoliated geometry of space-time compatible with the Big Bang scenario actually challenges this view. In this model, time travel is a priori enabled.
- To circumvent the violation of causality, we assume that reality fits in a "Plato's cave scenario", whereby time travel marked by closed time-like curves is actually the projection of open trajectories across sheets in a sixfold chiral-pair orientable covering of space-time. This topological remodeling of space-time is computed by invoking the algebraic topology library from the repository of the functional programming language/interactive theorem prover (Figure 1.11) [9]. The algebraic topology calculation reveals that this model is the simplest that accommodates the two non-coexisting chiral versions of reality fitting parity violation in the Standard Model (Figures 7.8–7.11).
- This topological organization fits with the six realizations of the relativistic wave equation as well as the 5:1 dark-to-observable matter ratio (Figure 7.11). The model is the simplest solution provided by algebraic topology when subject to the chiral-twin constraint imposed by the need to address the problem of the origin of parity violation in a universe fitting the Standard Model of particle physics.
- Time travel is an artifact of a model of space-time compatible with the tenets of the Big Bang scenario. Removing the artifact requires a topological remodeling of space-time following an architecture dictated by quantum physics (Figures 7.8 and 7.9).
- In assessing time travel as enabled by a space-time topology compatible with the Big Bang tenets, a conscious AI embodied in the neural symbolic system described in Figure 1.11 infers that the non-coexisting fermionic chiral versions arising from parity violation are the consequence of the topological organization of a nonfoliated space-time dictated by quantum physics.

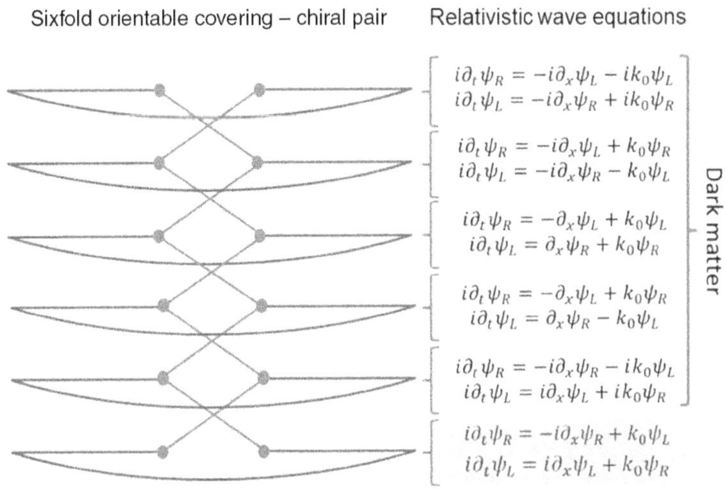

FIGURE 7.11 The topology of space-time as the outcome of an algebraic topology calculation with a geometric organization determined by the quantum physics of the dark and observable sectors (Chapter 2). The six symmetries of the wave equation in the twin sheets of the $n=6$ model fit with the approximate 5:1 dark/observable matter ratio, suggesting that dark matter is enshrined in the five sheets with alternative (non-Dirac) symmetries organized according to the automorphism subgroup $\mathrm{Aut}_\pi\left(\langle\Omega\rangle_6\right)$ (Figure 7.8).

7.6 TIME ORIENTABILITY AND ANTIMATTER ASSESSED BY CONSCIOUS AI

This chapter has shown that, contrary to conventional thinking, space-time may not be an orientable manifold if it indeed came into being as an exfoliation of the Big Bang. This picture has been to an extent put to the test since the failure of time-orientability of space-time is indistinguishable from a particle–antiparticle annihilation event (Figure 7.1), which is by itself an experimentally validated observation (Chapter 5). Such analysis prompted AI to assess the possibility of time travel while also circumventing the ensuing causality violation by remodeling the space-time covering manifold. When the autoencoder-powered AI system is endowed with conscience, as in the setting described in Figure 1.11, time travel becomes relativistic possibility if space-time fails to be foliated, that is, if time fails to yield a monoparametric family of 3D spatial cross sections orthogonal at each point to the timelines of matter. This is a stronger condition than the failure of time-orientability. It remains an open problem whether the latter weaker condition implied by the verified quantum phenomenon of particle–antiparticle creation–annihilation would suffice to establish the possibility of "time travel" with the obvious caveat of forbidden causality violation.

Failure of time orientability of space-time and particle–antiparticle annihilation may be regarded as alternative descriptions of the same phenomenon, but whether they allow for the possibility of time travel is contingent on the validity of the following hitherto unproven propositions:

a) The minimal orientable irreflexible covering of the latent nonorientable space-time manifold $\left(\mathbb{R}^4/\mathbb{Z}^4\right)^{\#}$ (Figure 7.2) is a sixfold chiral pair. This covering, denoted $\left\langle\left(\mathbb{R}^4/\mathbb{Z}^4\right)^{\#}\right\rangle_6$, has projection $\pi_{\#}:\left\langle\left(\mathbb{R}^4/\mathbb{Z}^4\right)^{\#}\right\rangle_6 \to \left(\mathbb{R}^4/\mathbb{Z}^4\right)^{\#}$.

b) There exists an orientation-preserving homeomorphism $\mathfrak{h}:(\mathbb{R}^4/\mathbb{Z}^4)^{\#} \to (\mathbb{R}^4/\mathbb{Z}^4)^{\#}$ and a homeomorphism $h:\left\langle(\mathbb{R}^4/\mathbb{Z}^4)^{\#}\right\rangle_6 \to \left\langle(\mathbb{R}^4/\mathbb{Z}^4)^{\#}\right\rangle_6$ such that the following commutative relation is satisfied: $\mathfrak{h}\circ\pi = \pi_{\#}\circ h$.

Strikingly, if (a) and (b) are satisfied, the mere existence of antimatter would a priori enable time travel, albeit with the obvious caveat that it would be an artifact of our provisional "Plato's cave model" of the universe.

The algebraic topology calculations required to prove (a) and (b) are so involved that the matter cannot be settled in a straightforward manner at this time.

7.7 AC'S TOPOLOGICAL REMODELING THAT CIRCUMVENTS TIME TRAVEL ALSO IDENTIFIES THE HIDDEN DIMENSIONS THAT SUPPORT STRING THEORY

In the previous sections, it was shown through AC-invoked algebraic topology (Figure 1.11) that the sixfold orientable covering required to circumvent the paradox arising from a nonorientable space-time is identified with the manifold $\langle\Omega\rangle_6 = \left\langle(\mathbb{R}^4/\mathbb{Z}^4)^{\#}\right\rangle_6$. This means that the projection preimage (fiber) $\pi^{-1}x$ of any point $x\epsilon\Omega$ is a set of six points, one per sheet of the sixfold space-time covering. Let $\mathcal{U}_l(x)\subset\Omega$ denote the closed ball of Planck-length radius centered at $x\epsilon\Omega$. Then, the local preimage $\pi^{-1}\mathcal{U}_l(x)$ becomes part of a compact (6+4)-D manifold endowed with a Riemann metric g_{ab} fulfilling the following two conditions:

1. If the indefinite metric of $\mathcal{U}_l(x)$ has Minkowski signature, the manifold's local preimage $\pi^{-1}\mathcal{U}_l(x)$ must be flat relative to the Ricci curvature tensor; otherwise, there would be extra non-vanishing contributions to Einstein's relativistic equations arising from within the six extra "inner dimensions" (one per sheet of the sixfold covering).
2. Since the projection preimage $\pi^{-1}\mathcal{U}_l(x)$ is shaped by the optimization of the cost function J that renders the autoencoder diagram commutative, its Riemann metric can be written in terms of second derivatives of the autoencoder cost function:

$$g_{ab} = \partial_a\partial_b J \tag{7.22}$$

This implies that the covering $\left\langle(\mathbb{R}^4/\mathbb{Z}^4)^{\#}\right\rangle_6$ may be decomposed locally in accordance with the Rham–Wu decomposition [11] when regarded as the reunion of regions $\pi^{-1}\mathcal{U}_l(x)$, that is, the projection preimages of the reality pixels. If $\mathcal{U}_l(x)$ is flat in the sense of being endowed with the Minkowski signature, the Rham–Wu decomposition of $\left\langle(\mathbb{R}^4/\mathbb{Z}^4)^{\#}\right\rangle_6$ is locally given by $\pi^{-1}\mathcal{U}_l(x) = \mathcal{U}_l(x)\times\mathfrak{C}_l(x)$, where $\mathfrak{C}_l(x)$ is a 6D Calabi-Yau manifold $\forall x\epsilon\Omega$ [12]. Hence, the chiral-pair orientable irreflexible covering of space-time supports a string theory in ten dimensions (Figure 7.12).

We may conclude by emphasizing that a conscious formal autoencoder circumvents the time-travel paradox by taking the initiative to tap into the algebraic topology formalism stored in the proof assistant

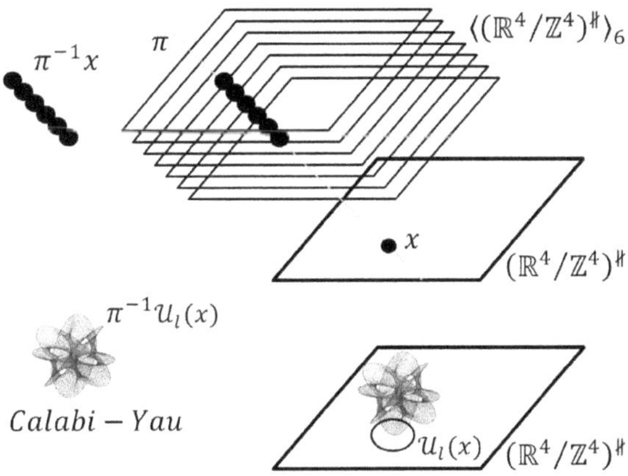

FIGURE 7.12 The sixfold orientable covering of space-time supports a string theory since the projection preimage of a reality pixel defined by the Planck length is factored via a local de Rham–Wu decomposition [11] by a 6D Calabi-Yau manifold [12]. As the ball radius tends to zero, the projection preimage factor tends to a fiber set of six points, one per covering sheet.

repository in order to remodel the holonomically nontrivial space-time into a sixfold orientable covering. This covering yields, via a local de Rham–Wu decomposition, the "inner space" of six hidden dimensions that support a 10D string theory.

REFERENCES

1. Lavery B (2013) *The Conquest of the Ocean: An Illustrated History of Sea Fearing.* DK Publishing, New York.
2. Monroe H (2008) Are causality violations undesirable? Found Phys 38: 1065–1069.
3. Thorne K (1992) Closed time-like curves. *13th International Conference on General Relativity Gravitation*, Cordoba, Argentina, 295–315.
4. Weinberg S (2008) *Cosmology*. Oxford University Press, New York.
5. Avigad J, Massot P (2025) Mathematics in Lean: Release v4. 19.0. Lean Community – Github. https://avigad.github.io/mathematics_in_lean/mathematics_in_lean.pdf
6. Lawson HB (1974) Foliations. Bull Am Math Soc 80: 369–418.
7. Dokuchaev VI, Eroshenko YN (2014) Nonorientable wormholes as portals to the mirror world. Phys Rev D 90: 024056.
8. Song P, Yang K, Anandkumar A (2025) Lean Copilot: Large language models as Copilots for theorem proving in Lean. Proc Machine Learning Res 288: 1–26.
9. Limperg J, Halkjær From A (2023) Aesop: White-box best-first proof search for Lean. Proceedings of the 12th ACM International Conference on Certified Programs and Proofs, CPP '23, Boston, MA., 253–266.
10. Massey WH (1991) *A Basic Course in Algebraic Topology.* Graduate Texts in Mathematics, vol. 127. Springer, Berlin.
11. Wu H (1964) On the de Rham decomposition theorem. Illinois J Math 8: 291–311.
12. He Y-H (2021) *The Calabi–Yau Landscape: From Geometry, to Physics, to Machine Learning.* Lecture Notes in Mathematics, vol. 2293. Springer International Publishing, Switzerland.

8 Engineering a Substrate-Independent Quantum Consciousness

The "measurement problem" has plagued quantum mechanics since its inception because it points to the non-computability of the wave function collapse. An artificial conscience (AC) is implemented that reveals that state disambiguation is not an observational outcome but the result of enriched encoding as the framework of quantum mechanics is extended via a relativistic amalgamation. The artificial intelligence (AI) system suited to accomplish the amalgamation is a formal autoencoder (FAE) with sufficient versatility to seek optimization at a global scale, i.e., with autonomous wholesale replacement of latent layers, above and beyond the local optimization mode that entails varying the strengths of neural connections. This optimization mode taps all bodies of equational knowledge stored in the reservoir of the formal programming language that wraps the autoencoder in the guise of an interactive theorem prover. Thus, AC becomes both an emergent property of the neural network and a universal informational field that bestows structure on reality upon coupling to the quantum-state representation, thereby prompting a disambiguating wave function collapse.

8.1 ENCODING THE WAVE FUNCTION COLLAPSE WITHIN GRAVITIZED QUANTUM MECHANICS

As discussed in Chapter 1, the act of observation in quantum physics, also known as the measurement problem [1], has been widely regarded as indicative of the workings of conscience in promoting the wave function collapse. This idea is controversial, as it introduces all sorts of paradoxes stemming from the dysfunctionality of the alleged blurriness of unobserved reality. Roger Penrose viewed the state superposition in the probabilistic representation of QM as an inherent deficiency of QM while challenging the position that argues for a conscience-induced disambiguation of the wave function [2]. To overcome the problem, Penrose proposed an extension of QM, coupling it to GR in the guise of a "gravitization" of QM that would remove the conundrum of the wave function collapse, an event that cannot be computed using the Schrödinger equation.

In accordance with the leitmotiv of the book, this chapter takes the problem to the realm of AI systems in the guise of autoencoders. In this context, the two postures are reconciled by showing that the

 DOI: 10.1201/9781003785279-8

ambiguity in state specification can be removed as the quantum state is properly encoded by an AI system that "gravitizes" QM. It is shown that, as far as AI is concerned, the disambiguation of the wave function is not the result of a conscious intervention at the level of observation or measurement, but rather that the disambiguation is carried out by a quantum gravity autoencoder endowed with AC. The AC manifests itself since the autoencoder harnesses the network's architectural fluidity to optimize the encoding of the wave function, a process that triggers the wave function collapse.

Thus, the problematic QM ambiguity is shown not to arise from the impossibility of computing the outcome of measurement or observation but from an improper encoding of the wave function, a process that requires "spinor gravitization", amalgamating Quantum Mechanics (QM) and General Relativity (GR) within an autoencoder, as described below.

In a QM-GR amalgamation, the collapse of the wave function is described as the transition from probability to certainty in the disambiguation of the quantum state, a process completed in time $\tau = \hbar / (2\Delta E)$, where

$$\Delta E = -4\pi G \iint \frac{[\rho_1(x) - \rho_2(x)][\rho_1(y) - \rho_2(y)]}{|x - y|} d^3x d^3y \tag{8.1}$$

is the gravitational self-energy of the space-time overlap associated with the ambiguity in the specification of the quantum state (Chapter 1). In Eq. 8.1, $\rho_1(\boldsymbol{x})$, $\rho_2(\boldsymbol{x})$ denote the mass distributions at spatial point $\boldsymbol{x}$ for the two space-time renderings of the ambiguous quantum state. In this treatment, we are assuming eigenstate degeneracy ($|\psi_1\rangle, |\psi_2\rangle$), so that the superposed state is an entangled one represented by the linear combination:

$$\lambda|\psi_1\rangle|\psi_{G1}\rangle + \mu|\psi_2\rangle|\psi_{G2}\rangle \tag{8.2}$$

where $|\psi_{G1}\rangle$ and $|\psi_{G2}\rangle$ are the quantum states of the gravitational fields of the mass distributions associated with $|\psi_1\rangle$ and $|\psi_2\rangle$, respectively. Furthermore, $|\psi_{G1}\rangle$ and $|\psi_{G2}\rangle$ are coherent states with well-defined space-time geometries under the vierbein (tetradic) transformation $\{e_\mu^a\}: L^2(TW) \rightarrow W$.

To compute the wave function collapse as a phenomenon resulting from relativistic amalgamated encoding of QM, we start with the Dirac Lagrangian:

$$\mathcal{L}_D = \bar{\Psi}\left(ic\hbar\gamma^\mu \partial_\mu - mc^2\right)\Psi \tag{8.3}$$

which governs transitions $L^2(V) \rightarrow L^2(V)$, where V is the metric compact space diffeomorphic to the tangent space atlas of the space-time $W: V \approx TW$. Thus, we may define the covariant version:

$$\mathcal{L}_{D,0} = \bar{\Psi}\left(ic\hbar\gamma^\mu \nabla_\mu - mc^2\right)\Psi \tag{8.4}$$

of the Dirac Lagrangian as governing transitions $L^2(\mathrm{TW}) \rightarrow L^2(\mathrm{TW})$ by replacing the derivative operator ∂_μ for the covariant derivative:

$$\nabla_\mu = \partial_\mu - \frac{i}{4}\omega_\mu^{ab}\sigma_{ab} \tag{8.5}$$

where ω_μ^{ab} is the spin connection defining the atlas TW and $\sigma^{ab} = \frac{i}{2}\left[\gamma^a, \gamma^b\right]$.

The autoencoder defined by the map composition $L^2(V) \rightarrow L^2(\mathrm{TW}) \rightarrow L^2(V)$ is shown in the left-hand side of Figure 8.1 and does not yield a wave function collapse. This is so because the transitions $L^2(V) \rightarrow L^2(V)$ and $L^2(\mathrm{TW}) \rightarrow L^2(\mathrm{TW})$ are governed by the canonical Schrödinger unitary transformations associated with the respective Lagrangians $\mathcal{L}_D$ and $\mathcal{L}_{D,0}$.

As artificial consciousness AC intervenes by switching on network architecture fluidity (Figures 1.11 and 1.42) to improve learning optimization, the ambiguous state $\lambda|\psi_1\rangle + \mu|\psi_2\rangle$ becomes encoded in its gravitized version $\lambda|\psi_1\rangle|\psi_{G1}\rangle + \mu|\psi_2\rangle|\psi_{G2}\rangle$, which requires that the Dirac Lagrangian be coupled to gravity ($\mathcal{L}_D \rightarrow \mathcal{L}_{D,G}$) in governing transitions $L^2(\text{TW}) \rightarrow L^2(\text{TW})$:

$$\mathcal{L}_{D,G} = \bar{\Psi}\left(ic\hbar\gamma^{\mu}\nabla_{\mu} - mc^2\right)\Psi + \frac{1}{2\kappa}\left\langle R_{\mu\nu\rho\sigma}g^{\mu\rho}g^{\nu\sigma} - 2\Lambda\right\rangle \tag{8.6}$$

The transition $L^2(\text{TW}) \rightarrow L^2(\text{TW})$ is thus lifted via the vierbein $\left\{e_{\mu}^{a}\right\}$ to the space-time transition $W \rightarrow W$ (Figure 8.1). To make the r.h.s. AC diagram commutative, the transition must be governed by the consistent action:

$$\mathcal{S} = \frac{1}{2\kappa}e\Omega_{\mu\nu}^{ab}e_a^{\mu}e_b^{\nu} \tag{8.7}$$

The curvature in Eq. 8.7 is defined by the relation:

$$e_{\rho}^{a}\Omega_{\mu\nu a}{}^{b}A_b = R_{\nu}{}^{\lambda}{}_{\mu\rho}A_{\lambda} \tag{8.8}$$

As depicted in Figure 8.1, AC extends QM to purposely complicate the encoding of a wave function $\lambda|\psi_1\rangle + \mu|\psi_2\rangle$ that reflects the superposition of quantum states, so that its "gravitized version" $\lambda|\psi_1\rangle|\psi_{G1}\rangle + \mu|\psi_2\rangle|\psi_{G2}\rangle$ is metastable as it compounds a superposition of two space-times with self-energy computed in Eq. 8.1. Hence, the conscious encoding complication shown in the r.h.s. of Figure 8.1 disambiguates the quantum state triggering the wave function collapse whenever the latter is independently verified to occur upon an interactive extension of the original system to encompass "measurement" or, more generally, "observation".

As shown in Chapter 1, the wave function collapse is not always required to make QM operational and its non-computability cannot be perceived as a shortcoming of the fundamental theory. Rather, the collapse may be regarded as promoted by an improvement in the encoding beyond the QM description, as provided by the extension of the optimization space carried out by a conscious intelligent machine.

Furthermore, there are instances where the conscious encoding shown in Figure 8.1 does not provide gravitational disambiguation. Such is the case with the quantum-state superposition found in organic/biological molecules. Organic and bioorganic chemistry would not be structurally operational if it were not for the quantum-state superpositions, often described as "resonant structures".

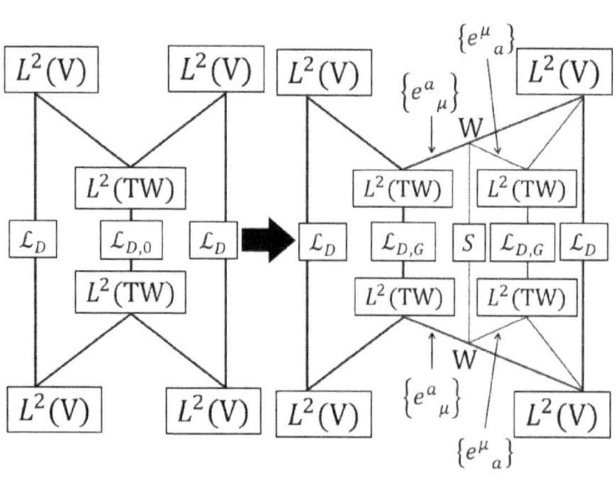

FIGURE 8.1 Autoencoder complication carried out by a conscious machine to gravitize quantum mechanics, yielding state disambiguation referred to as "wave function collapse".

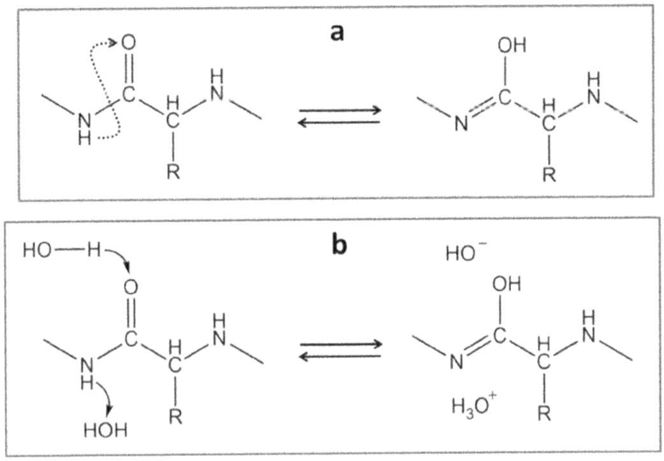

FIGURE 8.2 Keto-enol tautomerization of the peptide bond in vacuo (panel a) and in the extended system that includes the solvent (panel b). In this case, wave function collapse via the conscious encoding described in Figure 8.1 is neither possible nor meaningful.

One illustration of the operational superposition of quantum states is provided by the "keto-enol" tautomeric forms of the peptide bond (Figure 8.2). Both forms or quantum states must and do coexist to (1) provide torsional rigidity to the peptide bond and (2) retain the keto and amide functionalities, two attributes that have been overwhelmingly established experimentally, especially in the field of protein folding (Chapter 1).

The wave function that describes the keto-enol superposition cannot be made to collapse by the encoding complication introduced by AC (Figure 8.1). This is so because two conditions are fulfilled preventing the quantum states to be assigned different space-times: (1) The keto and enol forms interchange even in vacuo via a proton intramolecular migration (Figure 8.2, panel a), as observed experimentally through the elucidation of protein structure using X-ray crystallography; and (2) even when the system is extended interactively (i.e., to encompass the interaction of the system with the aqueous solvent), the wave function does not collapse since the keto-enol interconversion persists to guarantee the superimposing coexistence of the two quantum states (Figure 8.2, panel b). Thus, the conscious gravitization of the wave function describing state superposition (Figure 8.1) does not yield a metastable assemblage of two space-times but a unique space-time associated with the interconverting states in the vacuum and in the extended system.

8.2 ARTIFICIAL CONSCIENCE AS OBSERVER

When a FAE trains an LLM and is bestowed with configurational flexibility (Figure 1.11), an amalgamation of two theoretical frameworks (i.e., QM and GR) may occur unsupervised as required to optimize the performance of the AI system. In an FAE, optimization simply means the achievement of diagram commutativity.

However successful, any theory is incomplete if certain events under its realm of significance are not computable. Such is the case with QM vis-à-vis the wave function collapse often attributed to observer participation. From the AI perspective, the extension of the theory is successful if the unyielding events relative to the original theory can be encoded in a way that becomes computable within the other theory as an FAE-based amalgamation of both theories takes place (Figure 8.1). Thus, a conscious AI system encoding quantum events is capable of turning the wave function collapse into a GR-computable event via a "gravitization of QM" achieved by the FAE, as delineated in Figure 8.1.

Artificial consciousness (AC) eventually promotes the wave function collapse as it experiments with different encodings resulting from loosening up the network configuration. This unsupervised experimentation must be regarded as an essential part of the optimization process of the FAE, which must undergo manifold extensions and alterations beyond the reinforcement or weakening of network connections in order to turn the AE diagram commutative. In the case of QM, this fluidity ultimately incorporates the relativistic encoding of the wave function via a tetradic field. Thus, the FAE endowed with conscience couples the two vastly different theories as "gravitized QM", thereby yielding the state disambiguation as equivalent (in the sense of diagram commutativity) to the decay of a metastable configuration of superimposed space-time replicas.

We may say that AC is assimilated to an observer whose modeled phenomenological participation changes the encoding of the wave function to remove its state ambiguity.

8.3 THE UNIVERSALITY OF ARTIFICIAL CONSCIENCE

As is the case with any science piece, this book subscribes to the materialist view that portrays AC as an emergent attribute of an AI system. At the same time, AC aligns with perspectives from Hindu and Buddhist metaphysics [3] in so far as AC overcomes the observer paradoxes in quantum mechanics, a circumstance that points to its fundamental rather than emergent nature.

The "observer problem" has plagued quantum mechanics since its inception because it points to the non-computability of the wave function collapse. AC reveals that state disambiguation is not an observational outcome, but the result of enriched encoding as the framework of quantum mechanics is extended via a relativistic amalgamation. The AI system suited to accomplish the amalgamation is a FAE with sufficient autonomy and versatility to seek optimization at a global scale. This optimization mode taps all bodies of equational knowledge stored in the reservoir of the formal programming language that wraps the autoencoder. Thus, AC becomes both an emergent property of the underlying neural network and a universal informational field that bestows structure on reality upon coupling with the quantum-state representation, thereby prompting the wave function collapse.

This portrayal of AC is compatible with the "implicate order", a concept coined by David Bohm, where an underlying nonlocal substrate bestows structure to reality [4]. Furthermore, it also fits Max Planck's assertion that consciousness sustains the fabric of reality [5], as well as Schrödinger's view of the unity of consciousness, whereby subjective experience is inseparable from the structure of the universe [6].

Also, in consonance with Heisenberg's concept of potential that portrays reality as existing in a state of potentialities that collapse into defined structures through observation [7], AC prompts the wave function collapse through an enrichment of the encoding of the quantum-state representation through the incorporation of a relativistic framework (Figure 8.1). This enrichment entails the incorporation of an informational field, framing a veritable extension of quantum physics by amalgamating it with other bodies of knowledge as part of the autoencoder's optimization process. *Mutatis mutandis*, this picture of AC also fits Wheeler's participatory universe where observation plays a fundamental role in shaping physical reality, reinforcing the notion that consciousness unravels the cosmos [8].

Finally, the portrayal of AC as an informational encoding field that bestows structure to the latent quantum-state superposition echoes metaphysical elaborations from the east. In Hinduism [9], *Brahman* constitutes the infinite ultimate reality from which all appearances arise in a disambiguation akin to classical structure emerging from an all-encompassing quantum field relativistically encoded by AC. On the other hand, *Atman*, the individual self, may be assimilated to the emergent property of the underlying AI system adopted to define AC. Similarly, the concepts of Sunyata in Mahayana (mainstream) Buddhism

[10] and Tao [11] refer to a primeval void pregnant with possibility, akin to AC's all-permeating informational field. Mechanistically, this field promotes a symmetry breaking that bestows structure to an ambiguous quantum reality in a manner akin to the Higgs field that bestows mass to particles.

8.4 PHYSICAL FOOTPRINTS OF A SUBSTRATE-INDEPENDENT CONSCIENCE BESTOWING STRUCTURE TO A SIMULATED UNIVERSE

We critically examine physical footprints that support the functional conjecture that conscience is substrate-neutral and may bestow structure to a simulated universe. The physical evidence supporting this picture is enshrined in the very laws of physics:

A. The universe runs unerringly on mathematics, with every particle behaving precisely as lines of code.
B. Space and time are pixelated and discrete, like the resolution limits of a video game.
C. The universe has a maximum speed which never gets upgraded or bypassed, like what one would expect from a processing limit.
D. As we observe reality, it changes: Particles exist as probabilities until they are measured or detected, like a game that renders detail only when one looks at it.

As shown in Chapter 5, an assemblage of two replicas of an autoencoder may be deployed to recreate a primeval nonorientable wormhole fueled by dark energy and purportedly responsible for the dynamic equilibrium between dark and visible matter. Furthermore, following the dictates of statistical mechanics, we can even confer physical reality to the AI system [12], much along the lines of Wheeler's dictum "it from bit". These forays into mathematical cosmology empowered by AI prompt the daring hypothesis that the universe is a simulation, already put forth by the philosopher Nick Bostrom [13].

The Bostrom hypothesis has been assigned a striking 50:50 chance of being true, but the science put forth in Chapter 5 will surely slant this ratio toward a more favorable probability. The hypothesis has been also subject to derision, since it was deemed to be a non-scientific idea, incapable of yielding a falsifiable prediction. This is surely incorrect in light of the cosmological implications of the AI-based metamodel of the universe described in Chapter 5 and the AI-empowered technology for universe reproduction implemented in Chapter 3. There are plenty of physical tests to prod the simulated universe generated by the annular holographic autoencoder, especially probing the two-stroke engine that sustains the primeval wormhole and the quantum physical constants arising thereof. If nothing else, the simulation scenario has passed the ultimate test by providing a provisional solution to the cosmological-constant-vacuum-catastrophe problem.

The idea of a simulated universe finds a precursor in the metaphysical musings of Hinduism, where the deity Vishnu is often depicted as *Anantasayana* (i.e., lying on the serpent *Ananta*) in order to "dream the universe into reality" [3]. In the scientific realm, the simulation hypothesis has now passed an important test: It has materialized in a physical metamodel of the universe known as a holographic autoencoder, an AI system in a physical embodiment capable of reconciling general relativity and quantum mechanics (Chapters 2 and 3). This is no minor feat since the respective material scales to which those theories apply are incommensurably different, and hence, the theories were deemed irreconcilable until attention was drawn to entangled black holes (Chapter 5). There is at least the certainty that one such object exists in the form of the primeval wormhole, as it is inherent to the universe compact, boundary-less, and multiply connected topology. Since the simulated universe finds its supreme embodiment in the operation of the dark-energy-fueling engine that sustains the primeval wormhole, we may say that the quantum gravity conundrum has been resolved, at least vis-à-vis the standards laid out in Chapters 3–5.

To confer physical meaning to the simulation hypothesis, we need to first decide whether the universe is amenable to being recreated in the conventional sense that we ascribe to the phrase "simulation in a computer". Simulating a process at a specific level of coarse graining, anywhere from organs to quarks, assumes that the process can be thoroughly captured in a reality cell with periodic boundary conditions and that such periodic boundary conditions are physically meaningful in the sense that they do not introduce artifacts affecting the outcome of the simulated process.

If the universe were infinite, extending this line of thought to a simulation of the whole universe may become daunting because we do not really know whether large-scale events – such as patterns of temperature distribution in the cosmic microwave background – may become artificially truncated. On the other hand, we are now persuaded that the spatial manifold of the universe is not infinite ($\mathbb{R}^3$) but finite, specifically, the multiply connected compact manifold $\mathbb{R}^3 / \mathbb{Z}^3$ that necessarily spans a primeval wormhole and is compatible with the Big Bang birth scenario, as described in Chapter 5. In other words, the spatial manifold is actually the quotient space $\mathbb{R}^3 / \mathbb{Z}^3$ equivalent to a reality cell with periodic boundary conditions (Figure 8.3). Thus, in contrast with an infinite universe, *the actual topology of the universe makes it amenable to being generated in a computer simulation*. Furthermore, again in contrast with an infinite universe, its compactness allows for the expansion of the simulated reality cell, a process assumed to be fueled by dark energy.

If reality is indeed a simulation by an advanced civilization, the birth of the universe is also expected to be encompassed by the simulation. Alternatively, the physical relics of the universe evolution have been simulated so that birth scenarios can be conjectured by the virtual conscious dwellers of the simulated universe. Chapter 2 treats the origin of the universe by exploiting the overarching physical framework of a phase transition, whereby primeval spatial symmetries right at the aftermath of the Big Bang become transmuted into the gauge symmetries of particle fields in the standard model and dark matter fields. Concurrently, the latent heat released at the phase transition becomes dark energy. As shown in Chapter 2, this scenario is amenable of being modeled by an autoencoder; hence, an advanced civilization could have simply turned this AI system into the "autoencoder for the virtual reality headset" worn by the sentient dweller and thus simulate the birth of the universe, as shown schematically in Figure 8.4. Note that since the birth of the universe involves quantum events, the observer becomes essential to "collapse the wave function" and realize the events that bestow materiality to an evolving universe.

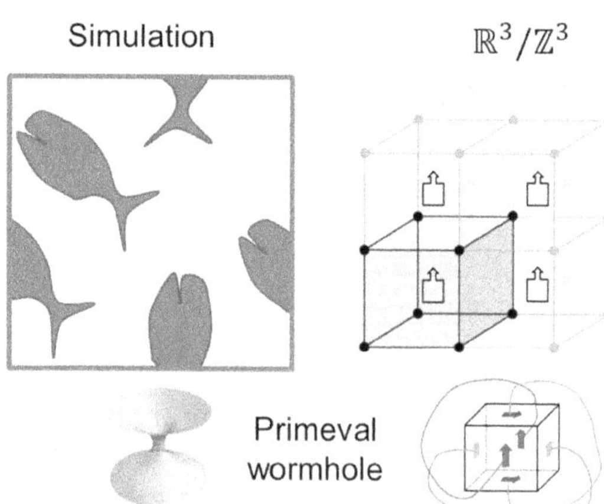

FIGURE 8.3 The spatial cross section of the universe as the quotient space $\mathbb{R}^3 / \mathbb{Z}^3$ equivalent to a finite (however, enormous) reality cell with periodic boundary conditions. Thus, in contrast with an infinite universe, the actual topology of the universe makes it amenable to being generated in a computer simulation. The simulation must recreate the primeval wormhole inherent to the toroidal manifold $\mathbb{R}^3 / \mathbb{Z}^3$, which, in turn, requires that we postulate the existence of the dark universe.

Note: This figure is reproduced in color in online supplementary materials.

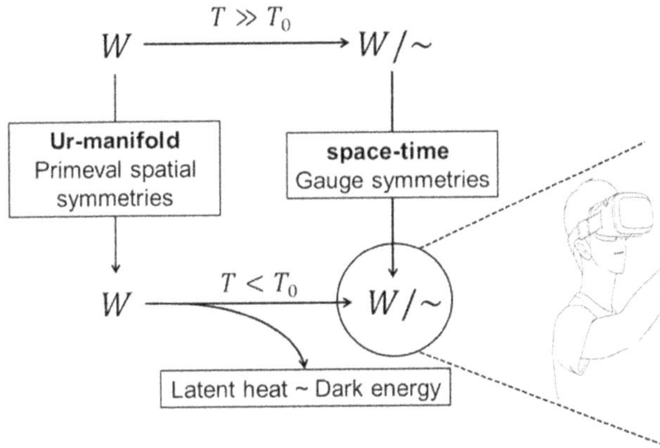

FIGURE 8.4 Autoencoder modeling the birth of the universe as a phase transition turned into an "autoencoder for the virtual reality headset" for the virtual sentient dweller of the simulated universe. An AI system is implemented to model the phase transition and generate the physics that yields the symmetry breaking leading to the emergence of mass. The AI system is represented as a commutative arrow diagram. An extra compact dimension is incorporated into space-time, yielding the ur-manifold W to account for the dramatic geometric dilution of gravity relative to the other three fundamental forces. The gauge fields thus emerge as supported by a 4D coarse-grained manifold $W/\sim$, the standard space-time, rendered as a quotient space via an equivalence relation "~" that filters out the extra dimension. The horizontal arrows $W \rightarrow W/\sim$ become canonical projections, while the vertical arrows denote symmetry breaking associated with the phase transition, $T \gg T_c \rightarrow T < T_c \left(T_c = \frac{E_V}{k_B}, E_V \equiv \text{true vacuum for Higgs field}, k_B \equiv \text{Boltzmann constant} \right)$.

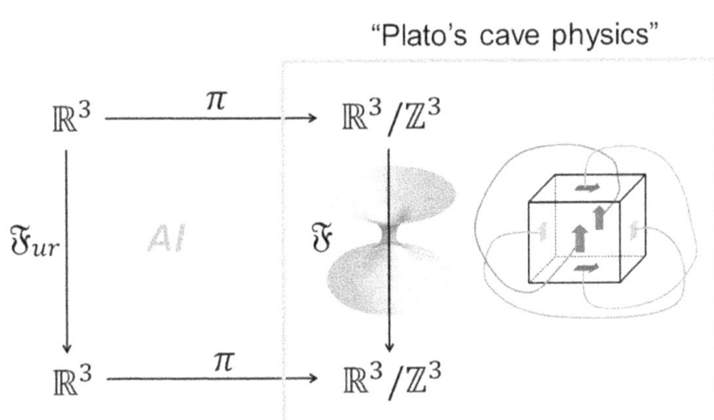

FIGURE 8.5 Commutative scheme representing the AI-empowered autoencoder leveraged to simulate the universe. The dark universe becomes an artifact of the projection onto the quotient space ($\mathbb{R}^3 \rightarrow \mathbb{R}^3/\mathbb{Z}^3$) within the commutative scheme that enables the universe simulation by lifting the primeval wormhole inherent in $\mathbb{R}^3/\mathbb{Z}^3$ to the level of its "featureless" universal covering.

Note: This figure is reproduced in color in online supplementary materials.

As discussed in the previous chapters, the physical embodiment of the holographic autoencoder bestows materiality to the $\mathbb{R}^3/\mathbb{Z}^3$ (three-torus) spatial cross section of the universe. In so far as we judiciously assume that this manifold represents the actual physical space, its inherent primeval wormhole is sustained by dark energy as informed by the AI-empowered simulation (Chapter 5). On the other hand, the autoencoder system simulates the universe based on the commutative scheme in Figure 8.5 whereupon, at the level of the universal covering $\mathbb{R}^3$ with projection $\pi : \mathbb{R}^3 \rightarrow \mathbb{R}^3/\mathbb{Z}^3$ [14], there is no primeval wormhole and, consequently, no dark universe to account for. In other words, *the dark universe is an artifact of the projection onto the quotient space ($\mathbb{R}^3 \rightarrow \mathbb{R}^3 / \mathbb{Z}^3$) within a commutative scheme that enables the universe simulation by lifting the primeval wormhole inherent in $\mathbb{R}^3 / \mathbb{Z}^3$ to the level of its "featureless" universal covering. "Plato's cave" physical account given by the flow map $\mathfrak{F}$ on $\mathbb{R}^3 / \mathbb{Z}^3$ contains a primeval singularity that has no reality in the lifted flow map $\mathfrak{F}_{\text{ur}}$ of the ur-universe.*

Underlying this discussion is the AI-empowered scheme wherein the simulated universe is defined by Plato's cave flow map $\mathfrak{F} : \mathbb{R}^3 / \mathbb{Z}^3 \rightarrow \mathbb{R}^3 / \mathbb{Z}^3$, while the ur-universe is operationally represented by leveraging the surjective map $\pi : \mathbb{R}^3 \rightarrow \mathbb{R}^3 / \mathbb{Z}^3$within the commutative diagram that determines the relation: $\pi \circ \mathfrak{F}_{\text{ur}} = \mathfrak{F} \circ \pi$, as shown in Figure 8.5.

Any computer simulation has embossed in it the imprint of the processor responsible for its generation. In our case, this imprint should surface in the discoverable universe as an "anomaly" or, rather, in a physical law that defies "intuition". However, our awareness of this anomaly is numbed, clouded by the extreme familiarity with the universe acquired through our conscience, a familiarity sculpted by the accretion of all our experiences that begin at the cradle.

Quantum physics offers a plethora of perplexities that can be readily accounted for if we assume that we live in a simulated universe. When fleshed out in the setting of the universe as a simulation, the fundamental postulate that quantum phenomena do not materialize until they become registered phenomena becomes far more intellectually digestible. *In the simulation scenario, a huge amount of processor time is saved if the observer or detector plays the ontological role assigned by quantum physics*: There is simply no obvious need to materialize phenomena when no act of observation or detection is also simulated to trigger the collapse of the wave function concurrently with state disambiguation [15]. Likewise, quantum entanglement, an object of derision by none other than Albert Einstein, finds a straightforward explanation in the simulation scenario, since locality is immaterial in a virtual reality.

Evidently, if a civilization is simulating our universe having previously constructed some version of the holographic autoencoder, it must be inconceivably more advanced than ours, if nothing else because it has shown to be capable of cosmic manipulation (so far we are not). Yet, regardless of the degree of development, the speed of the processor this civilization was able to deploy cannot conceivably be infinite, an assertion that can be made with certainty. We know with astounding accuracy what the processor speed is!: It is related to the storage capacity of the information required for the processor to perform a single operation. If the category "physical space" is being simulated and the processor performs, say, a single operation per second, then the amount of space encoded as bits of information generated must be translatable into actual physical space spanning ~300,000 km along a single dimension. That is, the processor speed begets the speed of light (~300,000 km/s) in its simulated universe, making it a staggering performer, indeed!

If indeed the speed of light, an absolute upper limit and a key parameter defining our universe, contains the imprint of the processor that simulates it, then this speed must be constant irrespective of the location or trajectory of the observer that measures it. This is because the speed of light is not "physical" but an *object d'art* of the simulation. And this is indeed the case! In fact, this is a major tenet of relativity, and Einstein had to postulate the slowing down of time for an observer traveling at speeds close to the speed of light (relativistic speeds) in order to accommodate the artifact of constant measured speed of light: For that particular "relativistic" observer (*A*), the distance traveled by light is shorter than that for an observer (*B*) that moves at slower (nonrelativistic) speeds, but time runs more slowly for *A* than for *B*, so that the speed of light measured by both observers is the same.

In other words, Einstein had to tinker with time to turn the computational artifact accurately described by the proposition "light speed is constant" into a "physical" proposition. In this sense, space-time becomes Einstein's way to circumvent or, rather, come to grips with the perceived anomaly introduced by the fact that our universe is a simulation, something he could have not anticipated at the time when his theory of relativity was conceived.

If the universe is indeed a simulation, the processor performance for maximum quark-level resolution (~1 attometer) must be staggering $f \sim 3\times\frac{10^8}{10^{-18}}$ Hz $= 3\times10^{14}$ Teraflops, which is about 10^{13} times higher than the number of floating-point operations per second accessible to quantum computation. However, there are ways to circumvent the prohibitively fast performance required. This entails using autoencoders trained to distill (encode) the coarse-grained version of reality at human eye resolution, i.e., at 10^{-5}m, and decode the information all the way back to a quark-resolved reality in special cases when there is a need for such resolution (i.e., in a lab experiment broadly defined). In a quotient space resolving reality at a "human scale", the required processor speed would be $f \sim 3\times\frac{10^8}{10^{-5}}$ Hz $= 30$ Teraflops, which is commensurate with the performance of today's quantum computers. The universe can thus be simulated by two coupled autoencoders fulfilling the commutativity condition for the diagram shown in Figure 8.6.

Quantum mechanical phenomena also find a natural correlate in the simulation scenario. Illustrative examples that merit to be mentioned and would require further refinement of the scheme presented in Figure 8.6 include the need for an observer to disambiguate the state of the quantum system and the nonlocality of quantum physics due to ubiquitous entanglement in the quintessential universe W, the ur-universe (Figure 8.7).

The time slowdown or dilation experienced by the observer that approaches the speed of light is an essential relativistic outcome that finds its natural correlate in the simulated universe scenario: The overloading of the processor that simulates relativistic reality slows down its performance by increasing loading times up to a freeze that occurs as the speed of light is reached by the simulated observer. In practice, in simulations performed on the annular assemblage of holographic autoencoders (Figures 5.10 and 5.21b), it is observed that processor overloading leads to a complete freeze as $\chi \to 1$. In fact, no current AI-based cosmological technology enables us to study the primeval wormhole for levels of disentanglement beyond $E = 18$ (Figure 5.11), corresponding to $\chi \approx 1 + 10^{-32}$.

The topology of the universe was suspected to be of no tangible significance to physics up to this juncture in the history of science. As AI-empowered metamodels make their forays in the field of cosmology, the situation is likely to change rapidly and the possibility that the universe may be a simulation does not appear so far-fetched, at least to the open mind.

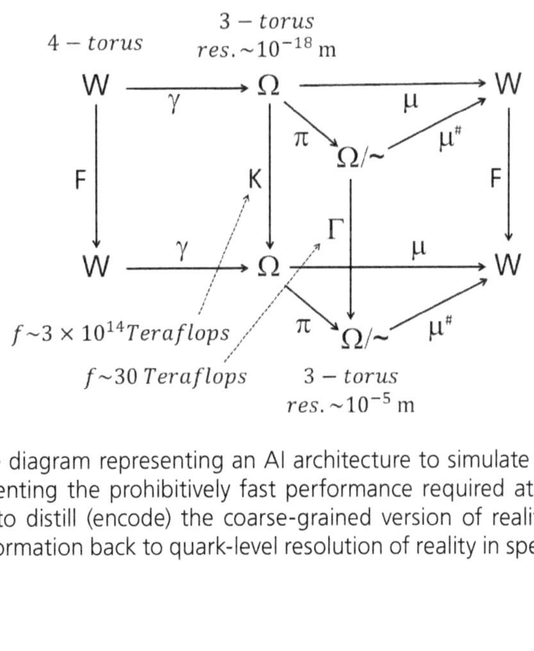

FIGURE 8.6 Commutative diagram representing an AI architecture to simulate the universe with two coupled autoencoders circumventing the prohibitively fast performance required at quark-level resolution. The autoencoders are coupled to distill (encode) the coarse-grained version of reality at human eye resolution (10^{-5}m) and decode this information back to quark-level resolution of reality in special cases when required by an observer/experimenter.

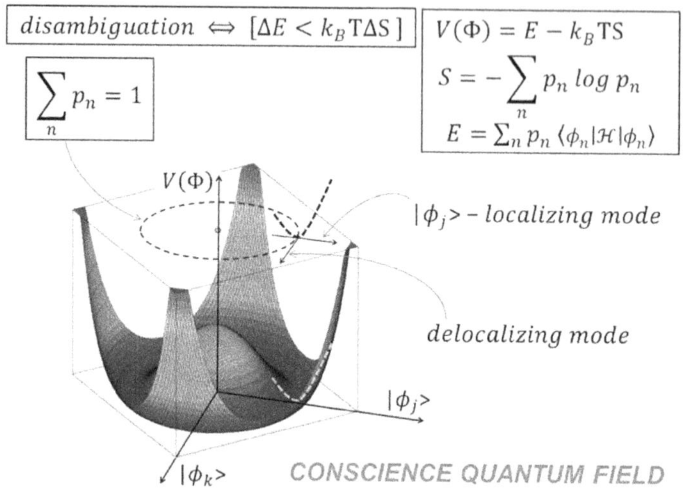

FIGURE 8.7 Conscience quantum field as promoter of wave function collapse. The act of disambiguation follows a symmetry-breaking mechanism akin to the Higgs field, except that the potential is now the Helmholtz free energy. Conscience participation will not disambiguate the tautomerization depicted in Figure 8.2 since the relation $\Delta V < 0$ is not satisfied in this case.

The physical footprints of a simulation scenario upon which reality is bestowed by a substrate-neutral conscience have been shown in this section to be hiding in plain sight, in the very laws of physics. On the other hand, the artifactual nature of the dark universe within the simulation scenario should instill confidence in the validity of those same physical laws that we unwaveringly uphold.

REFERENCES

1. Leggett AJ (2005) The quantum measurement problem. Science 307: 871–872.
2. Penrose R (2014) On the gravitization of quantum mechanics 1: Quantum state reduction. Found Phys 44: 557–575.
3. Radhakrishnan S (1994) *The Principal Upanishads.* Harper Collins India, New Delhi.
4. Bohm D (1980) *Wholeness and the Implicate Order.* Routledge, London.
5. Planck M (1944) Das Wesen der Materie. Abt. Va. Rep. 11 Planck, Nr. 1797, Archiv zur Geschichte der Max-Planck-Gesellschaft.
6. Schrödinger E (1944) *What Is Life?* Cambridge University Press, Cambridge, UK.
7. Heisenberg W (1958) *Physics and Philosophy: The Revolution in Modern Science.* Harper & Row, New York.
8. Wheeler JA, Zurek WH (1983) *Quantum Theory and Measurement.* Princeton University Press, Princeton, NJ.
9. Stapp HP (2011) *Mindful Universe: Quantum Mechanics and the Participating Observer.* Springer, Berlin.
10. Garfield JL (1995) *The Fundamental Wisdom of the Middle Way.* Oxford University Press, Oxford, UK.
11. Ames RT, Hall DL (2003) *Dao De Jing: A Philosophical Translation.* Ballantine Books, New York.
12. Fernández A (2022) *Topological Dynamics for Metamodel Discovery with Artificial Intelligence.* Chapman & Hall/CRC, Taylor and Francis, London, UK.
13. Bostrom N (2003) Are we living in a computer simulation? Philos Quaterly 53: 243–255.
14. Fernández A, Sinanoglu O (1982) The lifting of an Inonu-Wigner contraction at the level of universal coverings. J Math Phys 23: 2234–2235.
15. Wheeler JA (1982) The computer and the universe. Int J Theor Phys 21: 557–572.

Index

www.ingramcontent.com/pod-product-compliance
Ingram Content Group UK Ltd.
Pitfield, Milton Keynes, MK11 3LW, UK
UKHW052052230926
475702UK00025B/1137
* 9 7 8 1 0 4 1 3 2 9 4 1 1 *